Civil engineering construction contracts

Michael O'Reilly

Published by Thomas Telford Publishing, Thomas Telford Services Ltd,
1 Heron Quay, London E14 4JD

First published 1996

Distributors for Thomas Telford books are
USA: American Society of Civil Engineers, Publications Sales Department, 345 East 47th Street, New York, NY 10017-2398
Japan: Maruzen Co. Ltd, Book Department, 3–10 Nihonbashi 2-chome, Chuo-ku, Tokyo 103
Australia: DA Books and Journals, 648 Whitehorse Road, Mitcham 3132, Victoria

A catalogue record for this book is available from the British Library

ISBN: 0 7277 2530 0

Typeset in Great Britain by Alden, Oxford, Didcot and Northampton
Printed in Great Britain by Redwood Books, Trowbridge, Wiltshire

Contents

Table of cases xi

Table of statutes and secondary legislation xxvii

Chapter 1 Introduction 1

1. Brief historical introduction 1
2. Standard form contracts 3
3. Classification of civil engineering contracts 3
   - Classified according to the definition of the works 4
   - Classified according to the scope of the contractor's obligations 4
   - Classified according to the method of remuneration 5
   - Classified according to the project organisation 6
4. The procurement and legal environment 7
   - European procurement law 8
   - International contracts 8
   - General UK legislation 9
   - Housing Grants, Construction and Regeneration Act 1996 9

Chapter 2 Principles of contract law 11

1. The doctrine of privity 11
   - Supplementary agreements 12
   - Action in the tort of negligence 13
   - Assignment 13
   - Novation 14
   - Sub-letting consents and nomination 14
2. Formation and variation of contract 14
   - The elements of formation 15
   - Agreement 15
   - Authority 18
   - Consideration and intention to create legal relations 19
3. The terms of the contract 19
   - Mere representations 20
   - Construing contract terms 21
   - Conditions and warranties 22
   - Rectification 23

4. Implied terms 24
Terms implied by statute 24
Terms which are necessary for business efficacy 24
Usual implied terms 25
5. Exclusion, limitation and indemnity clauses 25
The Unfair Contract Terms Act 1977 25
6. Contracts and formality 26
The need for writing 26
Contracts under seal 26
7. Ending contracts 27
Agreement to discharge or vary the contract 27
Accepted fundamental breach or accepted repudiation 27
Determination under the terms of the contract 29
Frustration 29
8. Legal personality and agency 30
Legal personality 30
Agency 31

Chapter 3 General features of civil engineering contracts 33

1. Civil engineering contract documents 33
Drawings and specifications 33
Bills of quantities, schedules of rates, etc. 34
Programmes and method statements 35
2. Commonly implied terms and established constructions in civil engineering contracts 37
Payment 37
Time and progress 37
Quality of work and materials 38
The scope of the work 39
Cooperation between the employer and contractor 40
3. Performance 40
4. Instructions and variation orders 42
Agent's authority where the contract does not provide for variations 42
Terms entitling the employer to vary the works 43
Variations to be in a prescribed format 43
Variations requested by the contractor 44
Pricing of variations 44
5. Contract administration: decisions, certificates and adjudication 44
The Engineer and the ICE Conditions of Contract 47
Adjudication 47

6. Vicarious performance, assignment and novation 48
Vicarious performance and sub-contracting 48
Assignment 48
Novation 49
7. Insurance and indemnities 49
Construction of indemnity provisions in civil engineering contracts 50
Insurance 50
Third party rights against insurers 50
8. Determination clauses 51
The licence to occupy the site 51
Determination clauses 51
Wrongful determination and the court 52
Consequences of determination 52
Contractor's right to determine the contract 53
Passing of property on forfeiture 53
9. Guarantees and bonds 53
The use of guarantees and bonds 54
Construing the guarantee or bond 54
Injunction to restrain the creditor 54
Matters which release the surety under a guarantee 54
Bonds 55
The failure to obtain a guarantee or bond 56

Chapter 4 Extra-contractual entitlements **57**

1. Misrepresentation 57
Rescission for misrepresentation 58
Fraudulent misrepresentation 58
Negligent misstatement 58
Misrepresentation Act 1967 59
Where the misrepresentation has become a term 60
2. Claims for restitution 60
Where one person commences work in the expectation that a contract will eventuate 61
Where one person supplies work to another outside the scope of an existing contract 61
Where the contract ceases to exist 61
Where the plaintiff supplies services in an emergency 62
The quantification 62
3. Claims under a collateral or additional contract 62
Collateral contract 63
Compromise agreement 63

Chapter 5 Civil engineering professional services contracts **65**

1. Formation of design and professional services contracts 65
2. The terms of the agreement 66
   The standard of care 66
   Time 66
   Payment 66
   The extent of the work 67
   The duration of the consultant's obligation 67
   Termination 68
3. Delegation of design and consultancy services 68
4. Professional negligence: a failure to exercise reasonable skill and care 69
   Common professional practice 69
5. Examples of professional service obligations 71
   Collecting and investigating information 71
   The preparation of reports 72
   Supervision/inspection of work 73
   Safety 73

Chapter 6 Civil engineering claims: entitlements and evaluation **75**

1. Formulating claims 75
   The basis of valuation permitted 75
   The dispute resolution mechanisms provided for in the contract 75
   Notices 76
2. Statements of claim and response 76
   The content of the statement 76
   Other rules of pleading 77
3. Set-off and counterclaim 78
   Set-off 78
   Counterclaims 79
4. The evaluation of claims 79
   Damages 80
   Entitlements under the contract 83
   Quantum meruit 83
5. Interest and finance charges claims 83
   Interest as an entitlement under the contract 84
   Interest as damages flowing from breach 84
   Interest under statute 84
6. Extensions of time 85

The time for completion 85
Computing extensions of time 85
The opportunity to reprogramme 86
7. Prolongation, delay and disruption claims 87
Distinction between extensions of time and prolongation and delay claims 87
Prolongation, delay and disruption: heads of claim 88
8. Claims for defective work 90
9. Limitation 91
Contractual causes of action 91
Causes of action based on tortious negligence 92
Concealment 92

Chapter 7 Dispute resolution 93

1. Negotiated settlement 93
2. Alternative Dispute Resolution 94
Assisted negotiation 94
Non-binding tribunal 94
Interim binding tribunal 95
Binding tribunal 95
Procedure in ADR 96
3. Arbitration 96
The nature of arbitration 96
The arbitration legislation 100
The role of the courts in supervising arbitrations 100
The powers and responsibilities of the arbitrator 103
4. Litigation 104
Aspects of the procedures of the court 105
Short forms of procedure 106
Proposals for reform of the civil court procedure 107

Chapter 8 The ICE Conditions of Contract (6th Edition) 109

1. The scheme of the ICE Conditions of Contract 110
Personnel and administration 110
The Contract 110
Disputes 110
Time 110
Payment 111
Planning of operations 112
Unforeseen conditions 112
2. Commentary on the Conditions of Contract 112
Definitions and interpretation 112

Engineer and Engineer's Representative 117
Assignment and subcontracting 124
Contract documents 126
General obligations 131
Workmanship and materials 170
Commencement time and delays 175
Liquidated damages for delay 183
Certificate of substantial completion 188
Outstanding work and defects 190
Alterations, additions and omissions 192
Property in materials and Contractor's Equipment 199
Measurement 202
Provisional and Prime Cost Sums and Nominated Sub-Contracts 204
Certificates and payment 211
Remedies and powers 218
Frustration 224
War Clause 224
Settlement of disputes 228
Application to Scotland and Northern Ireland 234
Notices 235
Tax matters 236
The Construction (Design and Management) Regulations 1994 237
Special conditions 240
Short description of Works 241
Form of Tender 242
Form of Agreement 245
Form of Bond 246

Chapter 9 The FCEC Form of Sub-Contract (September 1991 Edition) 247

1. Introduction 247
The relationship between the Main Contract and the Sub-Contract 247
The relationship between the Main Works and Sub-Contract Works 248
2. Commentary on the Form of Sub-Contract 248

Chapter 10 Design and construct contracts and the ICE Design and Construct Conditions of Contract 289

1. Introduction to design and construct contracts 289
   Single point liability 289
   Constructability 290
   Simpler organisation from the employer's perspective 290
   The standard of design care 290
   Imposed design requirements 291
   Novation 292
   Collateral warranties 293
   The Construction (Design and Management) Regulations 1994 294
2. Introduction to the ICE Design and Construct Conditions of Contract 294
   Principal points of difference 295
   Terminology compared with the standard ICE Conditions of Contract 295
   The Contract 295
   The Employer's Representative 297
   The Contractor's obligations 299
   Variations 302
   Payment 304
   Dispute resolution 307

Chapter 11 The Engineering and Construction Contract (The NEC, Second Edition) 311

1. Introduction to the Engineering and Construction Contract 311
   The suite 311
   Style of drafting 312
   Core clauses 312
   Main Option clauses 312
   Secondary Option clauses 313
   Contract data 313
2. An outline of the contract 313
   The parties 313
   Administrators and decision-makers 313
   The Project Manager 313
   The obligations of the Contractor 314
   Time 314
   Payment 315

Compensation events 315
3. Commentary on selected core clauses 316
Early warning 316
Programme 319
Physical conditions 323
Disputes 326

Chapter 12 The Institution of Civil Engineers' Arbitration and Conciliation Procedures 331

1. Introduction to the ICE Arbitration Procedure (England and Wales) (1983) 331
New legislation and revised rules of procedure 331
Purpose of an arbitration procedure 332
The nature of the rules set out in the Arbitration Procedure 332
The binding effect of the Arbitration Procedure 332
2. Commentary on the ICE Arbitration Procedure (England and Wales) (1983) 333
Part A. Reference and appointment 333
Part B. Powers of the Arbitrator 339
Part C. Procedure before the Hearing 348
Part D. Procedure at the Hearing 356
Part E. After the Hearing 360
Part F. Short Procedure 365
Part G. Special Procedure for Experts 366
Part H. Interim Arbitration 368
Part J. Miscellaneous 369
3. Introduction to the ICE Conciliation Procedure (1994) 371
4. Commentary on the provisions of the Conciliation Procedure (1994) 372

Appendix 377
The Housing Grants, Construction and Regeneration Act 1996 377

Table of cases

Acsim (Southern) v. *Danish Contracting and Development Co. Ltd* (1989) 47 BLR 55 (CA) 79
Alan (W. J.) & Co. v. *El Nasr Export and Import Co.* [1972] 2 All ER 127 (CA) 41, 329
Alfred McAlpine Homes (North) Ltd v. *Property and Land Contractors Ltd* (1995) CILL 1130 89
Aluminium Industrie Vaasen v. *Romalpa Aluminium* [1976] 1 WLR 676 (CA) 53, 202
Amalgamated Building Contractors Ltd v. *Waltham Holy Cross U.D.C.* [1952] 2 All ER 452 (CA) 86
AMF International Ltd v. *Magnet Bowling Ltd* [1968] 1 WLR 1028 41, 271, 317
Anglian Water Authority v. *RDL Contracting* (1988) 43 BLR 98 45, 233
Antaios (The) (No. 2) [1984] 2 Lloyd's Rep. 235 102
Antaios Compania Naviera S.A. v. *Salen Rederierna* [1985] 1 AC 191 (HL) 21, 364
Arbitration Award No. 2 (1985), CLY 1994, 58 (Mr Hawker, arbitrator; on appeal to *Yorkshire Water Authority* v. *Sir Alfred McAlpine* (1985) 32 BLR 114 145
Arbitration Award No. 5 (1989) CLY 1994, 98 (Mr Uff QC, arbitrator; on appeal to the Court of Appeal: *Humber Oil Terminals Trustee* v. *Harbour and General* (1991) 59 BLR 1 (CA) 141
Armagas v. *Mundogas* [1986] 2 WLR 1063 (HL) 31
Armour v. *Thyssen Edalstahlwerke AG* [1991] 2 AC 339 (HL) (Scotland) 202
Ash v. *Buxted Poultry Ltd*, *The Times*, 29 November 1989 340
Ashville Investments v. *Elmer Contractors* [1989] QB 488 (CA) 24, 75, 98, 338
Ata Ul Haq v. *City Council of Nairobi* (1959) 28 BLR 76 (PC) 41
Atwell v. *Ministry of Public Buildings and Works* [1969] 1 WLR 1074 349
Aughton v. *M. F. Kent Services Ltd* (1991) 57 BLR 1 (CA) 98
Axel Johnson Petroleum AB v. *MG Mineral Group AG* [1992] 1 WLR 270 (CA) 78

Bacal v. *Northampton Development Corporation* (1975) 8 BLR 88 (CA) 140
Balcomb v. *Wards Construction (Medway) Ltd* (1980) 259 EG 765 72

Balfour Beatty Building Ltd v. *Chestermount Properties Ltd* (1993) 9 Constr. L.J. 117 — 21, 187
Balfour Beatty Construction Ltd v. *Kelston Sparkes Contractors Ltd*, CA 10th July 1996 — 269
Bank of India v. *Patel* [1983] 2 Lloyd's Rep. 298 (CA) — 54, 55
Bank of New Zealand v. *Simpson* [1900] AC 183 (PC) — 22
Barnard Pipeline Technology Ltd v. *Marston Construction Co. Ltd* (1992) CILL 743 — 26
Barnett v. *Chelsea and Kensington Hospital Management Committee* [1969] 1 QB 428 — 81
Barque Quilpué Ltd v. *Brown* [1904] 2 KB 261 (CA) — 40
Barrett (Ben) & Son (Brickwork) Ltd v. *Henry Boot Management Ltd* (1995) CILL 1026 — 98
Batty v. *Metropolitan Property Realisations Ltd* [1978] QB 554 (CA) — 72
Bell v. *Lever Brothers* [1932] AC 161 (HL) — 15
Beoco Ltd v. *Alfa Laval Co. Ltd* [1994] 3 WLR 1179 — 337
Bettini v. *Gye* (1876) 1 QBD 183 — 23
Bevan Investments Ltd v. *Brackhall & Struthers (No. 2)* [1973] 2 NZLR 45 (New Zealand Supreme Court) — 71
Bickerton v. *North West Metropolitan Regional Hospital Board* [1970] 1 WLR 607 (HL) — 293
Birch v. *Paramount Estates Ltd* (1956) 16 EG 396 — 20
Bisset v. *Wilkinson* [1927] AC 177 (PC) — 57
Blackpool and Fylde Aero Club v. *Blackpool Borough Council* [1990] 1 WLR 1195 (CA) — 16
Black Country Development Corporation v. *Kier Construction Ltd* QBD 10th July 1996 — 98
Blue Circle Industries v. *Holland Dredging Co. (UK) Ltd* (1987) 37 BLR 40 — 39, 43, 303
Blue Flame Mechanical Services Ltd v. *David Lord Engineering Ltd* (1992) CILL 760 — 101
Board of Governors of the Hospital for Sick Children v. *McLaughlin & Harvey plc* (1987) 6 Constr. L.J. 245 — 91
Bolam v. *Friern Hospital Management Committee* [1957] 1 WLR 582 — 69
Bonnell's Electrical Contractors v. *London Underground* (1995) CILL 1110 — 4
Bottoms v. *Mayor of York* (1892) *Hudson's Building Contracts* (4th edn) Vol. 2, p. 208 — 39
Bowen v. *Paramount Builders (Hamilton) Ltd* [1977] 1 NZLR 394 (New Zealand Court of Appeal) — 72
Bradley Egg Farm v. *Clifford* [1943] 2 All ER 378 (CA) — 30

Bremer Handelgesellschaft mbH v. *Westzucker GmbH* [1981] 2 Lloyd's Rep. 130 (CA) 363
Bremer Vulkan Schiffbau und Maschinenfabrik v. *South India Shipping Corporation* [1981] AC 909 (HL) 103, 332, 339
Bristol Corporation v. *Aird* [1913] AC 241 (HL) 125
British and Commonwealth Holdings v. *Quadrex Holdings* [1989] QB 842 (CA) 107, 356
British Steel Corporation v. *Cleveland Bridge and Engineering Co. Ltd* [1984] 1 All ER 504 17, 61, 248
British Waggon Co. v. *Lea & Co.* (1880) 5 QBD 149 47
British Westinghouse Electric and Manufacturing Co. v. *Underground Electric Railways of London* [1912] AC 673 (HL) 82
Brodie v. *Cardiff Corporation* [1919] AC 337 (HL) 21, 43, 120, 194
Brogden v. *Metropolitan Railway Co.* (1877) 2 App Cas 666 (HL) 18
Bryan (Allan) v. *Judith Anne Maloney* (1995) 74 BLR 35 (Australian High Court) 13
Butler Machine Tool Co. Ltd v. *Ex-Cell O Corporation (England) Ltd* [1979] 1 WLR 401 17

Cammell Laird & Co. Ltd v. *Manganese Bronze & Brass Co. Ltd* [1934] AC 402 (HL) 39
Campbell v. *Edwards* [1976] 1 All ER 785 95
Cana Construction v. *The Queen* (1973) 21 BLR 12 (Canadian Supreme Court) 16
Canada (The Queen in Right of) v. *Walter Cabott Construction Ltd* (1977) 21 BLR 46 40
Candler v. *Crane, Christmas & Co.* [1951] 2 KB 164 (CA) 73
Caparo Industries plc v. *Dickman* [1990] 2 AC 605 (HL) 13, 73
Carlisle Place Investments Ltd v. *Wimpey Construction (UK) Ltd* (1980) 15 BLR 109 104, 356
Carr v. *J. A. Berriman Pty Ltd* (1953) 27 ALJ 273 23, 28
Catherine L. (The) [1982] 1 Lloyd's Rep. 484 345
Cathery v. *Lithodmos Ltd* (1987) 41 BLR 76 (CA) 343
Central Provident Fund Board v. *Ho Bock Kee* (1981) 17 BLR 21 (Singapore) 52
Ceylon (Government of) v. *Chandris* [1963] 1 Lloyd's Rep. 214 361
Channel Tunnel Group v. *Balfour Beatty* [1993] AC 334 (HL) 101
Chatbrown v. *Alfred McAlpine Construction (Southern)* (1987) 35 BLR 44 (CA) 107
Chermar Productions Pty Ltd v. *Pretest Pty Ltd* (1991) 8 Constr. L.J. 44 (Supreme Court of Victoria) 51
Chester Grosvenor Hotel v. *Alfred McAlpine* (1991) 56 BLR 115 26, 251

Chichester Joinery v. *John Mowlem* (1987) 42 BLR 100 — 1
Christiani & Nielsen Ltd v. *Birmingham City Council* (1994) CILL 1014 — 334, 33
Comco Construction Pty Ltd v. *Westminster Property Pty Ltd* (1990) 8 Constr. L.J. 49 (Supreme Court of Western Australia) — 5
Comdel Commodities Ltd v. *Siporex Trade SA* [1990] 2 All ER 552 — 33
Commission for the New Towns v. *Cooper (Great Britain) Ltd* [1995] 2 WLR 677 (CA) — 2
Commission of the European Communities v. *Ireland* (1988) 44 BLR 1 —
Companie Interafricaine de Travaux (Comiat) v. *South African Transport Services* (1991) CLY 1994, 149 (South African Supreme Court) — 14
Cork v. *Kirby MacLean Ltd* [1952] 2 All ER 402 (CA) — 8
Costain Building and Civil Engineering Ltd v. *Scottish Rugby Union plc* (1993) 69 BLR 85 (Scottish Court of Session, Inner House) — 99, 21
Crabtree (B. J.) (Insulation) Ltd v *GPT Communication Systems Ltd* (1989) 59 BLR 43 (CA) — 34
Crawford v. *Charing Cross Hospital* (1953) *The Times*, 8 December — 71
Cropper v. *Smith* (1884) 26 ChD 700 — 78
Crosby (J.) & Sons v. *Portland Urban District Council* (1967) 5 BLR 121 (DC) — 77
Croshaw v. *Pritchard* (1899) 16 TLR 45 — 16
Croudace v. *London Borough of Lambeth* (1986) 33 BLR 20 (CA) — 45, 101, 107, 122
Crown Estate Commissioners v. *John Mowlem & Co.* (1994) 70 BLR 1 (CA) — 334
Currie v. *Misa* [1875] LR 10 Ex. 153 at 163 — 19

D. & C. Builders v. *Rees* [1966] 2 QB 617 (CA) — 27, 93
D. & F. Estates v. *Church Commissioners* [1989] AC 177 (HL) — 126
Dalton v. *Angus* (1881) 6 App Cas 740 (HL) — 126
Darlington Borough Council v. *Wiltshier Northern Ltd* [1995] 1 WLR 68 (CA) — 11
Davis Contractors Ltd v. *Fareham Urban District Council* [1956] AC 696 (HL) — 29
Davy Offshore Ltd v. *Emerald Field Contracting Ltd* (1991) 55 BLR 1 — 21
Dawber Williamson v. *Humberside County Council* (1979) 14 BLR 70 — 53

De Lassalle v. *Guildford* [1901] 2 KB 215 (CA) 63
Derry v. *Peek* (1889) 14 App Cas 337 (HL) 58
Dew & Co. Ltd v. *Tarmac Construction* (1978) 15 BLR 22 101
Dillingham Ltd v. *Downs* [1972] 2 NSWLR 49 (Supreme Court of New South Wales) 137
Donoghue v. *Stevenson* [1932] AC 562 (HL) 13
Douglas (R. M.) v. *Bass* (1990) 53 BLR 119 101
Dunlop Pneumatic Tyre Co. Ltd v. *New Garage and Motor Company* [1915] AC 79 (HL) 82
Dunlop Pneumatic Tyre Co. Ltd v. *Selfridge & Co. Ltd* [1915] AC 847 (HL) 12

Eagle Star Insurance v. *Yuval Insurance* [1978] 1 Lloyd's Rep. 357 101
East v. *Maurer* [1991] 1 WLR 461 (CA) 80
East Ham Borough Council v. *Bernard Sunley & Sons Ltd* [1966] AC 406 (HL) 41, 73, 90, 317
Eckersley and others v. *Binnie & Partners and others* (1988) CILL 388 (CA) 68, 69, 70, 71
Edgeworth Construction Ltd v. *N. D. Lea and Associates Ltd* (1993) (Supreme Court of Canada) 66 BLR 56 12
Edward Owen Engineering Ltd v. *Barclays Bank International Ltd* [1978] 1 QB 159 (CA) 55, 56
Edwin Hill & Partners v. *Leakcliffe Properties Ltd* (1984) 29 BLR 43 68
EEC Quarries Ltd v. *Merriman Ltd* (1988) 45 BLR 90 233
Ellis v. *Sheffield Gas Consumers Co.* (1853) 2 E & B 767 126
Ellis Mechanical Services Ltd v. *Wates Construction Ltd* (1976) 2 BLR 57 (CA) 46
Elissar (The) [1984] 2 Lloyd's Rep. 84 (CA) 102
Ellis-Don Ltd v. *The Parking Authority of Toronto* (1978) 28 BLR 98 (Supreme Court of Ontario) 40
Ellison v. *Bray* (1864) 9 LT 730 347
Elmes v. *Burgh Market Co.* (1891) *Hudson's Building Contracts* (4th edn) Vol. 2, p. 170 45
Enco Civil Engineering Ltd v. *Zeus International Development Ltd* (1991) 8 Constr. L.J. 164 101
English Industrial Estates v. *Kier* (1991) 56 BLR 93 194
English Industrial Estates v. *Wimpey* [1973] 1 Lloyd's Rep. 118 (CA) 127
Erith Contractor Ltd v. *Costain Civil Engineering Ltd* (1993) (unreported) 286
Esal (Commodities) Ltd v. *Oriental Credit Ltd* [1985] 2 Lloyd's Rep. 546 (CA) 55

Esso Petroleum Co. Ltd v. *Mardon* [1976] QB 801 (CA) 21, 5?
Evans, Davis and Caddick (Re) (1870) 22 LT 507 9?

Fairclough Building v. *Rhuddlan Borough Council* (1985) 30 BLR 26 (CA) 1?
Fairweather (H.) & Co. Ltd v. *London Borough of Wandsworth* (1987) 39 BLR 106 8?
Falcke v. *Scottish Imperial Insurance Co.* (1886) 34 ChD 235 6?
Felthouse v. *Bindley* (1862) 11 CB (NS) 869 1?
Fernbrook Trading Co. Ltd v. *Taggart* [1979] 1 NZLR 556 (New Zealand Supreme Court) 38, 85, 86 18?
Fidelitas Shipping Co. Ltd v. *V/O Exportchleb* [1966] 1 QB 630 (CA) 332, 35?
Finnegan v. *Sheffield City Council* (1988) 43 BLR 124 8?
Flender Werft v. *Aegean Maritime* [1990] 2 Lloyd's Rep. 27 342, 34?
Ford and Bemrose (Re) (1902), *Hudson's Building Contracts* (4th edn) Vol. 2, p. 324 (CA) 3?
Forman & Co. Proprietary Ltd v. *The Ship 'Liddlesdale'* [1900] AC 190 (PC) 39, 41, 4?
Fordgate Bingley Ltd v. *Argyll Stores Ltd* [1994] 39 EG 135 (2 EGLR 84) 97
Forsyth v. *Ruxley Electronics Ltd* [1995] 3 WLR 118 (HL) 82
Fox v. *Wellfair* [1981] 2 Lloyd's Rep. 514 (CA) 104, 347, 358
Freeman v. *Hensler* (1900) *Hudson's Building Contracts* (4th edn) Vol. 2, p. 292 (CA) 88
Frost v. *Moody Homes Ltd & others* (1989) CILL 504 91

Gilbert & Partners v. *Knight* [1968] 2 All ER 248 (CA) 67
Gilbert-Ash (Northern) v. *Modern Engineering (Bristol)* [1974] AC 689 (HL) 37, 46
Glasgow and South Western Railway v. *Boyd & Forrest* [1915] AC 526 (HL) 58
Gleeson (M. J.) Group plc v. *Wyatt of Snetterton Ltd* (1994) 72 BLR 15 (CA) 286
Glenlion Construction v. *Guinness Trust* (1988) 39 BLR 89 36, 40, 130
Gloucestershire County Council v. *Richardson* [1969] 1 AC 480 (HL) 14, 33, 38
Glynn v. *Margetson* [1893] AC 351 (HL) 127
GMTC Tools and Equipment Ltd v. *Yuasa Warwick Machinery Ltd* (1994) 73 BLR 102 (CA) 76
Gran Gelato v. *Richcliff* [1992] ChD 560 80
Gray v. *T. P. Bennett & Son* (1987) 43 BLR 63 92

Greater London Council v. *Cleveland Bridge Engineering Co. Ltd* (1984) 34 BLR 50 — 36, 38, 42, 150, 261
Greaves & Co (Contractors Ltd) v. *Baynham Meikle & Partners* [1975] 1 WLR 1095 (CA) — 39, 66, 290, 291

Hadley v. *Baxendale* (1854) 9 Ex 341 — 81, 255
Hanak v. *Green* [1958] 2 QB 9 (CA) — 79
Harbour Assurance v. *Kansa* [1993] QB 701 (CA) — 338
Hargreaves (B.) Ltd v. *Action 2000* (1992) 62 BLR 72 (CA) — 78
Harrison (M.) & Co. (Leeds) Ltd v. *Leeds City Council* (1980) 14 BLR 123 — 17
Harvela Investments Ltd v. *Royal Trust Company of Canada Trust* [1986] AC 207 (HL) — 16
Hawkins v. *Chrysler (UK) Ltd and Byrne Associates* (1986) 38 BLR 36 — 290
Hayter v. *Nelson* [1990] 2 Lloyd's Rep. 265 — 102, 107
Hedley Byrne v. *Heller and Partners* [1964] AC 465 (HL) — 13, 59
Helstan Securities Ltd v. *Hertfordshire County Council* [1978] 3 All ER 262 — 13, 49, 124
Henderson v. *Merrett* [1994] 3 WLR 761 (HL) — 13, 92
Henry Boot Construction Ltd v. *Central Lancashire Development Corporation* (1980) 15 BLR 8 — 87
Herkules Piling and Hercules Piling v. *Tilbury Construction* (1992) CILL 770 — 49
Heyman v. *Darwins Ltd* [1942] AC 365 (HL) — 98, 338
Hickman v. *Roberts* [1913] AC 229 (HL) — 45
Higgs & Hill Building Ltd v. *Campbell Davis Ltd* (1982) 28 BLR 47 — 97
Holland Dredging v. *Dredging and Construction Co.* (1987) 37 BLR 1 (CA) — 37, 133, 145
Holme v. *Guppy* (1838) 3 M & W 387 — 38
Honeywill and Stein Ltd v. *Larkin (London's Commercial Photographers) Ltd* [1933] All ER 77 (CA) — 126
Hong Kong Fir Shipping Co. Ltd v. *Kawasaki Kisen Kaisha* [1962] 2 QB 26 (CA) — 23
Hooker Constructions v. *Chris's Engineering Contracting Company* [1970] ALR 821 (Supreme Court of the Northern Territory of Australia) — 223, 262
Hounslow (London Borough of) v. *Twickenham Garden Developments Ltd* [1971] ChD 233 — 28, 42, 45, 51, 52, 223
Howard de Walden Estates Ltd v. *Costain Management Design Ltd* (1991) 55 BLR 124 — 44, 195
Humber Oil Terminals Trustee v. *Harbour and General* (1991) 59 BLR 1 (CA) — 133, 140, 141

Imodco v. *Wimpey and Taylor Woodrow* (1987) 40 BLR 1 (CA) 101, 10
Independent Broadcasting Authority v. *EMI Electronics Ltd and BICC Construction Ltd* (1980) 14 BLR 1 (HL) 12, 39, 71 291, 30
Investors in Industry v. *South Bedfordshire District Council* [1986] QB 1034 6

Jackson v. *Horizon Holidays Ltd* [1975] 3 All ER (CA) 1
Jacobs v. *Morton and Partners* (1994) 72 BLR 92 1
JMJ Contractors v. *Marples Ridgeway* (1985) 31 BLR 104
Joceleyne (The) [1977] 2 Lloyd's Rep. 121 33
Jones v. *Sherwood Computer Services* [1989] EGCS 172 9
Jordernson & Co. v. *Stara, Kopperbergs Bergslag Atkiebolag* [1931] 41 Lloyd's Rep. 201 10
Joscelyne v. *Nissen* [1970] 2 QB 86 (CA) 2

Kaye (P. and M.) v. *Hosier & Dickinson Ltd* [1972] 1 All ER 121 (HL) 9
Kelly Pipelines v. *British Gas* (1989) 48 BLR 126
Kingston-upon-Hull Corporation v. *Harding* [1892] 2 QB 494 (CA) 5
Kitsons Sheet Metal v. *Matthew Hall* (1989) 47 BLR 90 4
Kostas Melas (The) [1981] 1 Lloyd's Rep. 18 35
Kuwait Foreign Trading v. *Icori Estero SpA* [1993] ADRLJ 167 10
Lakers Mechanical Services v. *Boskalis Westminster* (1989) 5 Constr. L.J. 139 125
Leage (The) [1984] 2 Lloyd's Rep. 259 49
Lee v. *Showmen's Guild* [1952] 2 QB 329 (CA) 95
Leedsford Ltd v. *Bradford City Council* (1956) 24 BLR 45 (CA) 48, 153, 182, 253
Leggot v. *Barrett* (1880) 15 ChD 306 (CA) 249
Lep Air Services Ltd v. *Rolloswin* [1973] AC 331 (HL) 55
Leyland Shipping Co. v. *Norwich Union Fire Insurance Society* [1918] AC 350 (HL) 85
Linden Gardens Trust v. *Linesta Sludge Disposals Ltd and others* [1994] 1 AC 85 (HL) 13, 49, 124
Lindenburg v. *Joe Canning* (1992) 9 Constr. L.J. 43 4
Lloyds Bank v. *E. B. Savory & Co.* [1933] AC 201 (HL) 70
London, Chatham and Dover Railways v. *South Eastern Railway* [1893] AC 429 (HL) 84
Lorne Stewart Ltd v. *William Sindall plc and NW Thames Regional Health Authority* (1986) 35 BLR 109 98
LRE Engineering v. *Otto Simon Carves* (1981) 24 BLR 127 40
Lubenham Fidelities v. *South Pembrokeshire District Council* (1986) 33 BLR 39 (CA) 46

Luxor v. *Cooper* [1941] AC 108 (HL) 324

McAlpine Humberoak v. *McDermott International* (1992) 58 BLR 1 (CA) 29, 38, 187
McLaren Maycroft Co. v. *Fletcher Development Co. Ltd* [1973] 2 NZLR 100 (New Zealand Court of Appeal) 70, 73
McLaughlin & Harvey plc v. *P & O Developments* (1991) 55 BLR 101 232
McRae v. *Commonwealth Disposals Commission* (1950) 84 CLR 377 (High Court of Australia) 292
Martin Grant & Co. Ltd v. *Sir Lindsay Parkinson & Co. Ltd* (1984) 29 BLR 31 (CA) 40
Matto v. *Rodney Brown Associates* [1994] 41 EG 152 (2 EGLR 163) 72
May & Butcher v. *R.* [1934] 2 KB 17 (HL) 96, 293
Mayfield Holdings Ltd v. *Moana Reef Ltd* [1973] 1 NZLR 309 (New Zealand Supreme Court) 28
Maynard v. *West Midlands Regional Area Health Authority* [1984] 1 WLR 634 (HL) 70, 71
Mears Construction v. *Samuel Williams* (1977) 16 BLR 49 218
Mediterranean & Eastern Export Co. Ltd v. *Fortress Fabrics (Manchester) Ltd* [1948] 2 All ER 186 104, 347
Mercers v. *New Hampshire Insurance* (1992) 60 BLR 26 (CA) 55
Mersey Steel and Iron Co. v. *Naylor, Benzon & Co.* [1884] 9 App Cas 434 (HL) 28
Merton (London Borough of) v. *Stanley Hugh Leach* (1985) 32 BLR 51 36, 46, 47, 77, 120, 130, 299
Merton (London Borough of) v. *Lowe* (1982) 18 BLR 130 68, 91
Metropolitan Water Board v. *Dick, Kerr and Co. Ltd* [1918] AC 119 (HL) 29
Meyer v. *Gilmer* (1899) 19 NZLR 129 (New Zealand) 43
Mid Glamorgan County Council v. *J. Devonald Williams & Partners* (1991) 8 Constr. L.J. 61 77
Minter (F. G.) Ltd v *WHTSO* (1980) 13 BLR 1 (CA) 84, 117
Mitsui v. *AG of Hong Kong* (1986) 33 BLR 1 (PC) 324
Modern Engineering (Bristol) Ltd v. *C. Miskin & Son Ltd* [1981] 1 Lloyd's Rep. 135 (CA) 102
Modern Engineering (Bristol) v. *Gilbert-Ash* [1974] AC 689 (HL) 79
Mondel v. *Steel* (1841) 8 M & W 858 79
Monmouthshire County Council v. *Costelloe and Kemple* (1965) 5 BLR 83 (CA) 46, 234
Moneypenny v. *Hartland* (1824) 2 C & P 378 72, 137
Moorcock (The) (1889) 14 PD 64 24

Moresk Cleaners v. *Hicks* [1966] 2 Lloyd's Rep. 338 48, 69, 12(
Morrison-Knudsen International v. *Commonwealth of Australia* (1972) 13 BLR 114 (High Court of Australia) 13:
Mount Charlotte Investments plc v. *Prudential Assurance* [1995] 10 EG 129 (1 EGLR 15) 10·
Mowlem (John) & Co. v. *Eagle Star* (1995) CILL 1047 25{
Multi-Construction (Southern) Ltd v. *Stent Foundations Ltd* (1988) 41 BLR 98 9:
Murphy v. *Brentwood* [1991] 1 AC 398 (HL) 1:

National Coal Board v. *William Neill & Sons (St Helens) Ltd* (1983) 26 BLR 81 4
Neck v. *Taylor* [1893] 1 QB 560 (CA) 34:
Nema (The) [1981] 2 Lloyd's Rep. 239 10:
Nema (The) [1982] AC 724 (HL) 36·
Neodox v. *Swinton and Pendlebury Borough Council* (1958) 5 BLR 34 40, 13(
New Fenix Compagnie Anonyme v. *General Accident* [1911] 2 KB 619 (CA) 34:
Nicolene v. *Simmonds* [1953] 1 QB 543 (CA) 1{
Nimenia (The) [1986] QB 802 36:
Norjarl (K/S) v. *Hyundai Heavy Industries Co. Ltd* [1992] 1 QB 863 33(
Norta Wallpapers (Ireland) Ltd v. *John Sisk & Sons (Dublin) Ltd* (1978) 14 BLR 99 14
North Ocean Shipping Co. Ltd v. *Hyundai Construction Co. Ltd* [1979] QB 705 18, 93
North West Metropolitan Regional Hospital Board v. *T. A. Bickerton & Son* [1970] 1 WLR 607 (HL) 14
North West Water Authority v. *Binnie & Partners* (1989, December) QB unreported 69
Northern Regional Health Authority v. *Crouch* [1984] QB 644 (CA) 46, 98, 99, 102, 125, 309, 330
Nova (Jersey) Knit v. *Kammgarn Spinnerei GmbH* [1977] 1 WLR 713 (HL) 101
Nye Saunders & Partners v. *Alan E. Bristow* (1987) 37 BLR 92 69

Ogilvie Builders Ltd v. *Glasgow District Council* (1994) 68 BLR 122 (Scottish Court of Session) 84, 117
Owen v. *Nicholl* [1948] 1 All ER 707 (CA) 104, 360

Pacific Associates v. *Baxter* [1990] 1 QB 993 (CA) 12, 47
Panamena v. *Frederick Leyland & Co.* [1947] AC 428 (HL) 45, 46
Parkinson (Sir Lindsay) & Co. v. *Commissioners of Works* [1949] 2 KB 632 (CA) 42, 61, 248

Parkinson (Sir Lindsay) & Co. v. *Triplan* [1973] 1 QB 609 (CA) 341, 342
Patman and Fotheringham v. *Pilditch* (1904), *Hudson's Building Contracts* (4th edn) Vol. 2, p. 368 34
Payzu Ltd v. *Saunders* [1919] 2 KB 581 82
Peak Construction (Liverpool) v. *McKinney Foundations* (1970) 1 BLR 111 (CA) 38, 85, 180
Pearce (C. J.) & Co. Ltd v. *Hereford Corporation* (1968) 66 LGR 647 141
Pearson v. *Dublin Corporation* [1907] AC 351 (HL) 58
Perar BV v. *General Surety and Guarantee Co. Ltd* (1994) 66 BLR 72 (CA) 53
Percy Bilton v. *Greater London Council* [1982] 1 WLR 794 (HL) 14
Perini Pacific Ltd v. *Greater Vancouver Sewerage and Drainage District* (1966) 57 DLR (2d) 307 (British Columbia Court of Appeal) 85, 180
Philipps v. *Philipps* (1878) 4 QBD 127 77
Philips Hong Kong Ltd v. *AG of Hong Kong* (1993) 9 Constr. L.J. 202 (PC) 83
Photo Production v. *Securicor Transport* [1980] AC 827 (HL) 25, 26
Piggott Foundations Ltd v. *Shepherd Construction Ltd* (1993) 67 BLR 48 262
Pillings (C. M.) & Co. v. *Kent Investments Ltd* (1985) 30 BLR 80 46
Piper Double Glazing Limited v. *DC Contracts* [1994] 1 WLR 777 346
Porzelack K. G. v. *Porzelack UK Ltd* [1987] 1 WLR 420 342
Post Office v. *Norwich Union Fire Insurance Society Ltd* [1967] 2 QB 363 51
Pratt v. *Swanmore Builders Ltd and Baker* [1980] 2 Lloyd's Rep. 504 103, 371
Prenn v. *Simmonds* [1971] 1 WLR 1381 (HL) 22
President of India v. *Lips Maritime* [1988] AC 395 (HL) 84
Priestly v. *Stone* (1888), *Hudson's Building Contracts* (4th edn) Vol. 2, p. 134 34
Proctor v. *Bennis* [1887] ChD 740 62

R. v. *Demers* [1900] AC 103 (PC) 4
R. v. *Silverlock* [1894] 2 QB 766 (Crown Cases Reserved) 360
Radford v. *De Froberville* [1977] 1 WLR 1262 82
Ramsden v. *Dyson* (1866) LR 1 HL 129 62
Ranger v. *Great Western Railway* (1854) 5 HLC (HL) 52
Rayner (J. H.) (Mincing Lane) Ltd v. *Shaher Trading Co.* [1982] 1 Lloyd's Rep. 632 364
Rees & Kirby v. *Swansea City Council* (1985) 30 BLR 1 (CA) 84, 117
Rees Hough Ltd v. *Redland Reinforced Plastics Ltd* (1984) 27 BLR 136 26

Regalian Properties plc v. *London Docklands Development Corporation* [1995] 1 WLR 212 — 17, 61

Richard Roberts Holdings Ltd and Another v. *Douglas Smith Stimson Partnership* (1988) 46 BLR 50 — 67, 91

Richo International Ltd v. *Industrial Food Co. SAL, The Fayrouz III* [1989] 2 Lloyd's Rep. 10 — 34

Roberts v. *Bury Commissioners* (1870) LR 5 CP — 5

Roberts & Co. v. *Leicestershire County Council* [1961] ChD 555 — 2

Robinson v. *Harman* (1848) 1 Ex 850 — 8

Rolimpex Centrala Handlu Zagranicznego v. *Haji E. Dossa & Sons Ltd* [1971] 1 Lloyd's Rep. 380 — 361

Rose and Frank Co. v. *Crompton Brothers* [1925] AC 445 (HL) — 1

Rosehaugh Stanhope v. *Redpath Dorman Long* (1990) 50 BLR 75 (CA) — 2

Roxburghe v. *Cox* (1881) 17 ChD 520 (CA) — 13, 4

Royscot Trust Ltd v. *Rogerson* [1991] 2 QB 297 — 8

Ruxley Electronics v. *Forsyth* [1995] 2 WLR 118 (HL) — 4

Sainsbury (H. R. & S.) Ltd v. *Street* [1972] 3 All ER 1127 — 29

Salisbury v. *Woodland* [1970] 1 QB 324 (CA) — 12

Scammel v. *Ouston* [1941] AC 251 (HL) — 1

Schawel v. *Reade* [1913] IR 81 — 21

Schuler AG v. *Wickman Machine Tool Sales Ltd* [1974] AC 235 (HL) — 2

Scott v. *Avery* (1856) 5 HLC 811 (HL) — 101, 30

Scott v. *Mercantile Accident Insurance Co.* (1892) 8 TLR — 340

Scott Lithgow Ltd v. *Secretary of State for Defence* (1989) 45 BLR 6 (HL) — 258, 262

Sealand of the Pacific v. *Robert C. McHaffie Ltd* (1974) 51 DLR (3d) 702 (British Columbia Court of Appeal) — 72, 137

Secretary of State for Transport v. *Birse-Farr Joint Venture* (1993) 62 BLR 36 — 217

Selectmove Ltd (Re) [1995] 1 WLR 474 — 19

Shanklin Pier Ltd v. *Detel Products* [1951] 2 KB 854 — 12

Shanks & McEwan v. *Strathclyde Regional Council* (1994) CILL 916 (Scottish Court of Session) — 41, 194

Sharpe v. *E. T. Sweeting & Son Ltd* [1963] 1 WLR 665 — 13

Sharpe v. *San Paulo Railway Co.* (1873) LR 8 Ch App 597 — 34, 39, 42, 95

Shearson Lehman Brothers Inc. v. *Maclaine Watson & Co. Ltd* [1987] WLR 480 at 489 (CA) — 107

Sheldon v. *R. H. M. Outhwaite (Underwriting Agencies) Ltd* [1996] 1 AC 102 (HL) — 92

Sherry (Re) (1884) 25 ChD 692 (CA) 55
Sidney Kaye, Eric Firmin & Partners v. *Leon Joseph Bronesky* (1973) 4 BLR 1 65
Simaan General Contracting Co. v. *Pilkington Glass Ltd.* [1987] 1 WLR 516 342
Simplex Concrete Piles Ltd v. *St Pancras Borough Council* (1958) 14 BLR 80 44, 143, 195
Smallman Construction v. *Redpath Dorman Long* (1989) 47 BLR 41 (CA) 107
Smeaton Hanscombe & Co. v. *Sassoon I. Setty, Son & Co.* (No. 2) [1953] 1 WLR 1481 341
Smith (Re) (1896) LT 46 342
Solholt (The) [1983] 1 Lloyd's Rep. 605 82
South Shropshire District Council v. *Amos* [1986] 1 WLR 1271 (CA) 351
Southway Group Ltd v. *Wolff and Wolff* (1991) 57 BLR 33 18, 48
Spencer v. *Harding* (1870) LR 5 CP 561 16
Stevenson v. *McLean* (1880) 5 QBD 346 16
Stockport MBC v. *O'Reilly* [1978] 1 Lloyd's Rep. 595 42
Stockport MBC v. *O'Reilly* [1983] 2 Lloyd's Rep. 70 102
Sumpter v. *Hedges* [1898] 1 QB 673 37
Sutcliffe v. *Thackrah* [1974] AC 727 (HL) 45, 47, 121

Tai Hing Cotton Mill v. *Liu Chong Haing Bank* [1986] AC 80 (PC) 92
Tamplin v. *James* (1880) 15 ChD 215 (CA) 15
Tara Civil Engineering Ltd v. *Moorfield Developments Ltd* (1989) 46 BLR 72 28, 52, 222
Taunton-Collins v. *Cromie* [1964] 1 WLR 633 (CA) 101
Taylor v. *Caldwell* (1863) 32 LJQB 29
Temloc v. *Errill Properties Ltd* (1988) 39 BLR 30 (CA) 83, 181, 187
Themehelp Ltd v. *West and others* [1995] 3 WLR 751 (CA) 54
Thomas v. *Hammersmith Borough Council* [1938] 3 All ER 201 68
Thomas Feather & Co. (Bradford) Ltd v. *Keighley Corporation* [1953] 53 LGR 30 28
Thorn v. *Mayor and Commonalty of London* [1876] 1 AC 120 (HL) 39, 43
Toepfer v. *Warinco* [1978] 2 Lloyd's Rep. 569 31, 41, 121
Token Construction v. *Charlton Estates* (1973) 1 BLR 48 (CA) 46
Top Shop Estates v. *Danino* (1984) 273 EG 197 (1 EGLR 9) 104, 360
Town & City Properties (Developments) Ltd v. *Wiltshier Southern Ltd and Gilbert Powell* (1988) 44 BLR 109 103, 357, 360
Trade Fortitude (The) [1992] 1 Lloyd's Rep. 169 355

Trafalgar House Construction v. *General Surety & Guarantee Co. Ltd* [1996] 1 AC 199 (HL) 55
Trafalgar House Construction (Regions) Ltd v. *General Surety & Guarantee Co. Ltd* [1995] 3 WLR 204 (HL) 56, 135
Tramountana Armadora v. *Atlantic Shipping Co.* [1978] 1 Lloyd's Rep. 391 341
Trollope & Colls Ltd v. *Atomic Power Construction Ltd* [1963] 1 WLR 333 61
Tucker v. *Linger* (1883) 8 App Cas 508 (HL) 25
Turner v. *Fenton* [1982] 1 WLR 52 101
Turriff Ltd v. *Welsh National Water Development Authority* (1979) CLY 1994, 122 145
Tweddle v. *Atkinson* (1861) 1 B & S 393 11

Unione Stearinerie Lanza v. *Wiener* [1917] 2 KB 558 332
United Scientific Holdings v. *Burnley Council* [1978] AC 904 (HL) 135
Universal Petroleum Co. Ltd v. *Handels und Transportgesellschaft mbH* [1987] 2 All ER 737 (CA) 363
University of Warwick v. *Sir Robert McAlpine* (1988) 42 BLR 1 360

Vimeira (The) [1984] 2 Lloyd's Rep. 66 104
Vitol SA v. *Norelf Ltd* [1995] 3 WLR 549 (CA) 28, 223

Wadsworth v. *Lydall* [1981] 1 WLR 598 (CA) 84
Walters v. *Whessoe Ltd & Shell Refining Co. Ltd* (1960) 6 BLR 23 (CA) 50
Walton-on-the-Naze UDC (Re) v. *Moran* (1905) *Hudson's Building Contracts* (4th edn) Vol. 2, 376 35
Waring (F. R.) (UK) Ltd v. *Administracao Geral do Acucar e do Alcool* [1983] 1 Lloyd's Rep. 45 104, 347
Way v. *Latilla* [1937] 3 All ER 759 62
Wells v. *Army and Navy Co-operative Society* (1902) 86 LT 764 38
West Faulkner Associates v. *London Borough of Newham* (1994) 71 BLR 1 (CA) 223, 262
Westminster Corporation v. *Jarvis* [1970] 1 WLR 637 (HL) 189
Wharf Properties v. *Eric Cumine Associates* (1991) 52 BLR 1 (PC) 77
White & Carter (Councils) Ltd v. *McGregor* [1962] AC 413 (HL) 28, 42
Whitehouse v. *Jordan* [1981] 1 WLR 246 (HL) 360
Wigan Metropolitan Borough Council v. *Sharkey Bros Ltd* (1987) 4 Constr. L.J. 162 334
William Lacey (Hounslow) Ltd v. *Davis* [1957] 1 WLR 932 61

William Sindall plc v. *Cambridgeshire County Council* [1994] 1 WLR 1016 (CA) 80, 136
Williams v. *Fitzmaurice* (1858) 3 H & N 844 34, 39, 42, 132, 251, 301
Williams v. *Natural Life Health Foods*, *Independent*, 18 January 1996 30
Williams and Roffey Bros v. *Nicholls (Contractors) Ltd* [1991] 1 QB 1 19
Wilsher v. *Essex Area Health Authority* [1987] 2 WLR 425 (CA) 70
Wilson Smithett v. *Bangladesh Sugar* [1986] 1 Lloyd's Rep. 378 17
Winnipeg Condominium Corporation v. *Bird Construction Co. Ltd* (1995) 74 BLR 1 (Supreme Court of Canada) 13
Woodar Investment Development Ltd v. *Wimpey Construction UK Ltd* [1980] 1 WLR 571 (HL) 28
Workman, Clark & Co. v. *Lloyd Brazileno* [1908] 1 KB 968 (CA) 46

Yonge v. *Toynbee* [1910] 1 KB 215 (CA) 31
Yorkshire Water Authority v. *McAlpine* (1985) 32 BLR 114 37, 133, 145, 321
Young & Marten v. *McManus Childs* [1969] 1 AC 454 (HL) 33, 38, 173, 291, 301

Z Ltd v. *A–Z* [1982] QB 558 340

Table of statutes and secondary legislation

Law of Property Act 1925

s. 53(1) 93
s. 136(1) 49

Third Party (Rights against Insurers) Act 1930

General 51

Law Reform (Frustrated Contracts) Act 1943

General 30

Arbitration Act 1950 84, 100, 338, 341

s. 4 101
s. 12 103, 339, 340, 351
s. 14 355
s. 19 361
s. 19A 279
s. 23 102
s. 27 232, 334
s. 32 98

Occupier's Liability Act 1957

General 117, 133, 156

Misrepresentation Act 1967 58

s. 2(1) 59, 80
s. 2(2) 59, 80

Health and Safety at Work etc. Act 1974

General 9, 73, 133, 145, 156

Control of Pollution Act 1974

General 9

Arbitration Act 1975 100

s. 1 102, 107

Unfair Contract Terms Act 1977

General 13, 25, 26, 73, 251

Arbitration Act 1979 100, 364

s. 1 102, 361, 364
s. 3 102, 348, 362, 364

Limitation Act 1980 26

s. 5 91
s. 8 91
s. 32 92

Highways Act 1980

General 9, 166

Supreme Court Act 1981

General 84
s. 35A 78

Supply of Goods and Services Act 1982 37, 66

ss. 13–15 24, 66

Occupier's Liability Act 1984

General 117, 133, 156

Latent Damage Act 1986

General 92

Contracts (Applicable Law) Act 1990

General 8

Courts and Legal Services Act 1990

s. 100 99, 100, 330

Environmental Protection Act 1990

General 9

Public Supply Contracts Regulations 1991, SI 1991, No. 2679

General 8

Public Works Contracts Regulations 1991, SI 1991, No. 2680

General 8

New Roads and Street Works Act 1991

General 9, 163, 164

Utilities Supply and Works Contracts Regulations 1992, SI 1991, No. 3279

General 8

Public Services Contracts Regulations 1993, SI 1993, No. 3228

General 8

Construction (Design and Management) Regulations 1994

General 9, 109, 133, 145, 149, 156, 237, 238, 239, 259, 294, 321
Reg. 13 73, 239, 294
Reg. 15 150, 239
Reg. 16 156, 239
Reg. 21 74, 156, 294

Arbitration Act 1996

s. 5 98
s. 9 100
(4) 102
s. 12 232, 334
s. 30 338
s. 34 331, 339
(1)(g) 103
(2)(d) 351
(f) 354
(g) 360
ss. 35, 36 103, 331, 339
s. 37 103, 331, 339
(1)(a)(i) 347
(ii) 346

s. 38 103, 331, 339, 340
(3) 332, 341
s. 39 103, 331, 339, 355
(4) 332
ss. 40, 41 103, 331, 339
s. 47 355
s. 52 362
s. 56 361
s. 68 102
s. 69 102, 362, 364
s. 70(3) 361
s. 84 100
(2) 331
s. 86 101
s. 87 102, 362, 364

Housing Grants, Construction and Regeneration Act 1996

General 1, 109
ss. 104–107 9
s. 108 45, 48, 95, 329
(2)–(4)(6) 10
s. 109 10, 37
ss. 110–112 10
s. 113 10, 277

1

Introduction

A civil engineering construction contract is one in which the principal obligation is the construction of works of civil engineering. Civil engineering construction contracts are commercial contracts and are concluded, interpreted and enforced in the same way as any other such contract.[1]

1. Brief historical introduction

In the eighteenth and nineteenth centuries, many works of civil engineering, such as canal and railway construction, were promoted by private companies. The forms of contract used reflected the organisation of such works. The Engineer occupied a central position; he controlled the design, construction and commissioning of the works. In the second half of the nineteenth century, major schemes were promoted by public authorities, but the forms of contract used continued to bear the hallmark of the original paradigm with the Engineer as contract administrator. These public authority contracts were frequently drawn in standard formats. The Institution of Civil Engineers (ICE) and the Association of Consulting Engineers (ACE) drew up the first edition of the ICE Conditions of Contract based

1 When the Housing Grants, Construction and Regeneration Act 1996 comes into force (probably in 1997 or 1998), provisions as to adjudication and payment which comply with the Act will automatically be required or implied into construction contracts, thus distinguishing them to some degree from other commercial contracts. See the Appendix for the text of the Act.

largely on these public authority contracts.[2] The Federation of Civil Engineering Contractors (FCEC) joined the drafting committee for subsequent editions.[3]

Recent developments in contracting policy have prompted a reconsideration of the arrangements presupposed by forms such as the ICE Conditions of Contract. One such consideration relates to the growing practice of executing civil engineering works on a design and construct basis, rather than the traditional arrangement where the Engineer provides a design. In addition, developments in construction management and the increasing political and financial sensitivity of projects have caused commentators to question whether the primacy of the Engineer in these contracts is essential or, indeed, desirable. The ICE has been at the forefront of new contract development. In 1992 it published its Design and Construct Contract, which provides a vehicle for the contractor to carry the design obligations of the contract within a framework that is recognisable from the traditional ICE Conditions of Contract. But even more innovative was its publication in 1993 of the New Engineering Contract (NEC), in which the contract is designed to create a collaborative relationship between all parties to the works and to enhance the role of good management practice. This form of contract, now in its second edition,[4] has received much praise and the support of the Latham Report on procurement in the construction industry.[5]

In recent years, innovative forms of construction finance and procurement have been suggested. For instance, the UK Government has been keen to advance the concept of "private finance" for quasi-public schemes (including roads, railways, hospitals, schools, etc.) in which private sector finance is used to develop these works in return for the ability to collect direct or "shadow" income or tolls.[6]

---

2 The publication of the ICE Conditions of Contract was prompted by a paper published by E. J. Rimmer QC, The conditions of engineering contracts, *Proc. Instn Civ. Engrs*, February 1939. Rimmer was a civil engineer and the co-editor of the 8th Edition of *Hudson's Building and Engineering Contracts* (1959).

3 The FCEC also publishes a sub-contract standard form designed to be used in conjunction with the ICE Conditions of Contract.

4 Renamed the Engineering and Construction Contract to emphasise its suitability for all construction schemes.

5 *Constructing the team*, Sir Michael Latham, HMSO, 1994.

6 A shadow toll is not collected direct from the user, but from a funding body.

2. Standard form contracts

In recent decades there have been many calls for the construction industry to adopt a single standard form of contract.[7] The advantages of using standard form contracts are that they avoid much of the cost of a bespoke contract, they reduce the cost of tendering, enhance the contractor's confidence in the arrangement and contain fewer unforeseen anomalies.

Today, however, the tendency is for employers to experiment with new forms of contract and this has led to a wide range of standard forms as well as many instances of "home-made" contracts. The reason is that each employer has a different set of objectives, budgetary constraints and risk attitude[8] and hence demands a form to suit his particular needs.[9] Despite this experimentation, the vast majority of civil engineering contracts continue to be based on well-known standard form contracts published by independent bodies such as the ICE and FIDIC.[10] Nevertheless, it is very unusual for parties to contract on the basis of an unamended standard form. The printed form is used as a basic "template"; clauses are added, omitted, substituted or revised to suit the particular requirements of the parties (and principally those of the employer). Thus the ICE Conditions of Contract—which are designed as a traditional measure and value contract—frequently metamorphose into a lump sum, design and construct contract, or other species of contract.

3. Classification of civil engineering contracts

Since the parties are free to contract on any basis which to them seems convenient, and because the construction scenarios that can be conceived are infinitely various, a wide range of contractual arrangements and contract types have been developed. They may be classified in a number of ways including the following. Each focuses on a different aspect of the employer–contractor relationship.

---

7 The Banwell Report: *The Placing and Management of Contracts for Building and Civil Engineering Work*, HMSO, 1964; The Latham Report: *Constructing the team*, HMSO, 1994.

8 See O'Reilly M., Risk, construction contracts and construction law, 11 Constr. L.J. 343, 1995.

9 The publication of the NEC may, however, stimulate the use of a single standard form; the NEC is not a single contract, but a set of standard clauses which can be assembled to cater for the individual employer's objectives and risk attitude.

10 Féderation Internationale des Ingenieurs-Conseils.

Classified according to the definition of the works

Fixed specification Here the works are specified without provision for the employer to change the specification or for the contractor to claim for any additional costs. It amounts to the purchase of a defined product for a fixed price.

Defined specification with variations provisions[11] Here the works are specified, but the employer retains the right to change the specification if it suits him to do so, or if it is necessary for the proper completion or functioning of the works.

Term contracts A term contract is an agreement between the parties that work of a certain type[12] will be undertaken by the contractor at agreed rates for a stated period. The employer does not usually guarantee any particular amount of work,[13] but may agree that all work of the specified type will be awarded to the contractor.[14] Where the parties continue to operate the arrangement after the term has expired, either party may discontinue it upon reasonable notice.[15]

Classified according to the scope of the contractor's obligations

Construct-only contracts[16] This is the traditional arrangement in which the contractor is provided with a design and must build exactly what is shown. He assumes no responsibility for the design of the works.[17]

11 All standard form civil engineering contracts fall into this class.

12 Term contracts are widely used for civil engineering maintenance contracts for local authorities. They are sometimes referred to as maintenance contracts or standing orders.

13 See e.g. *R.* v. *Demers* [1900] AC 103 (PC).

14 *Kelly Pipelines* v. *British Gas* (1989) 48 BLR 126. But for a term contract which was held not to be exclusive, see *Bonnell's Electrical Contractors* v. *London Underground* (1995) CILL 1110.

15 *Bonnell's Electrical Contractors* v. *London Underground* (1995) CILL 1110.

16 This is the general scheme of traditional civil engineering contracts such as the ICE Conditions of Contract and FIDIC Conditions of Contract. The 6th Edition of the ICE Conditions of Contract now recognises that the contractor frequently undertakes elements of design—see Clause 8(2).

17 But see *Lindenburg* v. *Joe Canning* (1992) 9 Constr. L.J. 43 where a contractor was held to be 25% liable for failing to notice that walls shown on drawings for demolition were load-bearing.

Design and construct contracts[18] The contractor is engaged not only to construct the works but also to design them in accordance with the employer's general requirements.

Build, operate and transfer (and derivatives)[19] The key feature which distinguishes a BOT contract from other species of contract is the incorporation of an "operate" phase following construction and prior to transferring possession to the employer. The operate phase may be required for tuning operations to achieve optimum performance, proving that performance criteria have been met prior to handover, transferring skills to the employer's staff, or recovering the construction investment (private finance schemes only). Since some of these objectives (e.g. demonstrating that the works function properly) may be accomplished rapidly, while others (e.g. the realisation by the contractor of the fruit of his investment) may take decades, the "operate" phase may last as little as a few days or as much as a century.

Classified according to the method of remuneration

Lump sum contracts[20] The contractor is paid a fixed price for a fixed piece of work. The price may be altered from the agreed price if variations or other matters at the employer's risk occur. A "bill of quantities" or a "schedule of rates" may be included in the contract for the purpose of valuing variations.

---

18 For example, the ICE Design and Construct Conditions of Contract (1992). Design and build contracts are sometimes called "turnkey" contracts, though the latter term has no settled meaning.

19 See Stein S. W., Build–operate–transfer (BOT): a re-evaluation (1994) ICLR 101; Merna A., Payne H. and Smith N. J., Benefits of a structured concession agreement for build–own–operate–transfer (BOOT) projects (1993) ICLR 32.

20 Where the contractor agrees to construct works of description W for £*x*, a lump sum agreement is ordinarily generated. Many standard form contracts, e.g. the JCT Building Contracts, are lump sum contracts. Civil engineering contracts have tended to use measure and value schemes, though lump sum arrangements are becoming more common: see e.g. the Engineering and Construction Contract which has a lump sum option (Option A) and the ICE Design and Construct Conditions of Contract (1992).

Measure and value contracts[21] The price is computed by multiplying units of work required by agreed unit rates. The rates are usually set out in bills of quantities. The components of work to be included in the rates may be closely defined by reference to a "method of measurement".[22] Such contracts often provide that where there are significant changes in quantities or working conditions, the rates are to be modified to account for this.

Cost reimbursable contracts[23] The contractor is reimbursed for all his reasonable direct expenses and outgoings in connection with the scheme together with an additional sum representing the contractor's fee. The use of "targets" is common on such projects, with a target out-turn price and a target completion date. If the contractor exceeds the cost or time target he pays a penalty on his fee. If he performs the work at a keener price or faster than the target he receives a bonus on his fee.[24]

Classified according to the project organisation

Traditional[25] The employer commissions an engineer to design the works and engages a contractor to construct the works. During the construction, the engineer inspects the work in progress and is responsible for administering the contract, for example giving decisions on interim payments, variations and extensions of time, etc.

21 The traditional civil engineering contracts, e.g. the ICE Conditions of Contract 6th Edition and FIDIC Conditions of Contract 4th Edition, are measure and value contracts.

22 A number of standard methods of measurement are published: for instance see Civil Engineering Standard Method of Measurement, 3rd Edition (CESSM 3), Thomas Telford.

23 Cost reimbursement contracts are frequently used where the scope of the work is ill-defined. They enable construction to commence before the design is finalised. They are also used frequently for management contracts; see for example the Engineering and Construction Contract Option E.

24 See for example the Engineering and Construction Contract Option C and D for target contracts generally.

25 This is the general pattern adopted on most civil engineering projects, e.g. ICE Conditions of Contract, FIDIC Conditions of Contract, ICE Minor Works Conditions of Contract, as well as most building projects, e.g. JCT 80.

Management contracting[26] The work is let to the "management contractor" who then sub-contracts all of it to "trade contractors". The management contractor's primary role is to manage the project on behalf of the employer rather than to construct the works even though the management contractor's position *vis-à-vis* the employer is, in purely contractual terms, similar to that of a contractor under a traditional construction contract. In order to promote efficient management by the management contractor a risk-sharing agreement may be operated; a target price cost-plus-remuneration scheme is a typical method of achieving this.

Construction management[27] The employer employs the contractor who will construct the works directly but the work is controlled by a specialist construction management firm, also directly employed by the employer. This is similar in basic arrangement to the traditional construction system where a certifier appointed by the employer administers the contract. When a construction management team is employed, however, more than one contractor is often employed by the employer, the contractual and management systems tend to be significantly more sophisticated and there will often be a much greater interventionist coordinating role for the construction manager than for the typical certifier under a traditional construction contract.

4. The procurement and legal environment

It is important to recognise that civil engineering contracts exist within a procurement and legal environment which affects the way in which they are let, interpreted and operated. Three examples suffice to demonstrate this point.

---

26 The Engineering and Construction Contract Option F is a management contracting option.

27 The suite of Engineering and Construction Contract options produce construction management contracts which correspond to a greater or lesser extent to this model.

European procurement law

European directives have been published aimed at enhancing competition across the European Union.[28] A central theme is that public sector contracts should be open to all EU members on a common basis and that such contracts should not stipulate specifications which inherently favour goods or supplies from one locality or state.[29] Public sector contracts valued in excess of a prescribed limit[30] must comply with advertising and tendering procedures in accordance with the directives. Contractors who have been disadvantaged may seek damages.[31]

International contracts

Where a contract contains an international connection, English law does not necessarily apply.[32] Broadly speaking, the applicable law is that which the parties have chosen;[33] where no express choice is made, the applicable law is that of the country with which the contract is most closely connected.[34]

28 See The Public Supply Contracts Regulations 1991, SI 1991, No. 2679; The Public Works Contracts Regulations 1991, SI 1991, No. 2680; The Public Services Contracts Regulations 1993, SI 1993, No. 3228; The Utilities Supply and Works Contracts Regulations 1992, SI 1991, No. 3279.

29 See e.g. *The Commission of the European Communities* v. *Ireland* (1988) 44 BLR 1.

30 Currently 5 million ECU: Commission Notice in the Official Journal [1993] OJ C341/10.

31 Reg. 30 of The Public Supply Contracts Regulations 1991, SI 1991, No. 2679; Reg. 31 of The Public Works Contracts Regulations 1991, SI 1991, No. 2680. The procedure involves application to the High Court. See Bowsher M., Prospects for establishing an effective tender challenge regime: enforcing rights under EC procurement law in the English courts (1994), *Public Procurement Law Review* 30.

32 See generally the Contracts (Applicable Law) Act 1990.

33 In the vast majority of cases, the applicable law is expressly stated in the contract and no difficulty arises.

34 See e.g. *JMJ Contractors* v. *Marples Ridgeway* (1985) 31 BLR 104. Here the law applicable to a sub-contract made in the FCEC form was held to be Iraqi as this was the law applicable to the main contract. This case was decided prior to the enactment of the 1990 Act but remains indicative of the reasoning process which will be employed.

General UK legislation

A welter of legislation affects civil engineering construction. For example, highway engineers must consider the Highways Act 1980 and the New Roads and Street Works Act 1991. Pollution and noise are controlled by the Control of Pollution Act 1974 and The Environmental Protection Act 1990. Safety is to be managed in accordance with the Health and Safety at Work etc. Act 1974 and the Construction (Design and Management) Regulations 1994.

Housing Grants, Construction and Regeneration Act 1996

The Housing Grants, Construction and Regeneration Act 1996 was given Royal Assent in July 1996 but is not yet in force. Part II of the Act deals with construction contracts and is potentially very significant for the construction industry. Before it can be brought into force a document known as the "Scheme for Construction Contracts" must be produced by the relevant minister. At the time of writing, this document is still at its first draft stage and consultation is proceeding. A realistic timescale for the Act coming into force is late 1997 or in 1998. Section 104(6) provides that the Act only applies to contracts entered into after the date on which the Part II is brought into force, and so many of the provisions will not be tested until the turn of the century.

The text of the Act is set out in the Appendix, although its full impact will depend to a degree upon the wording and scope of the Scheme for Construction Contracts.

The Act defines a "construction contract" in Section 104 by reference to "construction operations". "Construction operations" are defined in Section 105(1). However, in Section 105(2) there are significant operations which are defined as not being construction operations and hence not subject to the Act. These include operations associated with mineral extraction, and process and power plants. Section 106 also excludes contracts where the employer is a residential occupier. Section 107 provides that the Act only applies to contracts in writing, and sets out in detail what is to be considered an agreement in writing. Section 106 allows the Government to alter these exclusions and it is thought that an amendment excluding Private Finance Initiative contracts will be made.

Section 108 sets out the major innovation of the Act, namely the introduction of compulsory adjudication for construction contracts to which the Act applies. The Section provides that the contract must contain adjudication provisions which comply with a specification set out in Sections 108(2), (3) and (4). Where there are no adjudication provisions or where they do not comply with the specification, the adjudication provisions of the Scheme for Construction Contracts will be implied into the contract: Section 108(6).

Sections 109 to 113 set out requirements as to payment provisions. There is a requirement in Section 109 that a "party to a construction contract is entitled to payment by instalments, stage payment or other periodic payments" for work lasting more than 45 days. Section 111 requires that where there is an intention to withhold payment, proper notice must be given. By Section 113 "conditional payment provisions", whereby payment to a party is made conditional upon payment by some other person, are ineffective.

2

Principles of contract law

A contract is an agreement which is enforceable at law. In principle, a civil engineering contract is formed, interpreted and enforced as any other contract.

1. The doctrine of privity

The doctrine of privity states that the parties to a contract are the only persons who have rights and/or liabilities under it. Thus, third parties have no enforceable interest or responsibility under a contract, even though it may exist partially or substantially for their benefit.[1] This doctrine is generally applicable,[2] though it has been criticised[3] and exceptions have been developed.[4]

Privity is important in the context of civil engineering projects, where there are typically many parties and many contracts. Contractual responsibility exists only where there is a contractual relationship. Contractual relationships frequently differ from the management and

1 *Tweddle* v. *Atkinson* (1861) 1 B & S 393.

2 *Darlington Borough Council* v. *Wiltshier Northern Ltd* [1995] 1 WLR 68 (CA): "the doctrine of privity of contract persists in all its artificial technicality" per Steyn LJ.

3 The Law Commission has published a draft Contracts (Rights of Third Parties) Bill 1996 which, if enacted, will entitle a third party to enforce a contract made for his benefit in defined circumstances.

4 In *Jackson* v. *Horizon Holidays Ltd* [1975] 3 All ER (CA) it was apparently held that where a person orders services on behalf of a group (in this case a father booked a family holiday) he may recover damages on behalf of them all. In *Darlington Borough Council* v. *Wiltshier Northern Ltd* [1995] 1 WLR 68 (CA), a constructive trust argument prevailed.

administrative relationships. For instance, the main communication during a civil engineering project may be between the contractor and the engineer; however, where there is no contract between them they will not be contractually liable one to the other.[5] Likewise, where an employer contracts with a contractor and the contractor sub-lets work to a sub-contractor, the employer may not ordinarily bring a contractual action against the sub-contractor for defective work;[6] nor may the sub-contractor bring an action in contract against the employer for payment.

A number of devices are used to circumvent or override the doctrine, as below.

Supplementary agreements

The normal chain of contracts may be supplemented by additional agreements, which create contractual relationships between parties who would not ordinarily be in such a relationship. Frequently, the construction contract states that the contractor may only sub-let work if the sub-contractor agrees to enter into a supplementary agreement with the employer. These supplementary contracts are often termed "direct warranties" or, where they are collateral to another contract, "collateral warranties". Such agreements are usually express. They may also arise by implication; where a prospective sub-contractor makes assurances to the employer to secure the employer's agreement to the sub-contract, a contract may arise by implication.[7]

5 So that where the contractor sued the engineer for his alleged failure to certify sums under a civil engineering contract, there was no option but to identify a non-contractual potential cause of action: see *Pacific Associates* v. *Baxter* [1990] 1 QB 993 (CA) where the action was framed in negligence and failed. But see also *Edgeworth Construction Ltd* v. *N. D. Lea and Associates Ltd* (1993) 66 BLR 56 (Supreme Court of Canada) where the contractor relied upon the engineer's drawings for tender purposes; the engineer owed a duty of care to the contractor.

6 *Dunlop Pneumatic Tyre Co. Ltd* v. *Selfridge & Co. Ltd* [1915] AC 847 (HL).

7 See *Shanklin Pier Ltd* v. *Detel Products* [1951] 2 KB 854 where an agreement was created between the employer and the sub-contractor. See also *Independent Broadcasting Authority* v. *EMI Electronics Ltd* and *BICC Construction Ltd* (1980) 14 BLR 1 (HL) where no agreement arose.

Action in the tort of negligence[8]

Where A injures B or damages his property, B may have an action in the tort of negligence against A even though there is no contract between A and B.[9] Where B's loss is purely economic,[10] however, B is unlikely to recover against A,[11] unless B's loss is a result of his reliance upon A's negligent misstatement.[12]

Assignment

A party is entitled to assign the benefits of his contract to any third person, unless the contract prohibits this.[13] Where the assignment is made, the assignee (i.e. the person to whom the benefits are assigned) is entitled to recover those benefits from the debtor (i.e. the person who owes the benefits). The debtor can use any defence against the assignee which would be available against the original party to whom the benefit was owed.[14] Most construction contracts contain restrictions on the right to assign, which can provide an effective defence to any action by an assignee.[15]

---

8 Where a term in a contract between A and B excludes a tortious remedy the intention of the parties will be enforced (provided the reasonableness test in the Unfair Contract Terms Act 1977 is satisfied). Otherwise, the common law is not antipathetic to concurrent liability; the plaintiff may choose a contractual or tortious action or remedy as he sees fit: *Henderson* v. *Merrett* [1994] 3 WLR 761 per Lord Goff at 788–789.

9 *Donoghue* v. *Stevenson* [1932] AC 562 (HL); *Sharpe* v. *E. T. Sweeting & Son Ltd* [1963] 1 WLR 665.

10 Defects in work provided under contract are classified as purely economic loss: *Murphy* v. *Brentwood* [1991] 1 AC 398 (HL). But "the complex structure exception" suggests that where defects cause damage to another distinct structure in the same building this may be treated as property damage and hence recoverable under the principle in *Donoghue* v. *Stevenson* [1932] AC 562; *Jacobs* v. *Morton and Partners* (1994) 72 BLR 92.

11 *Murphy* v. *Brentwood* [1991] 1 AC 398 (HL). This lead by the House of Lords has not been followed in Commonwealth jurisdictions. See for example *Winnipeg Condominium Corporation* v. *Bird Construction Co. Ltd* (1995) 74 BLR 1 (Supreme Court of Canada); *Allan Bryan* v. *Judith Anne Maloney* (1995) 74 BLR 35 (Australian High Court).

12 *Hedley Byrne* v. *Heller and Partners* [1964] AC 465 (HL); *Caparo Industries plc* v. *Dickman* [1990] 2 AC 605 (HL).

13 *Linden Gardens Trust* v. *Linesta Sludge Disposals Ltd and others* [1994] 1 AC 85 (HL).

14 *Roxburghe* v. *Cox* (1881) 17 ChD 520 (CA).

15 *Linden Gardens Trust* v. *Linesta Sludge Disposals Ltd and others* [1994] 1 AC 85 (HL); *Helstan Securities Ltd* v. *Hertfordshire County Council* [1978] 3 All ER 262.

Novation

A novation occurs where a contract subsisting between A and B is converted to one between B and C, with the agreement of all three. This involves no avoidance of the privity principle since the subsequent agreement is a valid contract which supersedes the original contract.

Sub-letting consents and nomination

Where an employer has no privity with a sub-contractor, he may, nevertheless, attempt to control the sub-contractor's identity and behaviour in a variety of ways. A term in the contract may require the contractor to obtain the employer's consent before sub-letting any work. A "nomination"[16] provision in the contract entitles the employer to instruct the contractor to contract with named sub-contractors; this has caused difficulty for the nominating employer on a number of occasions where the sub-contractor fails to complete and the employer has been held partly responsible for this.[17] Other means of controlling parties with whom one has no privity includes exclusion for misbehaviour or incompetence.

2. Formation and variation of contract

Most civil engineering work is performed under contract. The contract sets out the rights and duties of the parties. Sometimes the contract will have detailed written terms; sometimes it will be contained in a short letter or conversation. The existence of a contract is normally beyond dispute and both parties are frequently in agreement about the terms it contains. On occasions, however, the very existence of a contract and/or the identity of the terms it contains are disputed. In such cases, the process of formation of the contract must be considered.

16 See e.g. Clause 59 of the ICE Conditions of Contract.

17 See e.g. *North West Metropolitan Regional Hospital Board* v. *T. A. Bickerton & Son* [1970] 1 WLR 607 (HL); *Gloucestershire County Council* v. *Richardson* [1969] 1 AC 480 (HL); *Fairclough Building* v. *Rhuddlan Borough Council* (1985) 30 BLR 26 (CA); *Percy Bilton* v. *Greater London Council* [1982] 1 WLR 794 (HL). See also the Irish case of *Norta Wallpapers (Ireland) Ltd* v. *John Sisk & Sons (Dublin) Ltd* (1978) 14 BLR 99 where the nominated sub-contractor had a design responsibility.

At any stage after making a contract, the parties may agree to modify or vary the terms or, indeed, to discharge the contract altogether. In order to vary the contract[18] the same legal elements are required as are required to form a contract in the first instance.

The elements of formation

The principal requirements for the formation or variation of a contract are (*a*) agreement; (*b*) authority to contract; and (*c*) consideration and intention to create legal relations.[19]

Agreement

The question of whether or not an "agreement" has been reached is one of fact. It involves an objective assessment.[20] The test is whether or not an independent observer, appraised of the background facts known to the parties, would consider there to have been agreement, and if so, what agreement. The subjective state of the negotiators' minds is irrelevant. Where there is a mutual mistake concerning an important element of the proposed agreement this may prevent agreement.[21]

Often, the existence of agreement is clear, as where there is a signed document. But where there is a series of negotiations, it is useful to use a two-stage analysis: offer and acceptance. Once an offer has been made it may be accepted by the party to whom it is addressed. If it is unconditionally accepted there is agreement and a contract has come into existence.

An offer may only be accepted by the person to whom it is addressed (the offeree). No acceptance is possible after an offer ceases to exist. The offer may be destroyed by (*a*) being withdrawn by the offeror;

18 Variation of the contract must be distinguished from a variation under the contract; the former involves a revision to the basis of the agreement, the latter involves the operation of a mechanism which has been agreed between the parties.

19 *Rose and Frank Co.* v. *Crompton Brothers* [1925] AC 445 (HL).

20 *Tamplin* v. *James* (1880) 15 ChD 215 (CA).

21 *Bell* v. *Lever Brothers* [1932] AC 161 (HL).

(*b*) lapsing at a time specified by the offeror or at a reasonable time after being made; (*c*) being superseded by a subsequent offer (by the offeror or by the offeree);[22] or (*d*) being rejected by the offeree.

The formation of a straightforward civil engineering contract may thus proceed in the following stages.

Invitation to tender This is generally a pre-offer invitation,[23] in which the employer invites one or more contractors to tender quotations for a specified piece of work. There is generally no obligation for the person inviting tenders to accept the most attractive or any offer, though there may be an obligation to consider all tenders properly submitted.[24] An invitation to tender may rank as an offer if the invitation clearly states that a contract will be awarded to the person submitting the tender which most fully meets a stated criterion.[25]

Tender or quotation This is generally an offer by the contractor to undertake the work for the price or rates quoted and upon the conditions contained in his offer. Where the document is styled an "estimate" it may, nevertheless, amount to an offer;[26] it is a matter of construing the document.

Letter of intent Upon receipt of tenders, the employer will often write to a tenderer in terms such as "we intend to place a firm order with you shortly" and it may continue "please commence the works as

---

22 A request for clarification or further information may not amount to a counteroffer: *Stevenson* v. *McLean* (1880) 5 QBD 346.

23 *Spencer* v. *Harding* (1870) LR 5 CP 561. This is frequently termed an "invitation to treat".

24 *Blackpool and Fylde Aero Club* v. *Blackpool Borough Council* [1990] 1 WLR 1195 (CA).

25 Such as the lowest tender. See e.g. *Harvela Investments Ltd* v. *Royal Trust Company of Canada Trust* [1986] AC 207 (HL).

26 *Croshaw* v. *Pritchard* (1899) 16 TLR 45; see also *Cana Construction* v. *The Queen* (1973) 21 BLR 12 (Canadian Supreme Court) where an employer was responsible for inaccuracies in a document styled an estimate.

soon as possible". Such a communication does not generally rank as an acceptance as it is not unconditional.[27] Where the contractor commences work in pursuance of such a letter of intent, he will be entitled to be paid a reasonable sum for the work he has actually performed in the event that no contract eventuates.[28]

Acceptance This is the unconditional acceptance of the tender.[29] Civil engineering contracts tend to involve complex technical specifications and logistical arrangements. Frequently the formation of the contract is prefaced by lengthy and detailed negotiations. It is not uncommon for the negotiations to appear inconclusive. The question arises whether or not a contract then comes into existence at all. In practice, the law does not require absolute agreement on every detail. Provided that the parties have indicated a willingness to contract[30] and are in substantial agreement on the main issues,[31] a contract will often be created. Where the parties each put forward their own preferred terms, the question of whether a contract has in fact come into existence and if so, upon what terms, is a question to be decided in the light of all the circumstances.[32] Likewise where the parties commence work in accordance with draft terms without subsequent communication on the terms, the question of whether those terms govern the relationship between the parties is to be decided by

27 *British Steel Corporation* v. *Cleveland Bridge and Engineering Co. Ltd* [1984] 1 All ER 504; see also *Wilson Smithett* v. *Bangladesh Sugar* [1986] 1 Lloyd's Rep. 378 for a case where a contract came into existence.

28 See Chapter 4, section 2.

29 Note that the term "acceptance" is often used loosely to mean "acceptance pending final contract": see *M. Harrison & Co. (Leeds) Ltd* v. *Leeds City Council* (1980) 14 BLR 123.

30 This will normally be so where they have entered into negotiations about important matters such as price, quality and/or time. It will, however, be negatived where they set up some barrier, such as using the phrase "subject to contract" which normally has the effect of putting the contract on hold: *Regalian Properties plc* v. *London Docklands Development Corporation* [1995] 1 WLR 212.

31 In many cases, the price will be an essential term; but this will not always be so: see *British Steel Corporation* v. *Cleveland Bridge and Engineering Co. Ltd* [1984] 1 All ER 504.

32 *Butler Machine Tool Co. Ltd* v. *Ex-Cell O Corporation (England) Ltd* [1979] 1 WLR 401. This case is frequently said to be authority for the proposition that the "last shot wins". This was a potential test, but no ruling of principle was laid down. See also *Chichester Joinery* v. *John Mowlem* (1987) 42 BLR 100 for an example of the possible stages of negotiation through which a construction contract may progress.

reference to all the circumstances; but in appropriate circumstances, acceptance may be inferred from conduct alone.[33] Acceptance cannot, however, be inferred from mere silence;[34] where one party indicates that he will assume his terms to be accepted unless he hears to the contrary, and he hears nothing, this will not suffice.

The agreement must be genuine and not obtained by duress. Where a contract has been part-performed and one party uses a leverage it has obtained by that part-performance to extract a variation to the contract terms which are to his benefit, this will render the revised agreement voidable. In such cases the "innocent party" may either affirm or avoid the new arrangement created by the variation.[35]

The agreement must be reasonably clear and certain; otherwise there is, in fact, no agreement. Where a term which is at the very heart of the agreement contended for has no ascertainable meaning, there is no agreement.[36] Where, for example, the specification in an informal contract is so vague that it is not possible to say even in broad terms what the agreement is, no contract will be formed.[37] If the uncertain term can be severed from the agreement without destroying its basis, then this will be done.[38]

Authority

It is a logical prerequisite of any contract coming into existence between A and B that it is agreed between persons who are authorised to conclude a contract on behalf of A and B. The question of authority is dealt with below.[39] It is presumed that a person appointed to administer a construction contract is not authorised to vary its terms, unless a contrary intention is shown in the agreement. Thus an

33 *Brogden* v. *Metropolitan Railway Co.* (1877) 2 App Cas 666 (HL).

34 *Felthouse* v. *Bindley* (1862) 11 CB (NS) 869.

35 *North Ocean Shipping Co. Ltd* v. *Hyundai Construction Co. Ltd* [1979] QB 705.

36 *Scammel* v. *Ouston* [1941] AC 251 (HL).

37 *Southway Group Ltd* v. *Wolff and Wolff* (1991) 57 BLR 33.

38 *Nicolene* v. *Simmonds* [1953] 1 QB 543 (CA).

39 In section 8 of this chapter.

Engineer under the ICE family of contracts may vary the works, but not the contract.

Consideration and intention to create legal relations

The law distinguishes between two principal types of agreement; those which give rise to legally enforceable duties and those which do not. The former agreements are contracts; the latter are merely unenforceable promises. The underlying test concerns whether or not the parties intended to create legal relations. This test, however, is vague and raises the subjective intention of the parties; therefore, an alternative objective test is used. This focuses on the provision of "consideration". Consideration is the contribution which a party makes to support the agreement.[40] Since the law is concerned with ensuring not that commercial contracts are fair, but that they were intended to be binding, the question of whether or not the consideration is commercially adequate does not arise. In a civil engineering contract the contractor agrees to the works and the employer agrees to pay for it; each is providing consideration. It seems that a party can show sufficient consideration even if that party suffers no forbearance, detriment, loss or responsibility, as long as the other party benefits from the agreement.[41]

3. The terms of the contract

Where the agreement is written in a document, the terms of the agreement are usually readily established. A party is, however, entitled to claim that the agreement was made partly orally and partly in writing. He may claim also that the document does not accurately represent the agreement, in which case he may seek to have the agreement rectified or to show that it is supplemented by a collateral contract.

40 "A valuable consideration, in the sense of the law, may consist either in some right, interest, profit, or benefit accruing to the one party, or some forbearance, detriment, loss or responsibility, given suffered or undertaken by the other": *Currie* v. *Misa* [1875] LR 10 Ex. 153 at 163.

41 *Williams and Roffey Bros* v. *Nicholls (Contractors) Ltd* [1991] 1 QB 1; *Re Selectmove Ltd* [1995] 1 WLR 474.

The terms of a contract which is not reduced to formal writing are commonly more extensive than the express terms of the accepted offer, particularly where there has been a series of negotiations. An objective test is used to determine which terms of previously-made offers are to be carried over into any subsequent offer. Clearly, terms in a later offer destroy earlier inconsistent terms. But terms advanced during early negotiations may never have been contradicted and the question may arise as to whether or not they have survived the negotiations to become contract terms.

Furthermore, terms may be imported by reference to a document; second- and third-remove incorporation is common. If a document A is incorporated into the contract and A specifically incorporates document B, which in turn specifically incorporates document C, then the terms of C will form part of the contract, as well as those of A and B. It frequently happens that there are inconsistencies in these various documents, in which case the general rule is that terms in an incorporating document take precedence over those in an incorporated document.

Mere representations

During negotiations statements may be made which induce the other party to enter the contract on the terms beneficial to the person making the statement. Such statements may become contract terms but frequently they do not, in which case they are said to be "mere representations". In civil engineering it is common for site investigation information to be supplied to the contractor "for information only"; hence it is unlikely to become a term of the contract.

Whether or not statements become terms of the contract depends on the objective intention of the parties; all the surrounding circumstances must be taken into account. A number of "indicators" have been advocated. First, if the contract is put in writing and the statement is not included[42] this suggests that it is a mere representation.[43] Second, the earlier in negotiations a statement is

42 Where it was (objectively) intended to have been included, either party may claim a rectification.

43 *Birch* v. *Paramount Estates Ltd* (1956) 16 EG 396.

made the less likely it is to become a term.[44] Third, if one party is in a position of superior knowledge and he makes a statement in circumstances where he must be taken to know that the other will be relying upon that statement, that suggests that the statement will become a term.[45]

A mere representation which is inaccurate is termed a "misrepresentation". If one party misrepresents that a certain factual state of affairs exists and the representee has been induced to enter into the contract by that misrepresentation, then the plaintiff has an action for misrepresentation.[46]

Construing contract terms

Construction is the determination of the true objective meaning at law of terms in a contract. The subjective intentions or views of the parties either before or after making the contract are irrelevant. It becomes necessary to construe the contract when, for instance, there is a dispute about the meaning of a term or where mutually contradictory terms exist.

The main principle The main principle is that words in the contract are to be given their ordinary meaning, unless the parties are using commonly accepted trade or other specialist terms; in the latter case the law will take a view as to what the parties are presumed to know. The contract is to be construed as a whole[47] and a commonsense construction is to be preferred against constructions which appear to subvert the aim of the agreement.[48] When attempting to discern the commonsense construction it is proper to examine the purpose of a contract term and its underlying function in allocating risk.[49]

---

44 *Schawel* v. *Reade* [1913] IR 81, for a case where an early statement survived into the eventual contract.

45 *Esso Petroleum Co. Ltd* v. *Mardon* [1976] QB 801 (CA).

46 See Chapter 4.

47 *Brodie* v. *Cardiff Corporation* [1919] AC 337 (HL).

48 *Antaios Compania Naviera SA* v. *Salen Rederierna* [1985] 1 AC 191 (HL); *Davy Offshore Ltd* v. *Emerald Field Contracting Ltd* (1991) 55 BLR 1.

49 See e.g. *Balfour Beatty Building Ltd* v. *Chestermount Properties Ltd* (1993) 9 Constr. L.J. 117, per Coleman J. at 127.

Evidence of negotiations Evidence of negotiations prior to the formation of the agreement is inadmissible to prove the meaning of the contract, though the factual matrix which is established during negotiations may be admissible.[50]

The contra proferentem rule Where one party proffers his own form of contract, ambiguities upon which he relies are to be construed against him;[51] where the contract is in a standard form accepted in the industry there is some doubt concerning the applicability of the contra proferentem rule.

Priority The general rule of construction is that all terms are to be construed in the context of the contract as a whole; this gives rise to a presumption that no term is to carry greater weight than any other. However, where there is a plain contradiction, priority must be established. Many contracts give internal rules as to the order of priority which settle the matter.[52] Where this is not the case, determining priority between competing terms is assisted by a number of rules. All other things being equal, manuscript words prevail over typed words, which in turn prevail over printed words. The terms of incorporating documents prevail over terms in those which they incorporate.

Conditions and warranties

The law of contract has traditionally categorised terms into conditions and warranties. Conditions are terms whose breach undermines the purpose of the contract and entitles the innocent party to elect whether to continue with the contract or to treat it as being at an end. A breach of warranty merely entitles the innocent party to sue for damages. The classification into conditions and warranties depends upon a true

50 *Prenn* v. *Simmonds* [1971] 1 WLR 1381 (HL); *The Bank of New Zealand* v. *Simpson* [1900] AC 183 (PC). The explanation for this rule is as follows. Where proposed terms are offered and the offeree rejects them, the rejected terms add nothing to an understanding of the terms which are eventually accepted. However, where a fact is established during negotiations, this assists in understanding the agreement which was eventually made.

51 See e.g. *Rosehaugh Stanhope* v. *Redpath Dorman Long* (1990) 50 BLR 75 (CA).

52 The ICE Conditions of Contract do not; in fact Clause 5 states that all the Contract documents are mutually explanatory of one another.

construction of the contract rather than on any descriptions of terms as conditions or warranties.[53] In recent decades, however, the courts have shown a marked reluctance to force terms into the rigid categories of conditions and warranties.[54] The difficulty of such a categorisation in civil engineering contracts is readily demonstrated; many terms may be breached in a very minor or a very severe way. Such terms are properly treated as terms whose precise classification depends on the circumstances in which they are breached.[55] If parties require an entitlement to terminate the contract upon the breach of a specifically mentioned term, this should be expressly agreed.

Time clauses appear to be subject to a peculiarity. Time clauses are generally not conditions, or, as it is often put, they are not "of the essence". But they can be made so by notice, for example, that the employer wishes the work to be completed within a reasonable time.[56]

Rectification

Rectification is a discretionary remedy whereby a written contract document is amended to reflect the agreement reached by the parties. It may be granted in two distinct situations: (*a*) where a party can show convincingly that the document as drawn up does not reflect the objective intention of the parties at the time of its execution;[57] (*b*) where one party is labouring under a mistake and the other knows of,[58] conceals and takes advantage of that mistake, rectification may be granted where it would be inequitable to enforce the agreement as contained in the executed document.[59] An arbitrator may in general

53 *Bettini* v. *Gye* (1876) 1 QBD 183; *Schuler AG* v. *Wickman Machine Tool Sales Ltd* [1974] AC 235 (HL).

54 *Hong Kong Fir Shipping Co. Ltd* v. *Kawasaki Kisen Kaisha* [1962] 2 QB 26 (CA).

55 Election for fundamental breach is dealt with in section 7.

56 *Carr* v. *J. A. Berriman Pty Ltd* (1953) 27 ALJ 273.

57 *Joscelyne* v. *Nissen* [1970] 2 QB 86 (CA).

58 There must normally be actual knowledge. Thus mistakes in tendered prices which slip through without being noticed will remain binding. Where, however, the defendant must very strongly have suspected the mistake of the other, rectification may be ordered: *Commission for the New Towns* v. *Cooper (Great Britain) Ltd* [1995] 2 WLR 677 (CA).

59 *Roberts & Co.* v. *Leicestershire County Council* [1961] ChD 555. Here the employer amended the completion time without advising the contractor; the evidence was that the employer knew that the contractor had not noticed and rectification was granted.

grant rectification providing the arbitration agreement is sufficiently wide;[60] it is thought that he will not be able to rectify an agreement where this has the effect either of clothing him with jurisdiction or of depriving him of it.

4. Implied terms

In addition to the expressly agreed terms, terms may be implied. Implied terms cannot contradict express terms; but where they are implied they carry the same weight as express terms and may found a claim for breach of contract. They may be implied by statute, or in order to give the contract business efficacy. Furthermore, they may be implied where the express terms do not cover the matter which eventually becomes the subject of dispute, and it is necessary to define "usual practice" so as to determine the supposed objective intention of the parties.

Terms implied by statute

A number of statutes affect the content of commercial contracts. The Supply of Goods and Services Act 1982, for instance, implies terms into all contracts for the supply of services such as civil engineering contracts, except where these terms are inconsistent with the agreed express terms. This statute implies obligations to perform the work with reasonable care and skill, to do the work in a reasonable time and to pay a reasonable remuneration for the work.[61]

Terms which are necessary for business efficacy

Terms will be implied if they are necessary to give the contract proper business efficacy.[62] For instance, it will be implied that the employer may not hinder or prevent the contractor's performance; it will not, however, be implied that third parties will not prevent performance. It will be implied that in so far as is necessary the employer will

60 *Ashville Investments* v. *Elmer Contractors* [1989] QB 488 (CA).

61 Supply of Goods and Services Act 1982, sections 13, 14 and 15 respectively. See Chapter 5.

62 *The Moorcock* (1889) 14 PD 64.

cooperate with the contractor; but it will not be implied that the employer will organise his work so as best to accommodate the contractor.

Usual implied terms

Civil engineering construction is a complex process and the contracts which regulate it tend to be detailed. In some cases, however, the contract may be a simple agreement with very few express terms and the question arises of whether, in such cases, there is a body of "usual terms" which can be implied into the contract. Where, for instance, a project of long duration is agreed without mention of stage payments, does this mean that the employer does not have to pay anything until the end of the works?[63] It does not seem, on the business efficacy view, that any such implication is necessary but several judicial statements appear to suggest that a term to this effect might, nevertheless, be implied. When deciding whether payment was to be made in stages or as a single sum at the end, the court would certainly have regard to the usual practice in the industry.[64]

5. Exclusion, limitation and indemnity clauses

A clause in a contract purporting to exclude a party's liability for any event is an exclusion clause. A clause which purports only to limit liability is termed a limitation clause. Generally exclusion clauses are construed strictly against the person relying upon them; limitation clauses receive less strict construction.[65]

The Unfair Contract Terms Act 1977

Exclusion and limitation clauses are subject to the provisions of The Unfair Contract Terms Act 1977. The Act's significance is limited in

63 Note that Section 109 of the Housing Grants, Construction and Regeneration Act 1996 provides that stage payment provisions be implied into construction contracts.

64 Where a custom is so well-known and widely-used that the parties must be presumed to have contracted on that basis, the custom may be implied as a term: *Tucker* v. *Linger* (1883) 8 App Cas 508 (HL).

65 Generally, see *Photo Production* v. *Securicor Transport* [1980] AC 827 (HL).

civil engineering practice,[66] apart from two points. First, it restricts a party's ability to exclude his liability for his own negligence; liability for death or personal injury cannot be restricted at all and other liability cannot be restricted "except in so far as the term or notice satisfies the requirement of reasonableness".[67] Secondly, it restricts a person's right to exclude his liability for breach of contract where he is contracting on his own standard terms; any such clause must satisfy the test of reasonableness. This may become an issue where an employer maintains a standard contract.[68]

6. Contracts and formality

The need for writing

Some contracts, notably those of guarantee and for interests in land, need to be evidenced in writing; otherwise they are unenforceable. Excepting these, contracts may be in any form. They are equally binding whether they are made orally or by a highly formal sealed document.

Contracts under seal

The principal effect of executing a contract under seal is that the limitation period (i.e. the time following a breach of contract in which the innocent party is entitled to commence proceedings to enforce the agreement) is extended from six to twelve years.[69] Before being sealed the contract is known as a "simple" contract; when sealed, the contract is said to be "a specialty", "a deed" or "under seal". Standard form civil engineering contracts frequently entitle one or both parties to require the other to execute the contract under seal.[70]

---

66 In *Photo Production* v. *Securicor Transport* [1980] AC 827 (HL), Lord Wilberforce was of the clear opinion that in commercial contracts the Act would have limited impact: see especially 843.

67 Subsection 2(2); *Rees Hough Ltd* v. *Redland Reinforced Plastics Ltd* (1984) 27 BLR 136; *Barnard Pipeline Technology Limited* v. *Marston Construction Co. Ltd* (1992) CILL 743.

68 *Chester Grosvenor* v. *Alfred McAlpine* (1991) 56 BLR 115.

69 Limitation Act 1980.

70 See Clause 9 of the ICE Conditions of Contract, 6th Edition.

7. Ending contracts

A contract may be ended in two distinct senses. In its hard sense the contract may be eradicated so that the parties may not rely on its terms in future. In its soft sense, neither party is entitled to perform further obligations under the contract, but the terms of the contract survive and govern the rights of the parties to damages or other entitlements. Termination in its hard sense occurs where the contract is avoided or frustrated, or if the parties vary the agreement so that all claims, present and future, under the prior agreement are compromised.[71] Termination in its soft sense is achieved by purported performance, termination in accordance with the terms of the contract or an accepted fundamental breach.

Agreement to discharge or vary the contract

An agreement to discharge or vary the contract operates to substitute a new agreement for the one formerly in existence. Any agreement to discharge or vary the contract must be made by the parties (or their authorised agents) and be supported by consideration. Where both parties have unperformed obligations, sufficient consideration is supplied where each forgoes the right to insist on the other's further performance. If one party has completed his obligations, the other's agreement to tender a lesser performance is no consideration.[72] Where a party accepts less than complete performance from the other in return for some undertaking from that other, this is often referred to as an "accord and satisfaction". Such agreements are typically made to avoid disputes being litigated or arbitrated and the consideration provided often consists of one party forgoing the right to pursue a bona fide claim.

Accepted fundamental breach or accepted repudiation

A party to a contract may indicate that he no longer intends to comply with or be bound by the contract in one of two ways; a serious breach which strikes at the purpose of the contract, or by renouncing the

71 Known as an "accord and satisfaction".

72 *D & C Builders* v. *Rees* [1966] 2 QB 617 (CA).

contract. In either case, the "innocent party" may elect whether or not to continue with his own performance of the contract. The contract is not automatically terminated, and will continue to exist if the innocent party affirms it or fails clearly to indicate that he accepts the repudiation.[73] The innocent party may be entitled to continue to perform the agreement, even if the other party has no further interest in the subject matter of the agreement,[74] though this rule does not apply when the work is to be performed on the land of the party in breach[75] and may be unsound generally.

A repudiation, often referred to as an "anticipatory breach", consists not of the failure to perform the obligation but, rather, of the renunciation of the agreement itself. The innocent party may elect to terminate the agreement immediately upon the repudiation and does not have to wait until the time for the performance of the obligation in question. The test is whether the allegedly repudiatory conduct amounts to a renunciation of the contract or an absolute refusal to perform it.[76] Thus where a party relies in good faith on his own construction of the contract to support his refusal to perform obligations in particular circumstances, this will not amount to a repudiation,[77] even if his construction turns out to be mistaken.

A number of reported cases consider which breaches are sufficiently grave to entitle the innocent party to elect to terminate the agreement. The grant of only partial possession of the site,[78] failure to pay for deliveries of construction materials[79] and sub-letting in breach of the contract[80] have each been considered in this context. However, each contract and each breach is unique; accordingly reliance on historical

73 *Vitol SA* v *Norelf Ltd* [1995] 3 WLR 549 (CA).

74 *White & Carter (Councils) Ltd* v. *McGregor* [1962] AC 413 (HL).

75 *London Borough of Hounslow* v. *Twickenham Garden Developments* [1971] ChD 233; *Mayfield Holdings Ltd* v. *Moana Reef Ltd* [1973] 1 NZLR 309 (New Zealand Supreme Court); *Tara Civil Engineering Ltd* v. *Moorfield Developments Ltd* (1989) 46 BLR 72.

76 *Mersey Steel and Iron Co.* v. *Naylor Benzon & Co* [1884] 9 App Cas 434 (HL).

77 *Woodar Investment Development Ltd* v. *Wimpey Construction UK Ltd* [1980] 1 WLR 571 (HL).

78 *Carr* v. *J. A. Berriman Pty Ltd* (1953) 27 ALJ 273.

79 *Mersey Steel and Iron Co.* v. *Naylor, Benzon & Co.* [1884] 9 App Cas 434 (HL).

80 *Thomas Feather & Co. (Bradford) Ltd* v. *Keighley Corporation* [1953] 53 LGR 30.

instances provides at best a poor guide and may be positively misleading. Basic principles should be used in preference to specific examples. The important question in every case is whether the breach is so serious that it strikes at the purpose of the contract and constitutes a renunciation of it.

Determination under the terms of the contract

Determination clauses enable one or both parties to treat the contract at an end upon the happening of some specified event or circumstance.[81] This event or circumstance is normally a serious breach, a serious delay in performance or the contractor's insolvency; however, it need not be directly related to the performance of the project. The determination clause may also provide for compensation or transfer of property upon termination; such provisions will be enforceable unless they amount to a penalty.

Frustration

A contract is frustrated when the circumstances of its performance change so radically that the contract becomes an agreement to do something wholly different from that envisaged at the time the agreement was originally made. The frustrating event must not be within the control of either party. If the allegedly frustrating events are provided for in the agreement there can be no frustration.[82] Instances considered by the courts have included cases where the subject matter of the project is destroyed[83] and where equipment to be used for performing the project is requisitioned by the state for military purposes.[84] If the work merely becomes more difficult to perform as a result of shortages of labour and materials it will not be considered as frustrated.[85] Unless otherwise stated in the contract,[86] the Law Reform

---

81 See further Chapter 3, section 8.

82 *McAlpine Humberoak* v. *McDermott International* (1992) 58 BLR 1 (CA).

83 *Taylor* v. *Caldwell* (1863) 32 LJQB.

84 *Metropolitan Water Board* v. *Dick, Kerr and Co. Ltd* [1918] AC 119 (HL).

85 *Davis Contractors Ltd* v. *Fareham Urban District Council* [1956] AC 696 (HL).

86 Clause 64 of the ICE Conditions of Contract provides its own rules for property passed during the subsistence of the contract.

(Frustrated Contracts) Act 1943 provides for the effect of frustration on property transferred during the subsistence of the contract.

8. Legal personality and agency

Legal personality

The law recognises two categories of legal person: natural persons (i.e. human beings) and corporations. Natural persons of the age of majority may enter freely into contracts. Corporations may be created by charter or by statute (e.g. local authorities) or as companies by registration under the Companies Acts. Corporations are legally distinct from their members. Thus while a company is owned by its shareholders, it is, in law, an independent person. The liabilities of the companies do not, therefore, devolve upon the shareholders; creditors must sue the company itself rather than the shareholders if they wish to seek redress against the company. The distinction here is particularly important when the corporation is a limited (limited liability) company in liquidation, for here the personal liability of the shareholders is limited to the value of the initial stake in the company allotted to them. The directors of the company also escape liability for the company debts provided they have not been reckless or fraudulent in their dealings.[87]

Non-incorporated associations, such as clubs, partnerships, unincorporated joint ventures, etc. have no legal personality and so cannot formally contract in their own right but as a collection of persons (natural or legal or both).[88] Special rules exist in many cases enabling, for example, a partnership to sue and be sued in the name of the partnership. And many joint ventures are set up using a limited liability company as a vehicle with participants holding shares in proportion to their stake in the venture.

87 In exceptional circumstances, a director may also be liable for negligent misstatement where the company is established to sell his personal services: *Williams* v. *Natural Life Health Foods, Independent* 18 January 1996.

88 *Bradley Egg Farm* v. *Clifford* [1943] 2 All ER 378 (CA).

Agency

Most commercial organisations operate through agents. The statement "Company X contracts with Company Y" means that authorised representatives of X and Y have agreed, on behalf of X and Y respectively, that X and Y would be bound by the contract. In the law of agency, the authorised representatives are known as agents and the companies they represent are the principals. The key question relates to the authority of agents.

Where a person A (for agent) purports to act on behalf of a principal (P) in contract negotiations, A will bind P where he has been expressly authorised to do so. In addition, where A holds a position of apparent authority in relation to P (e.g. where A is P's managing director) he will have authority to bind P whenever it would be usual for someone of his rank to have such authority.[89] Where A acts without actual authority he may become personally liable.[90] Where A fraudulently represents the scope of his authority, P's liability (if any) will depend on all the circumstances.[91]

Generally speaking, Engineers and other contract administrators will not be authorised to vary a contract on behalf of their clients.[92] Some civil engineering contracts deal specifically with the extent and limitations on the Engineer's authority.[93]

89 This is frequently termed ostensible or apparent authority.

90 See e.g. *Yonge* v. *Toynbee* [1910] 1 KB 215 (CA).

91 *Armagas* v. *Mundogas* [1986] 2 WLR 1063 (HL).

92 *Toepfer* v. *Warinco* [1978] 2 Lloyd's Rep. 569.

93 See e.g. Clause 2(i)(c) of the ICE Conditions of Contract, 6th Edition.

3

General features of civil engineering contracts

1. Civil engineering contract documents

A civil engineering contract generally contains a standard form set of conditions (amended or unamended). In addition, there will be a variety of documents, some of which will be standard documents and some of which will be unique to the project, setting out the details of the scope of the work to be done, the standard which is to be achieved, ancillary (e.g. safety) requirements and mechanisms for computing the sums payable at any stage. Furthermore, method statements and programmes are frequently produced; these may either form part of the contract or be produced as a management tool without direct contractual status.

Drawings and specifications

The scope of civil engineering work is usually defined using drawings and specifications. The former set out the positional interrelationship between the items of work, while the latter set out the quality required. Where no specification is provided, it will be implied into the contract that work is to be done with proper skill and care, using good quality materials which are reasonably fit for their purpose.[1] The specification documents tend also to contain a variety of requirements and stipulations as to the manner of working. A number of standard forms

1 *Young & Marten* v. *McManus Childs* [1969] 1 AC 454 (HL); *Gloucestershire County Council* v. *Richardson* [1969] 1 AC 480 (HL).

of specifications are published which relate to specific sectors of the civil engineering industry.[2] It is common for large public and quasi-public sector bodies to have standard specification terms which they use on their own projects.

Bills of quantities, schedules of rates, etc.

Bills of quantities are lists of items with associated quantities. The effect of the bill of quantities within the contract is a matter of construing the contract in each case.

In contracts for a lump sum price, items required to complete the works must generally be provided despite their being omitted from the bill;[3] if there is no mechanism in the contract for recovering payment for these extra items, the contractor will have to pay for them. There may also be a presumption[4] that the quantities do not form a term in lump sum contracts unless the contract states otherwise; accordingly the contractor will not receive an additional payment if the quantities required are greater than stated in the bill.

For measure and value contracts, estimated quantities are set out for each class of work. When tendering, the contractor quotes a rate for each class. The bill total is the sum of all the products of rates and estimated quantities; but the sum payable is the product of the actual quantities and rates. The process by which the quantity of each item is determined is called "measurement", which may be physical measurement on site or the computation of areal or volumetric quantities from survey data. If an item of work is to be done for which there is no agreed rate, nor agreed mechanism for calculating its value,

---

2 See e.g. the Civil Engineering Specification for the Water Industry, 4th Edition, 1993, Water Research Centre.

3 *Williams* v. *Fitzmaurice* (1858) 3 H & N 844 per Channel B. "It was a contract for the erection of a house and though the flooring was not mentioned in express terms, it was necessarily implied."

4 This cannot be stated with any certainty since the principal authorities are from the 19th and early 20th centuries and the decisions seem somewhat lacking in harmony. Generally see *Hudson's Building Contracts* (4th edn) Vol. 2: *Patman and Fotheringham* v. *Pilditch* (1904) p. 368; *Priestly* v. *Stone* (1888) p. 134; *Re Ford and Bemrose* (1902) p. 324 (CA). See also *Sharpe* v. *San Paulo Railway Co.* (1873) LR 8 Ch App 597.

the contractor is entitled to be paid a reasonable rate/sum.[5] In the ICE Conditions of Contract, the bills are deemed to be prepared in accordance with the Civil Engineering Standard Method of Measurement.[6] The quantities in the bill are expressed to be estimates,[7] any errors or omissions are to be corrected by the Engineer and any items required to be added in will be paid for in accordance with the contract.[8] Accordingly, where items have been accidentally omitted from the bill, the contractor is compensated.

In addition to documents described as bills of quantities, similar documents described as schedules of rates,[9] schedules of prices,[10] etc. are frequently used. None of these terms are terms of art and their effect is determined by construing the agreement in each case.

Programmes and method statements

Management tools such as programmes and method statements are frequently generated. The status of any programme or method statement is determined by construing the contract. The status may be as follows.

Programmes etc. provided solely for information The contractor may be required to submit a programme solely for the purpose of demonstrating competence at tender stage. Such programmes or method statements have no contractual significance or effect.

Programmes etc. provided in accordance with the terms of the contract Civil engineering contracts frequently call for the contractor

---

5 *Re Walton-on-the-Naze UDC* v. *Moran* (1905) *Hudson's Building Contracts* (4th edn) Vol. 2, p. 376.

6 Methods of Measurement are frequently used with bills of quantities. These contain lists of item descriptions and define what is included in each item. Many define items using codes so that there can be no confusion as to which item is being referred to.

7 Clause 55(1).

8 Clause 55(2).

9 This term is frequently used for term contracts since the work to be done and hence the quantities are a matter of great uncertainty; all that is known are the rates.

10 This term is frequently used for lump sum contracts where the items are listed without quantities.

to indicate how he proposes to execute the work.[11] The programme enables the employer or his agent (*a*) to coordinate his design staff to provide design details and/or the relevant parts of the site in good time,[12] (*b*) to monitor and review the contractor's progress, and (*c*) to assist him in the evaluation of extensions of time. Method statements provided under the terms of the contract enable the employer's staff or agents to form their own assessment of the methods being used and their effect on other parts of the works or on third parties. They also enable the employer to call for such information and calculations as he reasonably needs to satisfy himself that the work is being properly executed. The programme and method statements produced in accordance with the contract must, however, always be interpreted in the light of the basic contract obligations: thus, a programme which shows a completion date earlier than the contract completion date will neither oblige the contractor to finish earlier than the contract date nor require the employer to render such cooperation as will enable the contractor to do so.[13] The contractor may revise his programme unless the contract expressly provides that he may not do so.[14]

Programmes etc. which rank as contractual (i.e. as terms of the contract) Here the contractor is required and entitled to perform the work in accordance with the programme and method statement and in the event that he is prevented from so doing for reasons at the

11 For example as required by the ICE Conditions of Contract, Clause 14.

12 The ICE Conditions of Contract, for instance, require the Contractor to serve a notice when information is required: see Clause 7. A properly annotated programme might suffice as written notice of the requirement for design information: *London Borough of Merton* v. *Stanley Hugh Leach* (1985) 32 BLR 51.

13 *Glenlion Construction* v. *Guinness Trust* (1988) 39 BLR 89. But see also the commentary on Clause 14 of the ICE Conditions of Contract in Chapter 8: the Engineer may accept a programme which shows a completion date earlier than the contract completion date; it is suggested that here the Employer is obliged to supply information etc. to enable the Contractor to comply with the programme until the Contractor is given notice that the programme is not considered definitive for the purposes of the Employer's cooperation.

14 *Greater London Council* v. *Cleveland Bridge Engineering Co. Ltd* (1984) 34 BLR 50: "There is, moreover, a general principle applicable to building and engineering contracts that in the absence of any indication to the contrary, a contractor is entitled to plan and perform the work as he pleases . . ." per Staughton J.

employer's risk, and thereby suffers a loss, he will be entitled to claim damages.[15]

2. Commonly implied terms and established constructions in civil engineering contracts

Terms are frequently implied into civil engineering contracts and a number of established constructions have developed.

Payment

Where no express terms relating to payment are agreed, the law will imply an obligation for the employer to pay a reasonable price.[16] There is no direct authority on the question of whether the employer is obliged to make stage payments on long duration projects.[17] There are dicta which suggest that such an implication can be made[18] and it is submitted that in proper circumstances such an implication could be raised through the device of trade custom.[19]

Time and progress

If there are no express provisions in the contract relating to time, the contractor's obligation is to complete the work within a reasonable time. Where the contractor is delayed by a matter for which the employer bears responsibility and there is no provision in the contract for extending any agreed time for completion, the time will be put at

---

15 *Yorkshire Water Authority* v. *McAlpine* (1985) 32 BLR 114; *Holland Dredging* v. *Dredging and Construction Co.* (1987) 37 BLR 1 (CA).

16 Supply of Goods and Services Act 1982.

17 This may create a problem for the contractor since civil engineering contracts are prima facie entire contracts. This means that no entitlement to payment arises until substantial completion: see e.g. *Sumpter* v. *Hedges* [1898] 1 QB 673.

18 *Gilbert-Ash (Northern)* v. *Modern Engineering (Bristol)* [1974] AC 689 (HL).

19 When in force, the Housing Grants, Construction and Regeneration Act 1996, Section 109, will require or imply an entitlement to periodic payments for works lasting more than 45 days.

large,[20] which means that the contractor will have a reasonable time to complete in all the circumstances.[21]

In the absence of contrary indications, the contractor has freedom to plan his work within any specified time constraints.[22] He is not generally obliged to proceed at any particular rate.[23] It is submitted, however, that there will be an implied term that the contractor will not allow progress to be so slow that it would be impossible for him to complete by dates specified in the contract.

Quality of work and materials

The contractor is required to complete the contract works with proper skill and care and in a good and workmanlike manner.[24] Subject to any express specification, the contractor warrants that the materials are of good quality; he may be liable for their quality even where the employer has suggested or nominated the supplier.[25] Furthermore, the materials are to be reasonably fit for the purpose for which they are used; where the employer has nominated the supplier this term will normally be excluded.[26] Where the employer makes known the purpose of the works, and the contractor holds himself out as skilled in works of that type, so that the employer reasonably relies on his

---

20 *Peak Construction (Liverpool)* v. *McKinney Foundations* (1970) 1 BLR 111 (CA); *McAlpine Humberoak* v. *McDermott International* (1992) 58 BLR 1 (CA); *Fernbrook Trading Co. Ltd* v. *Taggart* [1979] 1 NZLR 556 (New Zealand Supreme Court).

21 The origin of this expression seems to be *Holme* v. *Guppy* (1838) 3 M & W 387. Parke B. said that as a result of the prevention by the employer "The Plaintiffs [contractors] were therefore left at large; and consequently they are not to forfeit anything for the delay".

22 *Wells* v. *Army and Navy Co-operative Society* (1902) 86 LT 764.

23 *Greater London Council* v. *Cleveland Bridge Engineering Co. Ltd* (1984) 34 BLR 50.

24 *Young & Marten* v. *McManus Childs* [1969] 1 AC 454 (HL).

25 *Young & Marten* v. *McManus Childs* [1969] 1 AC 454 (HL); *Gloucestershire County Council* v. *Richardson* [1969] AC 480 (HL).

26 *Young & Marten* v. *McManus Childs* [1969] 1 AC 454 (HL); *Gloucestershire County Council* v. *Richardson* [1969] AC 480 (HL).

skill, the contractor will be responsible for providing works which are reasonably fit for their purpose.[27]

The scope of the work

In general, the contractor must supply all items which are reasonably to be defined as necessary to complete his obligations.[28] The work will normally include all those things which are necessary for the execution of the work[29] even if they were not foreseen at the time of making the contract; thus, for example, no term will be implied that a contractor is entitled to extra payment if poor ground conditions are unexpectedly encountered.[30] Nor will it be implied that the employer warrants that the work may be accomplished in accordance with any particular technique.[31] It will not normally be implied that a performance which is different from that in the specification will be a satisfactory performance merely because it is "as good as" that specified.[32] When the employer is entitled to vary the work it will be implied that the scope of such variations will be such as not to change the nature of the work contracted for.[33]

---

27 *Greaves & Co. Ltd* v. *Baynham Meikle* [1975] 1 WLR 1095 (CA). This term will ordinarily apply to design and build contracts: see *Independent Broadcasting Authority* v. *EMI Electronics Ltd and BICC Construction Ltd* (1980) 14 BLR 1 (HL). It may apply only to specific elements of the work: *Cammell Laird & Co. Ltd* v. *Manganese Bronze & Brass Co. Ltd* [1934] AC 402 (HL).

28 *Williams* v. *Fitzmaurice* (1858) 3 H & N 844.

29 But many contracts include terms which entitle a contractor to recover additional sums if work is omitted from the billed or described work.

30 *Bottoms* v. *Mayor of York* (1892) *Hudson's Building Contracts* (4th edn) Vol. 2, p. 208; *Sharpe* v *San Paulo Railway Co.* (1873) LR 8 Ch App 597.

31 *Thorn* v. *Mayor and Commonalty of London* [1876] 1 AC 120 (HL). Most civil engineering contracts deal expressly with this possibility.

32 *Forman & Co. Proprietary Ltd* v. *The Ship 'Liddlesdale'* [1900] AC 190 (PC).

33 *Blue Circle Industries* v. *Holland Dredging Co. (UK) Ltd* (1987) 37 BLR 40.

Cooperation between the employer and contractor

The employer is obliged not to prevent performance of the contractor's work.[34] The employer must grant proper and reasonable access to the site,[35] though he need not prevent third parties over whom he has no control from denying the contractor access.[36] Normally it will not be implied that the employer has to provide work opportunities and workfronts so as best to suit the contractor.[37] Furthermore the employer's implied obligation to supply necessary information for construction will be to supply that information at reasonable times given the contract timing[38] rather than timing which is designed to suit the contractor or to assist him in completing earlier than the contract period.[39] It will normally be implied that permits will be obtained by the party best placed to obtain them or by the party who customarily obtains them.[40]

3. Performance

The scope of the work will ordinarily be defined in the drawings, specifications and bills of quantities. For smaller projects the scope may be less comprehensively defined. Where there is any ambiguity, the terms must be construed in the normal way to determine the true extent of the obligations which have been undertaken.

34 *Barque Quilpué Ltd* v. *Brown* [1904] 2 KB 261 (CA): "In this contract, as in every other, there is an implied contract by each party that he will not do anything to prevent the other party from performing the contract or to delay him in performing it . . . it must not, however, be supposed that the law readily implies any special affirmative contract", per Vaughan-Williams LJ.

35 *The Queen in Right of Canada* v. *Walter Cabott Construction Ltd* 21 BLR 46.

36 *LRE Engineering* v. *Otto Simon Carves* (1981) 24 BLR 127.

37 *Kitsons Sheet Metal* v. *Matthew Hall* (1989) 47 BLR 90; *Martin Grant & Co. Ltd* v. *Sir Lindsay Parkinson & Co. Ltd* (1984) 29 BLR 31 (CA).

38 *Neodox* v. *Swinton and Pendlebury Borough Council* (1958) 5 BLR 34.

39 *Glenlion Construction* v. *Guinness Trust* (1988) 39 BLR 89.

40 *Ellis-Don Ltd* v. *The Parking Authority of Toronto* (1978) 28 BLR 98 (Supreme Court of Ontario).

Where the works are not in strict accordance with the original specification and there is no properly authorised variation to cover them, this will be a breach of contract by the contractor. Any non-compliance is normally inadvertent. It may, however, be deliberate as where the contractor believes that what is supplied is "as good as" what is specified or he considers that the employer's agent has approved the work. Claims that the work is "as good as" are rarely meritorious[41] unless the contract, on a true construction, specifies a minimum standard and what is provided is clearly better. Wherever it is reasonable to do so, the employer will be entitled to have the work re-done; but where this is unreasonable he will be able to claim damages for the reduced performance or value of the construction.[42] To succeed in a claim that the employer's agent approved the non-compliant work, the employer's agent must have been authorised to vary the contract or to waive[43] the employer's contractual rights and he must have done so expressly. In most civil engineering situations, the employer's agent is rarely authorised to vary the contract or to waive the employer's rights under it.[44] Clearly, the fact that the engineer merely witnesses the non-compliant work cannot be sufficient to exculpate the contractor; for the engineer is employed to protect the employer's interests, not to protect the contractor.[45] However, an engineer is ordinarily entitled to vary the specification; where the engineer's variation falls within the compass of the variation provisions, disconformities may become compliant.[46]

41 *Forman & Co. Proprietary Ltd* v. *The Ship 'Liddlesdale'* [1900] AC 190 (PC); but see also *Ata Ul Haq* v. *City Council of Nairobi* (1959) 28 BLR 76 (PC).

42 *Ruxley Electronics* v. *Forsyth* [1995] 3 WLR 118 (HL).

43 *W. J. Alan & Co* v. *El Nasr Export and Import Co.* [1972] 2 All ER 127 (CA).

44 *Toepfer* v. *Warinco* [1978] 2 Lloyd's Rep. 569: "It is well established that an architect or engineer has no implied authority from the building owner by whom he is employed to vary or waive the terms of a building contract", per Brandon J at 577. This is so even where he stipulates that the work is to be done to a specification and also to the satisfaction of the engineer, and the engineer indicates his approval: *National Coal Board* v. *William Neill & Sons (St Helens) Ltd* (1983) 26 BLR 81.

45 *East Ham Borough Council* v. *Bernard Sunley & Sons Ltd* [1966] AC 406 (HL). See also *AMF International Ltd* v. *Magnet Bowling* [1968] 1 WLR 1028: ". . . in general an architect owes no duty to a builder to tell him promptly during the course of construction, even as regards permanent work, when he is going wrong; he may, if he wishes, leave that to the final stages notwithstanding that the correction of a fault then may be much more costly to the builder than had his error been pointed out earlier", per Mocatta J at 1053.

46 *Shanks & McEwan* v. *Strathclyde Regional Council* (1994) CILL 916 (Scottish Court of Session).

4. Instructions and variation orders

A simple agreement that a contractor should execute work of description W for a stated sum usually amounts to an agreement to provide a fixed product[47] for a fixed price. The employer will ordinarily have no right under the contract to specify how the work is done[48] or to order that the work be varied. In practice, of course, most contractors will allow the employer to order variations. In this case, the original work will be paid for in accordance with the contract rates and the work outside the contract will be paid at any agreed rate, or, if none is agreed, at a reasonable rate.[49] There is also the related situation where the employer wishes to omit work. In the absence of a contractual power to omit agreed work, the employer will be in breach of contract if he instructs the contractor to omit any work. An employer does not, of course, have to suffer a contractor fixing any work to his land, even work to which he has already agreed.[50] But if he refuses to allow the contractor to undertake the originally agreed work, this will entitle the contractor to damages.

Agent's authority where the contract does not provide for variations

Where the contractor acts on an engineer's instruction to provide additional work, the employer will not be responsible for paying for that additional work, unless the engineer was properly authorised or the employer knew that the work was in progress and took no steps to alert the contractor.[51] In such cases, the engineer may become personally liable for payment.

47 Including all things reasonably necessary to complete that work: *Williams* v. *Fitzmaurice* (1858) 3 H & N 844; *Sharpe* v. *San Paulo Railway Co.* (1873) LR 8 ChD App 597.

48 *Greater London Council* v. *Cleveland Bridge Engineering Co. Ltd* (1984) 34 BLR 50: "... in the absence of anything to the contrary, a contractor is entitled to plan and perform the work as he pleases" per Staughton J. The employer can, however, insist that the performance be reasonable, e.g. that it is done at a reasonable hour, without causing a nuisance, etc.

49 *Sir Lindsay Parkinson & Co. Ltd* v. *Commissioners of Works* [1949] 2 KB 632 (CA).

50 *London Borough of Hounslow* v. *Twickenham Garden Developments Ltd* [1971] ChD 233; but see also *White & Carter (Councils) Ltd* v. *McGregor* [1962] AC 413 (HL).

51 *Stockport MBC* v. *O'Reilly* [1978] 1 Lloyd's Rep. 595.

Terms entitling the employer to vary the works

Most civil engineering contracts of any sophistication provide that the employer can, through his agent, issue instructions[52]or variations.[53] This is coupled with a right for the contractor to claim additional payment.[54] Where variations are permitted but no limit is expressed as to the permitted value or extent of variations, the general rule is that reasonable variations connected with the works may be ordered; entirely new works cannot be imposed on the contractor.[55]

Variations to be in a prescribed format

Where the contract requires that instructions or variation orders be given in a prescribed format or manner, any purported instruction or variation order which is not so given may be ineffective, and the contractor may not be entitled to be paid for it,[56] unless the employer has waived his right to insist on compliance with the strict terms of the agreement.[57] However, even where the contract provides that variation orders must be in a particular format, the absence of that format may not be fatal where the primary question relates to whether or not the work instructed or ordered falls within the scope of the original contract. This is particularly so if the contract contains an arbitration clause entitling the arbitrator to review the decisions of the person vested with authority to order variations.[58]

---

52 For instance under the ICE Conditions of Contract, Clause 13(1) requires the Contractor to comply with the Engineer's instructions. Where the instruction requires the Contractor to incur cost, this is recoverable in accordance with Clause 13(3).

53 Under the ICE Conditions of Contract, Clause 51 requires the Contractor to execute ordered variations which are necessary for the completion or functioning of the works. The variation is to be valued in accordance with Clause 52(1).

54 If this were not the case, the employer would be entitled to change the work without change in price. This would render the contract uncertain and potentially voidable.

55 *Thorn* v. *Mayor and Commonalty of London* [1876] 1 AC 120 (HL); *Blue Circle Industries* v. *Holland Dredging Co. (UK) Ltd* (1987) 37 BLR 40.

56 *Forman & Co. Proprietary Ltd* v. *The Ship 'Liddlesdale'* [1900] AC 190 (PC).

57 *Meyer* v. *Gilmer* (1899) 19 NZLR 129 (New Zealand) where the requirement for written variation orders was held to be waived on account of the attendance of the employer at the meetings where the oral variation orders were given.

58 *Brodie* v. *Cardiff Corporation* [1919] AC 337 (HL).

Variations requested by the contractor

Where the variation mechanism within the contract is properly operated, it matters not, in principle, for whose benefit the variation was required. Accordingly, a contractor may, in appropriate circumstances, recover additional payment for a variation, where that variation is made at his request and for his benefit.[59] Where the contractor is put on notice that the option is his and he will not be paid additional sums, he will not be entitled to be paid.[60]

Pricing of variations

Civil engineering contracts typically provide rules for the valuation of variations. Where no rules are agreed, the valuation will depend on the construction of the agreement. Broadly speaking, if the varied work is of a type which is identical or similar to that already priced in the contract, then the contract prices will apply to the varied work. Where the work is of a different type or is undertaken in different circumstances, and a price may readily be determined by analogy with work for which prices are given, then this analogous price will be payable. Where there is little relation between the varied work and that originally to be supplied, a reasonable price will be payable. Contractors frequently submit claims based on daywork rates published by various bodies; these are only applicable where the contract expressly permits them or they are shown to represent reasonable rates and the construction method and time taken are each reasonable.

5. Contract administration: decisions, certificates and adjudication

The proper administration of a civil engineering contract requires decisions to be made from time to time which amount to an interim

59 *Simplex Piling* v. *St Pancras Borough Council* (1958) 14 BLR 80.

60 *Howard de Walden Estates Ltd* v. *Costain Management Design Ltd* (1991) 55 BLR 124. Some standard form contracts provide that where the variation is required as a result of the contractor's default, no account may be taken of it in the valuation (e.g. ICE Conditions of Contract, Clause 51(3)); even this will not, it is thought, prevent the Contractor recovering for a variation which is simply made to accommodate the Contractor and which is not necessitated as a result of his prior default.

resolution of a potential dispute. These decisions are different in nature to decisions such as issuing a variation, because they involve an interpretation of the contract and involve a statement as to the rights of the parties. Many civil engineering contracts contain devices for such decisions to be made in a specified format. The devices which have been developed include certification,[61] formal decision making,[62] valuation, expert decision making and adjudication.[63] Some are expressed to be advisory, while others are expressed to be binding. Some are expressed to be final and unchallengeable, while others may be challenged at specified times by specified means.

All such processes are contractual in nature and their effect is determined by construing the contract. A party seeking to rely and enforce a decision purportedly made under such provisions must show that the decision complies with the express and implied requirements of the contract. While each process must be construed individually, a number of broad generalisations may be suggested.

The decision must be made in the circumstances authorised by the contract. The decision must be made by the person authorised by the contract[64] though the authorised decision-maker may delegate some of the background work, providing he takes the final decision.[65] The decision-maker must make his decision independently, fairly and in good faith,[66] though he need not afford the parties a full hearing or anything like it.[67] The decision need not be made in writing or in any form unless required by the contract,[68] though it should be

61 Most construction contracts have certification provision in relation to payment, so that the employer pays when the decision-maker certifies that payment is due. Certificates are also issued to define stages in the Works; for instance, under the ICE Conditions of Contract, a certificate is issued to fix the date of Substantial Completion and the date of the completion of the Defects Correction Period.

62 See e.g. the Engineer's decision under Clause 66 of the ICE Conditions of Contract.

63 When in force the Housing Grants, Construction and Regeneration Act 1996, Section 108, will require or imply adjudication provisions into the contract.

64 *Croudace* v. *London Borough of Lambeth* (1986) 33 BLR 20 (CA).

65 *Anglian Water Authority* v. *RDL Contracting* (1988) 43 BLR 98 at 112.

66 *Sutcliffe* v. *Thackrah* [1974] AC 727 (HL); *Hickman* v. *Roberts* [1913] AC 229 (HL); *Panamena* v. *Frederick Leyland & Co.* [1947] AC 428 (HL).

67 *London Borough of Hounslow* v. *Twickenham Garden Developments Ltd* [1971] ChD 233.

68 *Elmes* v. *Burgh Market Co.* (1891) *Hudson's Building Contracts* (4th edn) Vol. 2, p. 170.

unambiguous.[69] The substance of the decision is crucial so that if, for example, the contract requires a "certificate" to be issued, the decision must be the clear result of a certifying process;[70] and where a "decision" is required, the Engineer's informal rejection of claims is insufficient.[71] The decision must be published.[72] A decision certifying monies due creates an assignable debt,[73] but is not, unless otherwise stated, equivalent to a cheque and is subject to set-offs,[74] though any purported set-off must be supported by a bona fide claim.[75] Any refusal by the decision-maker to issue a decision for reasons outside the agreement exempts a party from obtaining that decision as a necessary precondition to some other event such as payment.[76]

Any power of review is governed by the contract. Only decisions which purport to express the parties' rights under the contract are, of course, reviewable. Purely executive decisions may be wrong and a breach of contract entitling the contractor to damages, but they cannot be revised by an arbitrator.[77] Where the parties agree that decisions made during the currency of the works may be reviewed by an arbitrator (e.g. where the contract contains an "open up, review and revise" power), the arbitrator is operating a mechanism under the authority of the contract, which the courts cannot do without special agreement or where the contractual mechanisms break down. Save in these special circumstances, the role of the court is restricted to enforcing the agreement.[78]

69 *Token Construction* v. *Charlton Estates* (1973) 1 BLR 48 (CA).

70 *Token Construction* v. *Charlton Estates* (1973) 1 BLR 48 (CA).

71 *Monmouthshire County Council* v. *Costelloe and Kemple* (1965) 5 BLR 83 (CA).

72 *Token Construction* v. *Charlton Estates* (1973) 1 BLR 48 (CA).

73 *Workman, Clark & Co.* v. *Lloyd Brazileno* [1908] 1 KB 968 (CA).

74 *Gilbert-Ash (Northern)* v. *Modern Engineering (Bristol)* [1974] AC 689 (HL). But note also *Lubenham Fidelities* v. *South Pembrokeshire District Council* (1986) 33 BLR 39 (CA) where the contractor was held not to be entitled to be paid except at the face value of a wrongly computed certificate.

75 *C. M. Pillings & Co.* v. *Kent Investments Ltd* (1985) 30 BLR 80; *Ellis Mechanical Services Ltd* v. *Wates Construction Ltd* (1976) 2 BLR 57 (CA).

76 *Panamena* v. *Frederick Leyland & Co.* [1947] AC 428 (HL).

77 *London Borough of Merton* v. *Stanley Hugh Leach* (1985) 32 BLR 51.

78 *Northern Regional Health Authority* v. *Crouch* [1984] QB 644 (CA).

The Engineer and the ICE Conditions of Contract

A historical feature of civil engineering contracts is the appointment of an Engineer under the contract. The Engineer acts as both the Employer's agent and an impartial decision-maker. The two roles must be carefully distinguished. In his role as Employer's agent, he must take proper care for the Employer's interests. In his role as impartial decision-maker, however, he has discretionary powers which he must exercise fairly with due regard to the interests of both parties. As the Employer's agent he may put the Employer into a position of breach of contract. But as the impartial decision-maker, the Employer does not warrant that the Engineer will act reasonably; the Employer warrants only that he will not improperly interfere with the Engineer's discretion.[79] While the "agent" and "certification" functions must be kept separate, an Engineer owes a duty of care to his client not to certify negligently.[80] He does not owe such a duty to the contractor unless a direct agreement is made between the Engineer and contractor.[81] A full commentary on the Engineer's power is given in Chapter 8.

Adjudication

Adjudication is the term used for a rapid decision given by an independent person; the decision is binding on the parties unless and until reviewed, for example, by an arbitrator. Limited adjudication provisions have been used for some time in building sub-contracts, and more extensively in the NEC family of contracts. In 1994, the Latham Report[82] recommended the adoption of adjudication as standard in construction contracts. This proposal has been accepted by Parliament; Section 108 of the Housing Grants, Construction and Regeneration Act 1996, when in force, will require or imply adjudication provisions. Until then, the effect of an adjudication

---

79 This paragraph has been derived from dicta in *London Borough of Merton* v. *Stanley Hugh Leach* (1985) 32 BLR 51. This case involved a JCT Contract, but the position of the Architect under the JCT Contract is virtually identical to that of the Engineer under the ICE Contract.

80 *Sutcliffe* v. *Thackrah* [1974] AC 727 (HL).

81 *Pacific Associates* v. *Baxter* [1990] 1 QB 993 (CA).

82 Sir Michael Latham, *Constructing the team*, HMSO, 1994.

clause is determined by construing the term in the ordinary way. When the Act comes into force, the term adjudication may attain the status of a term of art, and be construed by the courts by reference to the legislative provisions and intent.

6. Vicarious performance, assignment and novation

Vicarious performance and sub-contracting

A contractor who agrees to undertake work which is objectively defined is, subject to any restriction in the contract, ordinarily entitled to arrange for others to perform it on his behalf.[83] This is usually referred to as vicarious performance. It is usually accomplished by sub-contracting the work out. Where the work involves personal skill and attention it may not be vicariously performed.[84] The objective intention of all parties is paramount in deciding whether an obligation may be vicariously performed.[85]

Many standard form contracts contain clauses relating to sub-contracting, normally restricting its use unless and until approval has been obtained from the employer's agent. Unless the contract provides otherwise, approval may be unreasonably withheld[86] save where to do so is tantamount to preventing the contractor's performance of the work.

Assignment

A party is entitled to assign the benefits of his contract to any third person provided (*a*) there is no contractual prohibition and (*b*) the person to whom the benefit is owed has no role to play, save in acquiring the benefit. Where the assignment is made, the assignee (i.e. the person to whom the benefits are assigned) is entitled to recover

83 *British Waggon Co.* v. *Lea & Co.* (1880) 5 QBD 149.

84 *Moresk Cleaners* v. *Hicks* [1966] 2 Lloyd's Rep. 338.

85 *Southway Group Ltd* v. *Wolff and Wolff* (1991) 57 BLR 33.

86 *Leedsford Ltd* v. *Bradford Corporation* (1956) 24 BLR 45 (CA).

those benefits from the de[...] [...]n who owes the benefits).
The debtor can use any defen[...] [...]signee which would be
available against the assignor (i.e. [...] [...]hom the benefit was
originally owed).[87] Most construction[...] [...]ain restrictions on
the right to assign, which can provide a[...] [...]defence to any
action by an assignee.[88]

The question of whether contractual dispute processe[...] benefits are assigned along with the benefit requires a cons[...] [...]vering whether the assignment is a legal or an equitable assignment.[...] [...]n of case of a legal assignment, the assignor transfers absolute title [...]e writing with notice to the debtor;[90] here the right, for instance, to recover monies owed by arbitration is passed to the assignee. An equitable assignment need not be in writing, nor need the debtor be given notice; here the benefit of contractual disputes procedures is not transferred.

Novation

A novation occurs where a contract subsisting between A and B is converted to one between B and C, with the agreement of all three. The subsequent agreement is a new contract which supersedes the original contract.

7. Insurance and indemnities

Civil engineering contracts frequently contain requirements for works to be insured and for indemnities to be given. An indemnity undertaking is an undertaking by one person to make good a loss experienced by another; an insurance agreement is an example of such

87 *Roxburghe* v. *Cox* (1881) 17 ChD 520 (CA).

88 *Linden Gardens Trust* v. *Linesta Sludge Disposals Ltd and others* [1994] 1 AC 85 (HL); *Helstan Securities Ltd* v. *Hertfordshire County Council* [1978] 3 All ER 262.

89 *Herkules Piling and Hercules Piling* v. *Tilbury Construction* (1992) CILL 770; *The Leage* [1984] 2 Lloyd's Rep. 259.

90 Law of Property Act 1925, s. 136(1).

an undertaking. An[illegible] who makes good a loss for another is subrogated to th[illegible]ts. Thus an insurance company which has paid for [illegible]erson's damage is entitled to pursue any claims that th[illegible] person would have against a third person in respect of [illegible] damage.

[illegible] of indemnity provisions in civil engineering contracts

Con[illegible]ovisions and the form, scope and extent of any cover to be S[illegible]ided are generally construed strictly. Thus an indemnity is not to [illegible]e construed so as to include the consequences of the indemnified party's own negligence unless those consequences are covered either expressly or by necessary implication.[91] Where, however, there are exceptions to the indemnity, these tend to be construed so as to maintain the effect of the indemnity.

Insurance

The contractor will normally be required to insure the works against damage and to maintain insurance against third party claims and to indemnify the employer against such claims. Likewise, professionals will normally be required to maintain appropriate insurance cover.[92] The details of insurance contracts lies beyond the scope of this book. It should be noted, however, that they are contracts of utmost good faith and so all material facts must be disclosed to the insurer. If there has been a material non-disclosure the insurance contract may be avoided by the insurer.

Third party rights against insurers

Where an insured person is liable to a third party, his insurer will (provided the claim falls within the agreed cover) indemnify him against the third party's claim. Where, however, the insured person is insolvent and cannot pay the money, there are limited rights for the

---

91 *Walters v. Whessoe Ltd & Shell Refining Co. Ltd* (1960) 6 BLR 23 (CA).

92 Traditionally, professional indemnity cover provides cover against claims made during the requisite period, rather than for breaches during that period. Accordingly, designers are frequently required to maintain a stated level of cover for 12 years.

third party to proceed directly against the insurer.[93] The conditions include that the insured person is obliged by law to pay. This latter condition may mean that a court order or its equivalent must be obtained before recovery against the insurance company is possible.[94]

8. Determination clauses

A determination clause is a clause entitling one or other or both parties to terminate the agreement. The term "forfeiture clause" is frequently used for clauses entitling the employer to determine the contractor's employment in specified circumstances.

The licence to occupy the site

A contractor who agrees to perform work on the employer's site is necessarily licensed to be on the site for purposes properly connected with the work.[95] Where the employer revokes this licence this normally ranks as a repudiation of the contract. Such a revocation will, however, be proper where the contractor is in fundamental breach or has himself repudiated the agreement, or where the contract entitles the employer to terminate the contract. Whatever the reason for the revocation of the licence, the contractor trespasses if he remains on site.[96]

Determination clauses

These clauses entitle the employer to terminate the contractor's employment upon the happening of specified events. The events may be breaches (e.g. not completing on time, not proceeding with due diligence, not complying with the engineer's instructions, etc.) or they may be mere events of greater or lesser commercial importance (e.g.

93 Third Party (Rights against Insurers) Act 1930.

94 *Post Office* v. *Norwich Union Fire Insurance Society Ltd* [1967] 2 QB 363.

95 *London Borough of Hounslow* v. *Twickenham Garden Developments Ltd* [1971] ChD 233.

96 *Chermar Productions Pty Ltd* v. *Pretest Pty Ltd* (1991) 8 Constr. L.J. 44 (Supreme Court of Victoria).

the insolvency of the contractor). A trivial or transitory default which establishes the condition entitling one party to determine the contract is sufficient; the party who determines the contract is not obliged to act reasonably.[97] But forfeiture clauses are construed strictly.[98] They must be exercised in accordance with any time limits or within a reasonable time and in accordance with any procedures stipulated in the contract. Where the breach of the employer or his agents has caused the event,[99] the determination is wrongful and is ordinarily a repudiation.

Wrongful determination and the court

Where there has been a wrongful determination, the court will not order the employer to allow the contractor to resume his licence. Damages are usually an adequate remedy.[100] However, where the contract entitles the employer to use the contractor's plant and materials following a proper determination[101] and the employer is doing so, it is thought that an injunction may be issued to restrain such use.

Consequences of determination

Where the employer properly determines the contractor's contract, his right to claim damages depends upon the terms of the agreement.[102] If

97 *Comco Construction Pty Ltd* v. *Westminster Properties Pty Ltd* (1990) 8 Constr. L.J. 49 (Supreme Court of Western Australia).

98 *Roberts* v. *Bury Commissioners* (1870) LR 5 CP; *Central Provident Fund Board* v. *Ho Bock Kee* (1981) 17 BLR 21 (Singapore).

99 For example where delay in issuing plans has delayed the contractor: *Roberts* v. *Bury Commissioners* (1870) LR 5 CP.

100 In *London Borough of Hounslow* v. *Twickenham Garden Developments Ltd* [1971] ChD 233 the employer was refused an injunction to remove the contractor from site. This decision has not been followed. See *Tara Civil Engineering Ltd* v. *Moorfield Developments Ltd* (1989) 46 BLR 72 which is to be preferred.

101 As in *Ranger* v. *Great Western Railway* (1854) 5 HLC (HL).

102 Where wholly disproportionate consequences flow, the provision may be a penalty clause: *Ranger* v. *Great Western Railway* (1854) 5 HLC (HL).

there is no express reference to the consequences, the employer will be able to claim damages if the event which triggered the determination was a breach by the contractor. But where the event is not of itself a breach he will not be able to do so.[103] In any event he may not be able to claim the additional cost of completing the works as damages if, in discharging the contractor and bringing in a substitute, he has not mitigated his loss. Where the employer wrongly determines the licence the contractor will be able to claim for damages, including loss of profit, subject to his mitigating his loss.

Contractor's right to determine the contract

Some contracts entitle the contractor to determine his own employment in specified circumstances, usually connected with failure by the employer to make payments.

Passing of property on forfeiture

Many civil engineering contracts contain provisions by which the contractor's plant and materials vest in the employer upon being brought on the site. The purpose is to enable the employer to complete the works in the event of the contract being determined. These are effective where clear words are used, providing that the plant or materials are the property of the contractor in the first instance.[104]

9. Guarantees and bonds

A contract of guarantee is one in which a surety promises to answer to the creditor for the default of the debtor. A guarantee is generally a contract of indemnity which entitles the surety to set up any defence

---

103 Thus in *Perar BV* v. *General Surety and Guarantee Co. Ltd* (1994) 66 BLR 72 (CA) it was held that the contractor was not in breach for going into administrative receivership.

104 This may not be where the plant is hired or belongs to a sub-contractor: *Dawber Williamson* v. *Humberside County Council* (1979) 14 BLR 70. Nor where there is an effective retention of title clause: see *Aluminium Industrie Vaasen* v. *Romalpa Aluminium* [1976] 1 WLR 676 (CA) and Williams, G. A., Reservation of title in the construction industry (1987) 3 Constr. L.J. 252.

against the creditor which is available to the debtor. A bond is an obligation made as a deed whereby the bondsman promises to pay the obligee a sum when the obligee calls for it and satisfies any conditions stipulated in the bond, including a bare demand. A bond may be a guarantee or an on demand bond.

The use of guarantees and bonds

In civil engineering projects, guarantees and bonds are used to secure the performance of contractors, sub-contractors, suppliers and professionals. The most common situation is where the employer is the creditor, the contractor is the debtor and the bank which provides the guarantee is the surety. In this section this matrix will be used for illustrative purposes and the terms contractor and employer will be used in the stead of debtor and creditor.

Construing the guarantee or bond

In the event of the contractor's default (as defined in the guarantee/bond) the liability of the surety to the employer is determined by the terms of the guarantee/bond.

Injunction to restrain the creditor

A guarantee or bond is an autonomous agreement. Where, however, the employer is clearly acting fraudulently, he may be restrained by injunction from calling the obligation.[105]

Matters which release the surety under a guarantee

"There is no general principle that 'irregular' conduct on the part of the employer, even if prejudicial to the interests of the surety, discharges the surety."[106] Where, however, the employer and contractor act in concert to the prejudice of the surety, the obligation of the

105 *Themehelp Ltd* v. *West and others* [1995] 3 WLR 751 (CA).

106 *Bank of India* v. *Patel* [1983] 2 Lloyd's Rep. 298 (CA), per Goff LJ at 302.

surety may be avoided.[107] The employer's mere failure to intervene when the contractor acts improperly will not suffice;[108] but where the employer "finances" the default[109] or connives at it,[110] this will release the surety. This will be all the more so where the collusion is fraudulent.[111] Where the secured contract obligation is materially altered after the guarantee contract was agreed, the surety may be discharged; thus, extending the starting date or otherwise changing the contract obligation may effect a discharge.[112] Where the contractor repudiates the agreement and the employer accepts that repudiation the surety will not be discharged;[113] but where it is the contractor who accepts the repudiation of the employer, it may be.

Bonds

A bond may be (*a*) conditional (i.e. it may be a guarantee and be conditional upon proof of the default of the contractor and/or proof of damage)[114] or (*b*) on demand (i.e. the right to call upon the bond is triggered by a bare statement by the employer that the bond is due).[115] The nature of the bond will be determined by construing its terms.[116] Where the bond is of the conditional type then (subject to the terms of the bond) the bondsman may take advantage of any defences which

107 *Re Sherry* (1884) 25 ChD 692 (CA). Lord Selbourne said at 703: "A surety is undoubtedly and not unjustly the object of some favour both at law and in equity and ... is not to be prejudiced by any dealings without his consent between the secured creditor and the principal debtor."

108 *Kingston-upon-Hull Corporation* v. *Harding* [1892] 2 QB 494 (CA).

109 For example where the employer voluntarily makes advance payments to finance the work.

110 *Bank of India* v. *Patel* [1983] 2 Lloyd's Rep. 298 (CA), expressly approving the language of Bingham J at first instance.

111 Where money is paid over, and the surety later discovers the collusion, he may readily recover it using the principles of restitution or deceit. He may also be able to recover it from those who were personally involved.

112 *Mercers* v. *New Hampshire Insurance* (1992) 60 BLR 26 (CA).

113 *Lep Air Services Ltd* v. *Rolloswin* [1973] AC 331 (HL).

114 *Esal (Commodities) Ltd* v. *Oriental Credit Ltd* [1985] 2 Lloyd's Rep. 546 (CA).

115 *Edward Owen Engineering Ltd* v. *Barclays Bank International Ltd* [1978] 1 QB 159 (CA).

116 See *Trafalgar House Construction* v. *General Surety & Guarantee Co.* [1996] 1 AC 199 (HL). In this case the Court of Appeal construed the standard form of bond issued in association with the ICE Conditions of Contract virtually as an on demand bond. The House of Lords, however, has re-established that it is a performance bond and its call requires more than the employer's demand.

the contractor may have had.[117] A bond which is by its terms "on demand" is just that; no terms concerning proof or reasonableness will be implied and the only grounds for refusal to pay is clear evidence of fraud.[118] Where an on demand bond is called and the debtor shows that it was wrongly called (i.e. there was no default on the obligation secured), the position is unclear. A contractor who has given an indemnity to the bondsman may be entitled to an account from the employer and to recover any unaccounted for sum paid over under the principles of restitution.

The failure to obtain a guarantee or bond

Where the civil engineering contract requires the contractor to obtain a guarantee or bond, this may be a condition or a warranty.

117 *Trafalgar House Construction (Regions) Ltd* v. *General Surety & Guarantee Co. Ltd* [1995] 3 WLR 204 (HL).

118 *Edward Owen Engineering Ltd* v. *Barclays Bank International Ltd* [1978] 1 QB 159 (CA).

4

Extra-contractual entitlements

Contract provides the principal vehicle for claims in relation to civil engineering projects; claims may be for an entitlement provided by the contract or for damages for breach of contract. However, other possibilities exist and it is not uncommon for a claimant to place his claim on several independent bases, not all of them contractual. The most commonly advanced extra-contractual options include:

(*a*) claims for misrepresentation
(*b*) claims for restitution in the absence of contract
(*c*) claims under a collateral or supplementary contract.

1. Misrepresentation

During contract negotiations, statements may be made by one party to induce the other to enter the contract on the terms beneficial to the person making the statement. Such statements may become contract terms; where they do not, they are said to be "mere representations".[1] If a representation of fact[2] is inaccurate and was a material factor in inducing the representee to enter into the contract, the representee has an action. The action may be framed in one or more of the following ways:

(*a*) a claim to be entitled to rescind the agreement
(*b*) an action for fraudulent misrepresentation (deceit)

---

1 For example, the employer may provide a site investigation report to the contractor; or the contractor may provide information relating to his previous experience.

2 A statement of honest opinion is insufficient: *Bisset* v. *Wilkinson* [1927] AC 177 (PC). Where, however, the opinion is tendered in circumstances where the representor has skill or experience in relation to the subject matter this may be actionable: *Esso Petroleum Co. Ltd* v. *Mardon* [1976] QB 801 (CA).

(*c*) an action for negligent misstatement
(*d*) an action under the Misrepresentation Act 1967.

Rescission for misrepresentation

A party who was induced to enter into an agreement by an innocent (or other) misrepresentation is entitled to rescind the agreement, provided that it is possible to restore any property which had passed in the transaction. Thus, for example, where the contractor has not yet entered onto the site he will be entitled to rescind the agreement. In many practical situations, of course, such as in part-performed civil engineering contracts, it is impossible to restore property passed during the transaction, thus rendering rescission impossible.[3]

Fraudulent misrepresentation

A person who is induced to enter into a contract by a fraudulent representation has an action in the tort of deceit against the representor.[4] The word "fraudulent" has been given wide scope and includes reckless statements.[5] Thus on a project involving underground work, the engineer's reckless representation that existing protective works were of adequate depth was fraudulent; it defeated a clause in the contract requiring the contractor to be responsible for satisfying himself as to dimensions.[6]

Negligent misstatement

A party who had been induced to enter into a contract by the negligent misrepresentation of the other party can sue for damages for negligent

---

3 The leading case is *Glasgow and South Western Railway* v. *Boyd & Forrest* [1915] AC 526 (HL). Here the contractor completed a railway with significant groundworks, and then claimed rescission and a *quantum meruit* on the grounds that the employer's engineer had innocently misrepresented the condition of the soil strata. Rescission was refused on the grounds that a restoration of property was impossible.

4 The person making the misrepresentation must, of course, be authorised to do so on behalf of his principal. In the normal course of events, a senior engineer who makes representations of engineering fact during negotiations will be authorised to do so unless such authority is expressly limited.

5 In other words a representation made knowing it to be untrue or reckless as to whether it be true or false: *Derry* v. *Peek* (1889) 14 App Cas 337 (HL).

6 *Pearson* v. *Dublin Corporation* [1907] AC 351 (HL).

misstatement. In other cases, where no contract exists between the representor and the representee, an action for damages may exist where the representor professes to possess a degree of skill or experience and makes his statement knowing that the other may act on it.[7]

Misrepresentation Act 1967

The provisions of the Misrepresentation Act 1967 are in addition to the misrepresentee's previously existing causes of action. The Act deals with two matters of present importance. In the context of negligent misrepresentations, where the misrepresentee actually enters a contract, section 2(1) provides a new and stronger remedy for negligent misrepresentation by entitling him to recover as if the negligent misrepresentation had been fraudulent. Section 2(2) deals with the situation where rescission is not feasible or equitable and provides a statutory right to damages in lieu of rescission for innocent or negligent misrepresentation.

Section 2 of the Act provides as follows:

> (1) Where a person has entered into a contract after a misrepresentation has been made to him by another party thereto and as a result thereof he has suffered loss, then, if the person making the misrepresentation would be liable to damages in respect thereof had the misrepresentation been made fraudulently, that person shall be so liable notwithstanding that the misrepresentation was not made fraudulently, unless he proves that he had reasonable grounds to believe and did believe up to the time the contract was made that the facts represented were true.
>
> (2) Where a person has entered into a contract after a misrepresentation has been made to him otherwise than fraudulently, and he would be entitled, by reason of the misrepresentation, to rescind the contract, then, if it is claimed, in any proceedings arising out of the contract, that the contract ought to be or has been rescinded, the court or arbitrator may declare the contract subsisting and award damages in lieu of rescission, if of opinion that it would be equitable to do so,

7 *Hedley Byrne* v. *Heller and Partners* [1964] AC 465 (HL).

> having regard to the nature of the misrepresentation and the loss that would be caused by it if the contract were upheld, as well as to the loss that rescission would cause to the other party.

Where the misrepresentation has become a term

Where, for instance, an inaccurate site survey issued at tender stage becomes a term of the agreement, the claimant may elect to advance a claim for misrepresentation or for breach of contract.

2. Claims for restitution

A claim in restitution is a claim by A that:

(*a*) B has received a benefit from A; and
(*b*) it would be inequitable for B not to restore it to A either by the return of goods or by paying for services rendered.

A claim for restitution is necessarily made in the absence of a contract governing the situation.

In the context of civil engineering, restitutionary claims are most frequently advanced in respect of services rendered where there is no contract.[8] This may be because (*a*) A commences work for B in the expectation that a contract will eventuate; or (*b*) A supplies work to B outside the scope of an existing contract; or (*c*) the contract between A and B ceases to exist; or (*d*) A supplies services to B in an emergency. In each of these cases, A claims a reasonable amount for services rendered. This is frequently termed a *quantum meruit*.[9] This expression is used indiscriminately to refer also to payments under a contract where no payment is fixed, or where the measure of payment is agreed to be a "reasonable amount".

8 Such a claim is frequently termed a claim in quasi-contract.

9 As much as it is worth. See Levine M. F. and Williams J. H., Restitutionary quantum meruit—the cross roads (1992) 8 Constr. L.J. 244.

Where one person commences work in the expectation that a contract will eventuate

It is not uncommon for a professional or contractor to commence work before the contract is finally agreed. Where no contract eventuates, the professional or contractor is entitled to be paid a reasonable amount by the employer provided that the work which was carried out was done for the benefit of the employer and at his request, implied or express.[10] Where, however, the contractor commences work at his own risk, and no contract eventuates, he is not entitled to any payment.[11] While the party supplying work is entitled to be paid for it, the other party is not entitled to counterclaim, for instance for later delivery, since there is no contractually agreed delivery date.[12] If a contract is eventually agreed, work provided prior to its agreement will be covered by it, unless otherwise stated in the contract.[13]

Where one person supplies work to another outside the scope of an existing contract

Where a contract exists, it is common for the employer to request the contractor to do additional work. Where this is covered by the terms of a variation clause, the additional work will form part of the contractual work and will be paid in accordance with the terms of the contract. However, where the work is outside the contract scope, the contractor is entitled to be paid a reasonable amount for that work.[14]

Where the contract ceases to exist

A contract may be set aside for mutual mistake or duress; it may be rescinded or frustrated. A person who has performed work in pursuance of a non-existent contract at the request of the other party

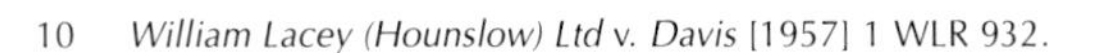

10 *William Lacey (Hounslow) Ltd* v. *Davis* [1957] 1 WLR 932.

11 *Regalian Properties plc* v. *London Docklands Development Corporation* [1995] 1 WLR 212.

12 *British Steel Corporation* v. *Cleveland Bridge and Engineering Co. Ltd* [1984] 1 All ER 504.

13 *Trollope & Colls Ltd* v. *Atomic Power Construction Ltd* [1963] 1 WLR 333.

14 *Sir Lindsay Parkinson & Co.* v. *Commissioner of Works* [1949] 2 KB 632 (CA).

will generally be paid a reasonable amount unless the contract states what is to happen in such an event.[15]

Where the plaintiff supplies services in an emergency

Work may be supplied in an emergency in order to preserve the integrity of the work, other property, or the safety of those in the vicinity. Immediate action is often called for and the person specifically responsible for the work is not always available, or in a position to carry out the preventative work himself. An intervener may perform the emergency operations. It may be that where the intervener acts wholly officiously he will not be entitled to be paid;[16] where however he acts in pursuit of some interest he has (e.g. if the collapse would damage his own works) he will be entitled to a reasonable amount.

The quantification

A *quantum meruit* means a reasonable sum in all the circumstances. It is not simply the cost of the work plus a reasonable addition for "profit". The computation may be informed by the nature and terms of an intended agreement.[17]

3. Claims under a collateral or additional contract

Several possibilities exist for action under a contract which is collateral to or substitutes for the primary contract.

15 For example, Clause 64 of the ICE Conditions of Contract provide rules for payment in the event of frustration.

16 *Falcke* v. *Scottish Imperial Insurance Co.* (1886) 34 ChD 235 at 248–249 per Bowen LJ. Where, however, the person who benefits from his intervention knows of it and does nothing to indicate that the intervener will not be paid, it seems that the intervener may claim: Lord Cranworth in *Ramsden* v. *Dyson* (1866) LR 1 HL 129 140–141 cited in *Proctor* v. *Bennis* [1887] ChD 740 by Cotton LJ.

17 *Way* v. *Latilla* [1937] 3 All ER 759.

Collateral contract

During negotiations one party may agree to enter into the contract in return for some advantage. For example, a prospective house purchaser may agree to sign the agreement provided some undertaking is given as to the state of the drains.[18] In some cases, this collateral undertaking can form the substance of a contract in its own right, and provides a right of action.

Compromise agreement

Where the parties agree to compromise their disputes, a new agreement comes into existence. Thus where a contractor accepts a promise of £x in full and final satisfaction of all claims, and the employer reneges, the contractor's action is under the compromise agreement.[19]

18 *De Lassalle* v. *Guildford* [1901] 2 KB 215 (CA).

19 It will be a question of construction whether or not the arbitration provisions (if any) in the primary contract will be incorporated into the accord and satisfaction. In practice, however, there may be little scope for dispute and summary judgment will be available.

5

Civil engineering professional services contracts

Contracts for professional services in civil engineering practice are variously styled "appointments", "retainers", "engagements", etc.; they include contracts in respect of design and engineering services, surveys, feasibility studies and the provision of supervision services.

1. Formation of design and professional services contracts

Contracts for professional services are formed in the same way as other contracts.[1] In practice, there tend to be fewer detailed negotiations than for a civil engineering construction contract.[2] The agreement is typically made in or evidenced by an exchange of letters referring to a standard form of conditions or a standard scale of fees such as the forms published by the Association of Consulting Engineers. While professionals frequently contract on the standard conditions published by their professional institutions, this general practice is, in itself, insufficient to make those conditions part of the contract without the agreement of the client, unless that particular client and consultant habitually contract under those conditions.[3]

1 See Chapter 2.

2 Although where competitive tendering is used, the process may involve qualification stages, tender bids, negotiations, etc.

3 *Sidney Kaye, Eric Firmin & Partners* v. *Leon Joseph Bronesky* (1973) 4 BLR 1.

2. The terms of the agreement

Any agreed terms prevail. Where the contract is made informally, terms will be implied by the Supply of Goods and Services Act 1982 and by the usual operation of law to give the agreement business efficacy.[4] Terms relating to the standard of care, remuneration, time and scope are of obvious importance.

The standard of care

Unless the parties expressly agree otherwise, the principal obligations of the consultant will normally be to exercise "reasonable care and skill"[5] in and about the performance of the work covered by the agreement. Where the agreement stipulates a higher standard, this will be enforced; thus where the circumstances indicate that the design must be fit for its purpose, such a term will be taken to apply.[6] The failure to comply with the standard of reasonable skill and care is commonly termed "professional negligence".

Time

Unless the agreement states otherwise the professional will be obliged to complete his work within a reasonable time.[7]

Payment

Unless the agreement states otherwise, the professional will be entitled to recover a reasonable fee for his work.[8] Conditions published by professional bodies contain scales of fees so that if the

---

4 See Chapters 2 and 3.

5 Section 13 of the Supply of Goods and Services Act 1982. In most circumstances, the words care and skill are transposed to read "reasonable skill and care".

6 *Greaves Contractors* v. *Baynham Meikle & Partners* [1975] 1 WLR 1095 (CA). The engineer was employed as sub-contractor to a contractor engaged to design and build a warehouse. A term was implied into the design agreement that the building would be reasonably fit for its purpose as the engineer knew that this was the standard to which the contractor was working and that the contractor intended the engineer's obligation to be on the same basis.

7 Section 14 of the Supply of Goods and Services Act 1982.

8 Section 15 of the Supply of Goods and Services Act 1982.

scope of the work increases or decreases, the remuneration will also be increased or decreased in line with the changes in the scope of the work. Where a lump sum fee is agreed and there is a possibility that the scope of works may increase during the currency of the project, the agreement should say how the remuneration is to be affected. Where this is not done and the professional voluntarily undertakes more work, he will not be entitled to be paid for it[9] unless it was reasonably clear to the client that the professional was expecting to receive additional payment.

The extent of the work

Where the consultant contracts on the basis of one of the standard forms of appointment published by one of the professional institutions, the scope and extent of the consultant's obligations will be as set out in that form. Where, however, the agreement is informal the parameters of the agreed obligation may be inferred only from the entire factual matrix; for instance, the duty to act as certifier may be inferred from the fact that the construction contract to be used contains certification provisions. Where specialist work is included in the project and that specialist work would normally fall outside the professional's area of competence, it is a question of fact whether or not the professional is responsible for its design. Thus on a project where there is no engineer employed and the architect becomes involved in seeking quotations for specialist engineering work, he may well become responsible for its design.[10]

The duration of the consultant's obligation

A professional's design duty continues throughout the design and construction phases. A design which was initially and justifiably thought to have been suitable may in the event turn out to be unsuitable; this will not render the designer negligent. However, the discovery of the design defect prior to the completion of construction reactivates the professional's duty in relation to the design and imposes on him a duty to take such steps as are necessary to correct

9 *Gilbert & Partners* v. *Knight* [1968] 2 All ER 248 (CA).

10 *Richard Roberts Holdings Ltd and another* v. *Douglas Smith Stimson Partnership* (1988) 46 BLR 50.

the results of the inadequate design.[11] It seems, however, that once a project is completed, a consultant need not continue to watch out for it.[12]

Termination

Most standard form contracts contain provisions for terminating the consultancy agreement. Where, however, a consultant has embarked upon a project and there are no agreed termination provisions, the client cannot terminate the consultant's appointment at will. The consultant is to be allowed to complete the project which he commences. Where he is prevented from so doing this will be a breach of contract entitling him to damages for the loss of profit.[13]

3. Delegation of design and consultancy services

A construction contractor is normally entitled to sub-let work. Since the employer seeks a defined result, the employer can look to the contractor to perform the contract or to pay for its non-performance in any event. In the case of design or other professional contracts different considerations generally apply. The professional is frequently chosen because of his skill, reputation or flair in a particular field and a client might be rightly aggrieved to find that the work which he had entrusted to a particular person had been delegated or sub-let. In addition, the consultant's obligation is not normally to achieve a specified result but to exercise reasonable skill and care; accordingly difficult questions of liability may arise where the designer sub-lets the work to an apparently competent person. For these reasons, it has been established that a consultant may not

11 *London Borough of Merton* v. *Lowe* (1982) 18 BLR 130.

12 It was suggested by the judge at first instance in *Eckersley and others* v. *Binnie & Partners and others* (1988) CILL 388 (CA) that a professional might have a continuing duty. This suggestion was disapproved by Bingham LJ in the Court of Appeal.

13 *Thomas* v. *Hammersmith Borough Council* [1938] 3 All ER 201; *Edwin Hill & Partners* v. *Leakcliffe Properties Ltd* (1984) 29 BLR 43.

ordinarily delegate design work,[14] including financial appraisals,[15] unless authorised by his client to do so.[16]

4. Professional negligence: a failure to exercise reasonable skill and care

The term "negligence" is ordinarily used to refer to the tort of negligence. However, the composite form "professional negligence" is used to indicate either a breach of a term in a contract requiring the professional to exercise reasonable skill and care, or a breach of a duty owed by a professional in tort. This dual use derives from the fact that the duties owed in the tort of negligence and under a professional contract are identical.[17]

A professional does not necessarily breach his duty by making a design error or by supplying bad advice. He will be in breach only if he fails to use reasonable skill and care.[18] The question of whether or not he has met this standard must be answered in the light of all the circumstances existing at the time of the alleged breach. Caution must be exercised when attempting to make assessments after the event.[19]

Common professional practice

A person who holds himself out to be skilled in a particular profession must exercise "the standard of the ordinary skilled man exercising and professing to have that special skill. A man need not possess the highest expert skill; it is well established law that it is sufficient if he exercises the ordinary skill of the ordinary competent man exercising that particular art."[20] Likewise, "... a professional man should command the corpus of knowledge which forms part of the

14 *Moresk Cleaners* v. *Hicks* [1966] 2 Lloyd's Rep. 338.

15 *Nye Saunders & Partners* v. *Alan E. Bristow* (1987) 37 BLR 92.

16 *Investors in Industry* v. *South Bedfordshire District Council* [1986] QB 1034.

17 *North West Water Authority* v. *Binnie & Partners* (1989) QB unreported, December 1989.

18 Unless of course the consultant has agreed to supply a design etc. which is fit for its purpose.

19 *Eckersley and others* v. *Binnie & Partners and others* (1988) CILL 388 (CA).

20 McNair J in *Bolam* v. *Friern Hospital Management Committee* [1957] 1 WLR 582.

professional equipment of the ordinary member of his profession. He should not lag behind other ordinarily assiduous members of his profession in knowledge of new advances, discoveries and developments in his field. He should have such awareness as an ordinarily competent practitioner would have of the deficiencies in his knowledge and the limitations in his skill. He should be alert to the hazards and risks inherent in the professional task he undertakes to the extent that other ordinarily competent members of his profession would be alert. He must bring to any professional task he undertakes no less expertise, skill and care than other ordinarily competent members of his profession would bring, but need bring no more. The standard is that of the reasonable average. The law does not require of a professional man that he be a paragon, combining the qualities of polymath and prophet .... In deciding whether a professional man had fallen short of the standards observed by ordinarily skilled and competent members of his profession, it is the standard prevailing at the time of the acts or omissions which provided the relevant yardstick. He is not to be judged by the wisdom of hindsight."[21] The standard of care in the context of professional practice is not to be judged according to the standard that a reasonable man would expect from such a professional but from the standard that other members of that profession would consider appropriate.[22] The claimant must therefore establish that what the consultant has done, or his failure to do it, falls below this standard.[23] Inexperienced professionals are required to work to the same standard as experienced professionals doing the same work.[24] Where there are several schools of thought and the defendant can show that he had followed a recognised school he will not be in breach.[25] Where professional work requires an extension

21 *Eckersley and others* v. *Binnie & Partners and others* (1988) CILL 388 (CA) per Bingham LJ (dissenting).

22 Unless the standard which other professionals apply is clearly unreasonable: *Lloyds Bank* v. *E. B. Savory & Co.* [1933] AC 201 (HL).

23 *McLaren Maycroft Co.* v. *Fletcher Development Co. Ltd* [1973] 2 NZLR 100 (New Zealand Court of Appeal).

24 *Wilsher* v. *Essex Area Health Authority* [1987] 2 WLR 425 (CA): "this notion of a duty tailored to the actor, rather than to the act which he elects to perform, has no place in the law of tort", per Mustill LJ at 440. It is submitted that the same applies to contractual liability.

25 *Maynard* v. *West Midlands Regional Health Authority* [1984] 1 WLR 634 (HL).

to existing practice, professionals discharge their burden by paying consideration to the potential problems as other members of their profession would have considered to be reasonably sufficient.[26]

Codes of practice, whether drawn up under the direction of the British Standards Institute, some equivalent foreign committee, or by the relevant professional bodies may be considered to comprise a formal statement of what is considered "good practice". A failure to undertake work in accordance with a design standard may be prima facie evidence of breach unless it can be demonstrated that it conforms with accepted engineering practice by rational analysis;[27] nevertheless, all the relevant circumstances must be taken into account.

Specialist skills are to be judged against the standards in that specialism.[28] The standard of "common professional practice" is problematical where unique work is being undertaken or where it is so specialist that no body of practice exists; nevertheless the court will determine the relevant level.

A professional should keep reasonably abreast of new developments.[29] He need not read every article appearing in the professional literature, and need not adopt techniques advocated in those articles until such techniques become accepted practice in the relevant discipline.[30]

5. Examples of professional service obligations

Collecting and investigating information

A person engaged to undertake a survey must consider what information is required. Thus a person undertaking a site survey for a development may be in breach if he fails to investigate the history of

26 *Independent Broadcasting Authority* v. *EMI Electronics Ltd and BICC Construction Ltd* (1980) 14 BLR 1 (HL).

27 *Bevan Investments Ltd* v. *Brackhall & Struthers (No. 2)* [1973] 2 NZLR 45 (New Zealand Supreme Court), per Beattie J at 66.

28 *Maynard* v. *West Midlands Regional Area Health Authority* [1984] 1 WLR 634 (HL).

29 *Eckersley and others* v. *Binnie & Partners and others* (1988) CILL 388 (CA).

30 *Crawford* v. *Charing Cross Hospital* (1953) *The Times*, 8 December.

the site[31] or to investigate adjacent sites for evidence of geological faults which may affect his development.[32] A surveyor must investigate the reliability of any information which is supplied to him.[33] He must not simply rely on the claims of a manufacturer of proprietary products.[34] A builder was held to be negligent for failing to take additional precautionary measures (such as obtaining expert advice) when the soil conditions which manifested themselves should have alerted him to the fact that his initial design was inadequate;[35] accordingly if, as they so often do, ground conditions on site indicate that additional site investigation or a revised design is required, the engineer should be prepared to recommend additional necessary work or alterations.

The preparation of reports

Where a professional prepares an inaccurate report, he will be liable to his client where his client suffers loss as a result of his failure to use reasonable skill and care. Where matters touched upon in the report genuinely cannot be forecast (e.g. future movement of foundations) the engineer must make this clear to the client.[36] In some cases a professional may owe a duty to third parties who rely upon his report. Thus, where accounts were prepared for a company in order to attract investors and were negligently prepared, investors who relied on them and suffered a loss as a result were able to recover from the accountant. They owed a duty "... to any third person to whom they themselves show the accounts, or to whom they know their employer is going to show the accounts, so as to induce him to invest money or take some other action on them. But I do not think the duty can be extended still further so as to include strangers of whom they have heard nothing and to whom their employer without their knowledge may choose to

31 *Balcomb* v. *Wards Construction (Medway) Ltd* (1980) 259 EG 765.

32 *Batty* v. *Metropolitan Property Realisations Ltd* [1978] QB 554 (CA).

33 *Moneypenny* v. *Hartland* (1824) 2 C & P 378.

34 *Sealand of the Pacific* v. *Robert C. McHaffie Ltd* (1974) 51 DLR (3d) 702 (British Columbia Court of Appeal).

35 *Bowen* v. *Paramount Builders (Hamilton) Ltd* [1977] 1 NZLR 394 (New Zealand Court of Appeal).

36 *Matto* v. *Rodney Brown Associates* [1994] 41 EG 152 (2 EGLR 163).

show their accounts."[37] Professionals frequently attempt to restrict their liability to third parties by the use of disclaimers. Such disclaimers are effective only in so far as they satisfy the reasonableness tests laid down in the Unfair Contract Terms Act 1977.

Supervision/inspection of work

An engineer's duty to supervise and/or inspect work in progress will depend on his contract with his client. An engineer engaged to supervise the work is not ordinarily obliged to watch every detail of the work undertaken by the contractor.[38] He is entitled to consider how best to employ the resources which are allocated to the project; thus where he has no reason to suspect that there is anything amiss, he need not spend a great deal of resources in supervising the contractor.[39]

Safety

In addition to the ordinary obligations imposed by the Health and Safety at Work etc. Act 1974, professionals must be aware of the Construction (Design and Management) Regulations 1994. By Regulation 13 the designer shall ensure that any design he prepares pays proper regard to risks during and after construction and gives priority to measures which will protect people. He must ensure that information is available as to the design and he must cooperate with the planning supervisor. In addition, a professional may be engaged as a planning supervisor under the regulation; if so he must exercise proper competence in this role. The regulations seem to impose no

---

37 *Candler* v. *Crane, Christmas and Co.* [1951] 2 KB 164 (CA); see also *Caparo Industries plc* v. *Dickman* [1990] 2 AC 605 (HL).

38 *McLaren Maycroft Co.* v. *Fletcher Development Co. Ltd* [1973] 2 NZLR 100 (New Zealand Court of Appeal).

39 *East Ham Corporation* v. *Bernard Sunley & Sons Ltd* [1966] AC 406 (HL).

new direct civil liability;[40] however, a consultant who acts in disregard of the regulations may be deemed to have failed to exercise the general skill and care required of a professional and thus attract liability.

40 Reg. 21. Note, however, that there are two exceptions. By one of these a client who allows the construction phase to commence without a health and safety plan may be liable, presumably to persons injured as a consequence. It may be that a professional adviser who fails to alert the client of this obligation may attract liability.

6

Civil engineering claims: entitlements and evaluation

1. Formulating claims

A claim may be based on the contract either (*a*) as an entitlement provided for under the contract; or (*b*) as damages for breach of a term of the contract. It may also be advanced as a claim related to, but outside, the contract[1] (*a*) as a claim for misrepresentation; (*b*) as a claim for a *quantum meruit* where the work claimed for does not form part of the contract work; (*c*) as a claim for negligence or negligent misstatement; or (*d*) under the terms of a collateral contract. Claims may be advanced on alternative bases. The following matters may be considered when deciding how best to frame a civil engineering claim.

The basis of valuation permitted

Frequently the claimant's entitlement under the contract will be more generous than the damages which he would recover for breach of contract.[2] Accordingly, given the choice, he may prefer to run his case primarily on the basis of an entitlement under the contract.

The dispute resolution mechanisms provided for in the contract

The applicability of such mechanisms to the dispute will depend upon the true construction of the contract provisions[3] and the factual basis

---

1 See Chapter 4.

2 For the measure of damages see below, section 6.

3 *Ashville Investments* v. *Elmer Contractors* [1989] QB 488 (CA).

alleged. For instance, a claim for restitution on the grounds that no contract eventuated deprives the claimant of the benefit of any dispute resolution mechanisms in the contract.[4] A claim for misrepresentation may fall within the terms of the dispute resolution clause, though this will depend on the construction of the contract. A claim under a collateral contract may be available if it can be shown that the dispute resolution provisions of the principal agreement are incorporated into the collateral agreement.

Notices

Civil engineering contracts frequently contain provisions which purport to deprive or limit[5] the right to claim unless proper notices are served. While a failure to comply with notice provisions may prevent the contractor from receiving benefits under the contract, breaches of contract or extra-contractual remedies will not normally be affected.

2. Statements of claim and response

The content of the statement

Subject to any requirements in the contract, the parties are entitled to formulate and set out their claims as they wish.[6] All claims, however, should contain a clear statement of (*a*) the nature of the claim, (*b*) the basis of the alleged entitlement and (*c*) the sums claimed and their computation.

A clear statement of the nature of the claim The claim should state whether it is a claim for breach of contract or alternatively a claim for an entitlement under the contract or some other type of claim. A party

4 This is not always a bad thing as far as the claimant is concerned, as he may then be entitled to serve a writ immediately rather than operating time-consuming dispute provisions in the contract.

5 For example, under the ICE Conditions of Contract, notices are required in a number of situations. They rarely operate as conditions precedent, but the claimant's right to claim is limited if his failure to serve a notice has prejudiced the Engineer: see Clause 52(4)(e).

6 *GMTC Tools and Equipment Ltd* v. *Yuasa Warwick Machinery Ltd* [1994] 73 BLR 102 (CA).

is entitled to advance alternative claims even if they are inconsistent providing one claim does not undermine another.

A clear statement of the basis of the entitlement Where a breach of contract is alleged, the term which is allegedly breached should be set out (or clearly referred to) and the facts of the alleged breach should be set out succinctly. The claim should include all material facts so that the other party is alerted to the case being advanced.[7] The case must be pleaded with sufficient particularity and the causation between any breaches of contract alleged and the consequences which it is alleged flow from them must be outlined.[8]

A clear statement of the amount and calculation of the sums claimed For example, where a breach of contract is alleged, the claim will be for damages and a calculation giving a proper breakdown of how the damages have been calculated must be submitted.

Other rules of pleading

Complex interactions Generally speaking, the causal link between any breach and damage[9] must be pleaded and demonstrated. In some civil engineering situations, however, the interactions between individual items are complex and the precise cause and effect for each individual loss is difficult to establish. While there can be no excuse for failing to state in proper particularity the basis of the claim,[10] the complexity which arises in civil engineering cases has been judicially recognised; a party is not, it seems, obliged to show the individual effect of each and every minor factor where the degree of interaction is great.[11]

---

7 *Philipps* v. *Philipps* (1878) 4 QBD 127 at 139 per Cotton LJ.

8 *Wharf Properties* v. *Eric Cumine Associates* (1991) 52 BLR 1 (PC).

9 Or, as the case may be, between the term giving an entitlement and the sums claimed.

10 *Wharf Properties* v. *Eric Cumine Associates* (1991) 52 BLR 1 (PC).

11 *J. Crosby & Sons* v. *Portland Urban District Council* (1967) 5 BLR 121 (DC); *London Borough of Merton* v. *Stanley Hugh Leach* (1985) 32 BLR 51; *Mid Glamorgan County Council* v. *J. Devonald Williams & Partners* (1991) 8 Constr. L.J. 61. See also Byrne, D., Total costs and global claims (1995) ICLR 531; Wilson M., Global claims at the cross roads (1995) 11 Constr. L.J. 15.

Applicability of rules of court to other claims In court proceedings the rules of pleading are reasonably narrowly defined; even here, however, the touchstone is whether or not the other party can be expected to understand the claim or response being advanced. These rules of pleading are not directly applicable to other claims or to arbitration proceedings. Nevertheless, some court rules are used habitually in all proceedings. Thus a claim for interest should be specifically pleaded.[12] And any matter which the party pleading will allege constitutes a release from the claim of the other party must specifically be pleaded.[13]

Amendment of pleadings Parties are generally entitled to amend their pleadings at any stage,[14] though they may be subject to a costs order on this account.[15]

3. Set-off and counterclaim

Set-off

Where A claims against B, B may reply that the sums claimed by A are to be diminished or extinguished because A owes B money. Here, a potential set-off arises. A set-off may be claimed at law or in equity. Furthermore, the contract may provide its own set-off provisions.

Set-off at law This species of set-off is rare in civil engineering cases. A prerequisite is that the sum to be set-off has been ascertained with certainty.[16] A surveyor's estimate does not render a claim ascertained.[17]

---

12 Order 18 Rule 8(4) of the Rules of the Supreme Court requires that "any claim for interest under Section 35A of the [Supreme Court] Act or otherwise" must be pleaded.

13 For example an accord and satisfaction or a time bar under the Limitation Acts.

14 They may not do so if this will cause unacceptable prejudice to the other party; but generally speaking any prejudice may be remedied by an appropriate order for costs.

15 *Cropper* v. *Smith* (1884) 26 ChD 700 at 710 per Bowen LJ.

16 *Axel Johnson Petroleum AB* v. *MG Mineral Group AG* [1992] 1 WLR 270 (CA).

17 *B. Hargreaves Ltd* v. *Action 2000* (1992) 62 BLR 72 (CA).

Set-off in equity B may set-off unliquidated cross-claims against A's claim providing they arise out of or in connection with the same contract as A's claim against B.[18] Equitable set-offs are generally effective, even where the contractor is in possession of a certificate ostensibly entitling him to immediate payment.[19] There are two distinct types of equitable set-off: (*a*) a cross-claim against the claimant arising out of or in connection with the same contract as the claim; (*b*) a claim by the respondent to be entitled to abate the claim because the work which the claimant claims to be paid for is defective and/or incomplete and hence has a value less than that claimed. Claims properly framed as (*a*) are true equitable set-offs while those properly framed as (*b*) are known as abatement claims.[20] Contracts frequently refer to and limit rights of "set-off". Terms purporting to exclude rights of set-off are construed strictly, so that a term excluding set-off may, in certain circumstances, not exclude an abatement claim.[21]

Contractual set-off Where the parties agree upon terms relating to set-off this will be enforced according to its construction. Thus rights of set-off in respect of other contracts between the two parties may be generated.

Counterclaims

Where A claims against B and B believes that he has a claim against A, he may counterclaim, providing the relevant tribunal has jurisdiction to entertain it. The tribunal will ordinarily have jurisdiction to hear any cross-claim which amounts to a set-off, since a set-off is a defence.

4. The evaluation of claims

Where a claim is made, and the respondent's liability is demonstrated, the amount due must be evaluated. The amount due may be recovered as one or more of the following.

18 *Hanak* v. *Green* [1958] 2 QB 9 (CA).

19 *Modern Engineering (Bristol)* v. *Gilbert-Ash* [1974] AC 689 (HL).

20 *Mondel* v. *Steel* (1841) 8 M & W 858; *Modern Engineering (Bristol)* v. *Gilbert-Ash* [1974] AC 689 (HL).

21 *Acsim (Southern)* v. *Danish Contracting and Development Co. Ltd* (1989) 47 BLR 55 (CA).

(*a*) Damages: compensation due as a result of the respondent's breach of duty.[22]
(*b*) Contractual entitlement: sums due in accordance with the contract terms.[23]
(*c*) *Quantum meruit*: a reasonable sum payable where the employer has requested the contractor to perform the work but no contract exists between them.
(*d*) Indemnity: a sum payable in order to reimburse a party for some payment which he is obliged to make.[24]

Damages

Damages are compensatory and are awarded to place the innocent party in the position in which he would have been had the breach not occurred.[25] Damages in contract are valued as the loss which the claimant has suffered as a result of the breach of contract, namely the difference between what he has received and what he would have received without the breach. Damages for misrepresentation are valued as the loss experienced by the claimant as a result of the misrepresentation.[26] Damages for breach of a tortious duty are valued as the loss suffered by the claimant as a result of the breach of duty to take care.

22 For example breach of contract, breach of duty not to misrepresent, breach of a tortious duty of care.

23 For example under Clause 12 of the ICE Conditions of Contract. Where adverse ground conditions are encountered which could not reasonably have been foreseen by an experienced contractor, this does not cause the employer to be in breach of contract. Nevertheless, the contractor is entitled to his cost plus an element for profit as an entitlement under Clause 12.

24 For example, where a surety guarantees to pay over a sum to an employer in the event of a contractor's default, he is entitled to an indemnity from the contractor.

25 In *Robinson* v. *Harman* (1848) 1 Ex 850 Parke B. said: "The rule of common law is, that where a party sustains a loss by reason of a breach of contract, he is, so far as money can do it, to be placed in the same situation, with respect to damages, as if the contract had been performed".

26 See *East* v. *Maurer* [1991] 1 WLR 461 (CA) (Fraudulent misrepresentation); on the measure of damages recoverable under sections 2(1), (2) of the Misrepresentation Act 1967 see *Royscot Trust Ltd* v. *Rogerson* [1991] 2 QB 297; *Gran Gelato* v. *Richcliff* [1992] ChD 560; *William Sindall plc* v. *Cambridgeshire County Council* [1994] 1 WLR 1016 (CA).

The claimant must demonstrate that he has suffered the loss[27] and that the damage is attributable to the respondent's breach. This involves demonstrating that: (*a*) the breach caused the loss; (*b*) the loss was a reasonably foreseeable result of the breach; (*c*) the evaluation proposed is reasonable in all the circumstances; and (*d*) the claimant is not claiming for any loss which he could have avoided.

Causation "If the damage would not have happened but for a particular fault, then that fault is the cause of the damage; if it would have happened just the same, fault or no fault, the fault is not the cause of the damage."[28] In some instances where more than one cause contributes to the plaintiff's damage there are complicating factors which must be considered.

Reasonable foreseeability The test of reasonable foreseeability is set out in *Hadley* v. *Baxendale*.[29] "Where two parties have made a contract which one of them has broken, the damage which the other party ought to receive in respect of such breach of contract should be [1] such as may fairly and reasonably be considered as either arising naturally i.e. according to the usual course of things, from such breach of contract itself, or [2] such as may reasonably be supposed to have been in the contemplation of both parties at the time they made the contract as the probable result of such breach of it." [Numbers added.] This passage refers to two distinct limbs as indicated. The first relates to the damage which may be recovered in the general case; while the second, often termed "special damages", is recoverable where the claimant notifies the respondent at the time of making the contract that these losses are likely in the event of a breach.

---

27 Where a claimant's damage is made good by a third party (e.g. a purchaser who buys a defective building at the full market rate) complications may arise. Where the loss is made good by an insurer, the insurance company is entitled to take over the plaintiff's claim under the principle of subrogation.

28 *Cork* v. *Kirby MacLean Ltd* [1952] 2 All ER 402 (CA); see also *Barnett* v. *Chelsea and Kensington Hospital Management Committee* [1969] 1 QB 428.

29 (1854) 9 Ex 341. Strictly speaking, this case relates to breaches of contract only. However, in tort, fairly similar principles apply.

Reasonable evaluation The decided cases show that there are no hard and fast rules.[30] The measure of damage cannot be determined in the abstract. The court does not disregard the hopes and aspirations or the individual predilections of the particular plaintiff in applying the basic principle.[31]

Mitigation of loss The claimant may not claim for any loss which he could reasonably have avoided. This is commonly termed "the duty to mitigate".[32] This terminology is slightly misleading as the plaintiff is under no duty to mitigate the loss. However, his entitlement to damages is limited to losses which could not reasonably be prevented by mitigation.[33]

Liquidated damages Liquidated damages are prospective damages which have been reduced to a fixed sum agreed in the contract. They may be contrasted with "unliquidated damages" which have not been reduced to a defined sum and which are the ordinary damages recoverable upon a breach of contract. Liquidated damages are commonly used in a variety of situations including (*a*) liquidated damages for delay[34] expressed as a certain amount of money per unit of time (e.g. per hour, per day, per month, etc.) and (*b*) liquidated damages for inefficiency in the completed works. A liquidated damages clause may not be used to penalise a breach excessively. Accordingly if it is clear that a liquidated damage clause is a penalty clause and cannot represent a genuine pre-estimate of the loss, then the courts will not enforce it but will treat the damages as unliquidated and will assess them.[35]

---

30 *British Westinghouse Electric and Manufacturing Co.* v. *Underground Electric Railways of London* [1912] AC 673 (HL) (at page 688): "The quantum of damages is a question of fact, and the only guidance which the law can give is to lay down general principles which afford at times but scanty assistance." See *Forsyth* v. *Ruxley Electronics Ltd* [1995] 3 WLR 118 (HL).

31 *Radford* v. *De Froberville* [1977] 1 WLR 1262.

32 *Payzu Ltd* v. *Saunders* [1919] 2 KB 581; Lord Haldane in *British Westinghouse Electric and Manufacturing Co.* v. *Underground Electric Railways of London* [1912] AC 673 (HL).

33 *The Solholt* [1983] 1 Lloyd's Rep. 605.

34 See Hosie J., Assessment of damages for delay in construction contracts: liquidated and unliquidated damages (1994) 10 Constr. L.J. 214.

35 *Dunlop Pneumatic Tyre Co. Ltd* v. *New Garage and Motor Company* [1915] AC 79 (HL).

However, the person attempting to show that the liquidated damages are a penalty has a difficult task.[36] Liquidated damages clauses may be used to limit the damages payable. Where, for example, the word "nil" had been inserted as the rate of liquidated damages in a standard printed form of contract, this meant that the rate was £0 and that this was "an exhaustive agreement as to the damages which are or are not to be payable to the contractor in the event of his failure to complete the works on time".[37]

Entitlements under the contract

Parties often include terms in their agreement designed to regulate entitlement in a variety of eventualities. These include payment for variations to work ordered in accordance with the contract, the occurrence of stated risks, etc. The entitlement which follows from such clauses is a question of construing the agreement. It is suggested that the following rules of construction are appropriate in the absence of anything to the contrary. First, where no method of computation is provided, the claimant must demonstrate that the sums claimed flow from the relevant event and that he has mitigated his loss. Second, where the claimant agrees to supply the respondent with a service under the agreement, he may claim a reasonable sum for profit in all the circumstances, unless the agreement provides otherwise.

Quantum meruit

A claim for *quantum meruit* is a claim for a reasonable sum in all the circumstances.[38]

5. Interest and finance charges claims

Due to the time which is often required to resolve civil engineering claims, interest frequently constitutes a major element of claims. Interest may be claimed as:

36 *Philips Hong Kong Ltd* v. *AG of Hong Kong* (1993) 9 Constr. L.J. 202 (PC).

37 *Temloc* v. *Errill Properties Ltd* (1988) 39 BLR 30 (CA) per Nourse LJ.

38 See Chapter 4, section 2.

(*a*) an entitlement under the contract;
(*b*) damages flowing from the breach; or
(*c*) under statute.

Interest as an entitlement under the contract

The parties may agree that late payment of sums or entitlements attracts interest, including compound interest.[39] Terms entitling the contractor to e.g. direct loss and expense may be construed to allow the recovery of "financing charges".[40]

Interest as damages flowing from breach

The common law maintains a fiction that a person suffers no loss by the late payment of damages owed to him; hence he cannot claim interest as damages for the delayed payment of those damages under the first limb of *Hadley* v. *Baxendale*.[41] Any claim for interest as damages must, therefore, fall within the second limb. This requires[42] that the claimant must have expressly advised the respondent at the time of making the contract that he would incur "special damages" (e.g. interest) if the claimant is not paid promptly.

Interest under statute

The court/arbitrator has discretionary powers under the Supreme Court Act 1981 and the Arbitration Act 1950 to award interest from the date when a cause of action arose.

---

39 See e.g. Clause 60(7) of the ICE Conditions of Contract.

40 For example in *F. G. Minter Ltd* v. *WHTSO* (1980) 13 BLR 1 (CA) the "direct loss and/or expense" provisions of the 1963 Standard Form Building Contract were construed to include financing charges. See also *Rees & Kirby* v. *Swansea City Council* (1985) 30 BLR 1 (CA); *Ogilvie Builders Ltd* v. *Glasgow District Council* (1994) 68 BLR 122 (Scottish Court of Session).

41 *London, Chatham and Dover Railways* v. *South Eastern Railway* [1893] AC 429 (HL); *President of India* v. *Lips Maritime* [1988] AC 395 (HL).

42 *Wadsworth* v. *Lydall* [1981] 1 WLR 598 (CA).

6. Extensions of time

The time for completion

Where a civil engineering contract is silent as to the time for completion, the contractor will have a reasonable time to complete the work. Where a completion time is agreed, but the contractor is delayed by some matter at the employer's risk, the completion time reverts to a reasonable time unless there is a term in the agreement enabling the time for completion to be extended. Where the time for completion reverts to a reasonable time, any liquidated damages provisions are rendered uncertain and ineffective.[43] In most civil engineering contracts, there are elaborate clauses for extensions of time, designed to avoid this consequence. These clauses, however, tend to focus on the circumstances which entitle the contractor to an extension of time, rather than the proper basis on which it might be computed.

Where the contractor's delay is caused by a number of concurrent factors, for some of which only the employer is at risk, difficult questions may arise. The dominant cause approach[44] is frequently used to provide assistance; this requires that the cause which may be said to be the dominant or fundamental cause of the delay is identified. If the employer is at risk for delays due to that cause, an extension is granted. Otherwise no extension is granted. However this test has been doubted, for where both parties are culpable to a degree, it seems just that each should bear an apportionment of the delay.[45]

Computing extensions of time

Where computer-based programming techniques are used to compute extensions of time, care should be exercised as these models lack transparency. Care must be taken to ensure that the assumptions take

---

43 *Peak Construction (Liverpool)* v. *McKinney Foundations* (1970) 1 BLR 111 (CA); *Fernbrook Trading Co. Ltd* v. *Taggart* [1979] 1 NZLR 556 (New Zealand Supreme Court; *Perini Pacific Ltd* v. *Greater Vancouver Sewerage and Drainage District* (1966) 57 DLR (2d) 307 (British Columbia Court of Appeal).

44 *Leyland Shipping Co.* v. *Norwich Union Fire Insurance Society* [1918] AC 350 (HL).

45 See e.g. the remarks of Judge Fox-Andrews QC in *H. Fairweather & Co. Ltd* v. *London Borough of Wandsworth* (1987) 39 BLR 106.

into account real construction practice. For example, the relationship between resource availability, resource costs and the activities which will be scheduled to run is often poor and unrealistic.

Delays are frequently calculated on the basis of an addition or a subtraction model. The addition model takes a credible initial programme and adds in the delay events to give a revised completion date; the difference in original and revised completion dates is the computed extension. The subtraction model requires that an as-constructed programme be devised, from which all the delaying events are subtracted to give the date by which the contractor would have completed but for the delays. Neither of these models takes into account the fact that the project is administered from day to day, without the benefit of hindsight as to what delays will eventually arise and what their effects will be.

The opportunity to reprogramme

It is important for the contractor to know what the completion date is to be. Under most civil engineering contracts, the decision-maker is required to grant extensions of time.[46] The contract may specify the time in which the grant is to be made, but generally this is not so and the decision-maker must grant extensions within a reasonable time. What amounts to a reasonable time will depend on the assistance provided by the contractor,[47] the complexity of the delays and the extent of the time consequences ascertainable at that stage.[48]

Where the decision-maker fails to grant a proper extension within a reasonable period, this may be a breach by the employer.[49] It may also provide the contractor with an entitlement under the contract. The contractor is entitled to aim to complete by the currently-set completion date. Accordingly, if he is obliged to accelerate his work to

46 For example under the ICE Conditions of Contract, the Engineer is empowered to grant extensions of time.

47 For example where the contractor holds relevant records and has access to project programming data he might reasonably be expected to assist the decision-maker.

48 *Amalgamated Building Contractors Ltd* v. *Waltham Holy Cross UDC* [1952] 2 All ER 452 (CA).

49 *Fernbrook Trading Co. Ltd* v. *Taggart* [1979] 1 NZLR 556 (New Zealand Supreme Court).

complete in the foreshortened period and he thereby suffers loss, he may be able to recover this loss as damages.

7. Prolongation, delay and disruption claims

Contractors frequently claim that their work has been delayed, prolonged or disrupted. They will be entitled to additional costs or damages only if they can bring their claim within one of the usual heads of claim. Thus, if they wish to claim for breach of contract, they must identify the term,[50] show that the employer is in breach of it and that his loss is caused by the breach. Alternatively, the employer may have misrepresented the state of facilities or quality of access or possession to be provided, in which case a claim for misrepresentation (or negligent misstatement) may lie. Alternatively (and most commonly) there will be provisions in the contract entitling the contractor to claim in a variety of circumstances.[51]

Distinction between extensions of time and prolongation and delay claims

A claim for an extension of time is essentially a claim that no damages for delay may be levied by the employer until some later date than that originally stated. A claim for prolongation etc. bears a resemblance to an extension of time claim; for example project programming techniques are frequently employed to evaluate both and indeed it is rare that a prolongation claim will succeed where an extension of time claim does not.[52] However, a claim for prolongation etc. is directed at

50 This may be an express or an implied term; see Chapter 3, section 5 on implied terms as to "cooperation between the employer and contractor".

51 In the ICE Conditions of Contract, for instance, Clause 31(2) entitles the Contractor to claim for interference from other contractors etc. where such interference could not reasonably have been foreseen. In addition, Clause 52(2) provides for the revision of contract rates as a result of variations which render the original rates unreasonable or inapplicable.

52 Although the delay may come within the extension of time clause, that of itself is not sufficient to allow the contractor to claim for prolongation. Thus, under the ICE Conditions of Contract, the contractor is entitled to an extension of time for exceptionally inclement weather; however, the effect of this is simply to stop the clock running on the contract period and does not rank as a matter for which the employer assumes any other responsibility. See also *Henry Boot Construction Ltd* v. *Central Lancashire Development Corporation* (1980) 15 BLR 8.

the loss which the contractor suffers as a result of disruptions for which the employer is responsible. In short, a claim for an extension of time is a defensive claim while a claim for delay, prolongation, etc. is an active claim.

Prolongation, delay and disruption: heads of claim

Claims for prolongation etc. frequently arise from reduced site access, cranage, storage facilities, etc., interference from other contractors or personnel, delays in issue of information, free-issue materials, preceding work or necessary reprogramming causing inefficient working (e.g. winter working). The heads of claim are frequently as follows. Note that these heads may overlap and it is important to ensure that no "double recovery" is sought.

On-site resources being employed for a longer period than was reasonably anticipated This may be due to the site facilities, such as offices, power, etc. being required for a longer period;[53] or specified items of plant being required for longer as where delay pushes work into the winter period.[54] In some civil engineering contracts, there are provisions to bill some items as time-related[55] so that the effects of prolongation associated with these items may readily be computed.

Increased allocation of on-site resources Where the efficiency of plant or operatives is reduced, additional machinery or personnel may be needed to achieve the same rates of production. Furthermore, additional or different measures may be required, such as additional cranes, safety measures, etc. In addition, disruption may cause wastage of or damage to materials, as where concrete or asphalt deliveries are held waiting or non-productivity of personnel or plant where work is delayed. The costs of additional resources on site are readily computed, and it is a question of fact and reasonableness

53 These items are frequently referred to as "preliminary items" as they are usually placed first in a traditional bill of quantities.

54 *Freeman* v. *Hensler* (1900) *Hudson's Building Contracts* (4th edn) Vol. 2, p. 292 (CA).

55 For example, where the contract incorporates CESMM 3, section 7 of which deals with time- and other method-related charges.

whether any item of site resource is allowable for any period. Consequential effects may involve complex interactions, and the contractor must demonstrate reasonably that the loss claimed did occur as a result of the delay of which he complains.[56]

Increased "head-office overheads" The costs of head-office overheads are usually claimed as a supplement to the direct on-site costs. Their computation is frequently controversial.[57] It is suggested that the factors which might properly be taken into account include (*a*) whether the proportion claimed for overheads is within the normal range for that type of contractor;[58] (*b*) whether the overhead percentage claimed is supported by historical accounts; and (*c*) the relationship between site work and the correlated head-office input.[59]

Profit Unless the contract provides otherwise,[60] profit may only be recovered, in principle, to the extent that it represents a loss. In a buoyant market it will not be difficult to show that the resources used on the project could have been profitably employed elsewhere. Where

---

56 See the discussion in section 2 above and the cases cited there.

57 A number of formulae have been advocated, including the Hudson Formula, the Emden Formula and the Eichlay Formula (USA). The use of formulae has received some qualified support, e.g. *Finnegan* v. *Sheffield City Council* (1988) 43 BLR 124. But they are based on unstated assumptions about the state of the market, the uniformity of application of overheads to site work, etc. It is submitted that they provide no advantage over a traditional calculation in which the overheads which are being claimed are set out with reasonable particularity. In this regard see also *Alfred McAlpine Homes (North) Ltd* v. *Property and Land Contractors Ltd* (1995) CILL 1130.

58 Some small specialist contractors may run with high head-office overheads due to the small scale of their operations, the need to supply engineering support on all projects, etc. Larger contractors working in areas involving highly repetitive low-technology work may run with significantly lower overheads.

59 For example for some low-technology projects, such as straightforward muck-haulage, a major increase in work on a well-established site may entail little head-office input. On the other hand, an apparently minor change which requires careful planning may result in significantly more than a proportionate increase in head-office input.

60 For example the ICE Conditions of Contract provide for profit for additional permanent or temporary work where there is interference from other contractors: Clause 31(2).

the market is slower it is suggested that there should still be a presumption that a modest level of profit be recoverable.[61]

Interest or financing charges These are dealt with in section 5 above.

8. Claims for defective work

A contractor whose work does not comply with the specification is in breach of contract. His employer is entitled to recover damages. Various bases for computing the quantum of damages have been suggested, including the diminution in value of the building and the cost of repair. Frequently the building experiences only a small diminution in value despite the defective work whereas the cost of remedying that work is very costly. In such a case the contractor will attempt to have damages calculated on a diminution basis while the employer will seek to recover the cost of the remedial work. It is clear that there are no hard and fast rules and that the court will take all the circumstances into account before deciding on the appropriate basis of computation.[62] Non-compliance with the specification may have a number of practical effects and it is suggested that it is not possible to decide how damages should be calculated without a consideration of these effects. Where reinstatement is the appropriate basis of calculation of quantum, the time at which those damages are to be calculated is usually at the time when those defects could reasonably have been detected.[63]

Due to the delay in resolving defects claims, an employer who finds defects in his structure frequently calls upon consultants to advise him as to remedial works. Where consultants advise remedial works the employer is, generally, entitled to rely upon their advice and recover

61 The levels of profit margin in the construction industry tend to seem small, since they are expressed on turnover. However, as a contractor's turnover is usually several times the capital he employs, the profit margin expressed as a proportion of capital employed is significantly higher.

62 See e.g. *East Ham Borough Council* v. *Bernard Sunley & Sons Ltd* [1966] AC 406 (HL).

63 *East Ham Borough Council* v. *Bernard Sunley & Sons Ltd* [1966] AC 406 (HL).

the cost from the party or parties liable for the defects. Where, however, the consultant gives advice which is itself negligent, the contractor is not obliged to pay for any additional or excessive work[64] or any work which results in a structure that is an improvement over the original specification.[65]

9. Limitation

An aggrieved party may take action to recover under a contract or other entitlement during the "limitation period".[66] Upon the expiry of the limitation period, any potential action is said to be "time barred". The limitation period in most civil actions related to civil engineering is six years and runs from the time at which the "cause of action" arises; the exception is an action on contracts "under seal" where the limitation period is twelve years. The appropriate proceedings must commence before the limitation period expires. In litigation the proceedings begin when the writ is issued; arbitration proceedings commence with the issue of the arbitration notice.

Contractual causes of action

In contract cases, the cause of action arises at the time of the breach of contract. For civil engineering design and/or construction works this is usually the date of completion on site, since a contractor who creates a defect is entitled to return to remedy it[67] and a designer has a continuing duty in respect of the design until the works are complete.[68] Breaches by the employer of contractual obligations to provide access, produce drawings, information, etc. occur at the time when the employer failed to provide or produce them; the cause of

---

64 *Frost* v. *Moody Homes Ltd & others* (1989) CILL 504; *The Board of Governors of the Hospital for Sick Children* v. *McLaughlin & Harvey plc* (1987) 6 Constr. L.J. 245.

65 This is termed "betterment". See for example *Richard Roberts Holdings Ltd and another* v. *Douglas Smith Stimson Partnership* (1988) 46 BLR 50.

66 Limitation Act 1980; see sections 5 and 8 respectively for simple contracts and deeds.

67 *P. and M. Kaye* v. *Hosier & Dickinson Ltd* [1972] 1 All ER 121 (HL) where Lord Diplock said at 139: "Provided that the contractor puts it right timeously I do not think that the parties intended that any temporary disconformity should of itself amount to a breach of contract by the contractor."

68 *London Borough of Merton* v. *Lowe* (1982) 18 BLR 130.

action for claims which arise out of a failure to certify sums probably arises at the date of the final certificate.

Causes of action based on tortious negligence

In cases based in negligence, the cause of action arises at the date of damage, which frequently leads to a longer limitation period. Employers who wish to sue engineers frequently do so in the tort of negligence to take advantage of this extended period.[69]

Concealment

Section 32 of the Limitation Act 1980 provides that where the defendant has been fraudulent, and this has affected the plaintiff's knowledge of his right to make a claim against the defendant, the period of limitation does not run until the plaintiff has, or could reasonably have had, knowledge of his cause of action[70] and may be reactivated by a subsequent act of concealment.[71]

---

69 At one stage it was thought that where the employer and engineer were in a contractual relationship, the limitation period would be that prescribed by the contract: *Tai Hing Cotton Mill* v. *Liu Chong Haing Bank* [1986] AC 80 (PC). However, it now seems clear that a claimant may take advantage of a tortious remedy even where a contractual remedy exists: see *Henderson* v. *Merrett* [1994] 3 WLR 761 (HL). Note also that the Latent Damage Act 1986 provides that no negligence claim may be brought more than 15 years after the breach of duty. The protection offered by this rule is likely, on occasions, to be of value to professionals.

70 See e.g. *Gray* v. *T. P. Bennett & Son* (1987) 43 BLR 63.

71 *Sheldon* v. *R. H. M. Outhwaite (Underwriting Agencies) Ltd* [1996] 1 AC 102 (HL).

7

Dispute resolution

1. Negotiated settlement

The vast majority of disputes and differences are settled by amicable negotiation leading to a binding agreement. Such agreements are frequently made at the conclusion of a project and include a full and final settlement of all claims by both parties.[1] Such settlements are variations to the existing contractual obligations and hence need to be agreed by properly authorised agents of the negotiating parties; likewise, the agreement must be entered into freely[2] and supported by consideration.[3] A subsequent failure on the part of one of the parties to honour the settlement will not revive the earlier contract; the appropriate cause of action is for breach of the settlement agreement.[4] In principle, there is normally no need for a negotiated settlement to be put into writing or indeed to be evidenced in any way whatever.[5] It is enforceable if made orally at a meeting.

1 Such an agreement is commonly referred to as an "accord and satisfaction".

2 *North Ocean Shipping Co. Ltd* v. *Hyundai Construction Co. Ltd* [1979] QB 705.

3 See *D & C Builders* v. *Rees* [1966] 2 QB 617 (CA).

4 Order 14 summary judgment will ordinarily be available.

5 If the settlement includes a transfer of an interest in land then evidence in writing is required. Also note that section 53(1) of the 1925 Law of Property Act requires that a "disposition of an equitable interest or trust subsisting at the time of the disposition must be in writing [and] signed . . .". This detailed technical provision may be important when the contents of funds held by one party on behalf of another—e.g. the retention fund held by an employer pending the successful conclusion of the maintenance period—are included in a settlement. If such a fund is to form part of the agreement, the settlement should always be in writing and signed to avoid a later claim that it is unenforceable for lack of formality.

2. Alternative Dispute Resolution

A range of techniques, known generically as Alternative Dispute Resolution (ADR), has been developed to facilitate dispute resolution. ADR techniques such as mediation and conciliation have recently been introduced into a number of standard form contracts, including civil engineering contracts.[6]

ADR techniques may be classified into four principal types: (*a*) assisted negotiation; (*b*) non-binding tribunal; (*c*) interim binding tribunal; and (*d*) binding tribunal.

Assisted negotiation

A third party neutral (often called a mediator or conciliator) attempts to bring the parties together either in open session or in private meetings to discuss and to facilitate a resolution. The assisted negotiation may be facilitative or evaluative: in the former case, the neutral is not entitled to suggest terms for settlement or to make a recommendation; in the latter the neutral is enjoined to make a recommendation. It is important that the neutral's terms of reference are clearly defined in advance, including his power, if any, to meet the parties in private.[7]

Non-binding tribunal

A panel is established to hear submissions from the parties. Typically the panel will include an employee of each party who is not directly involved with the dispute. There will also be a neutral on the panel. The parties may present their case in a strictly limited time and the

6 See for instance Clause 66 of the ICE Conditions of Contract which contains a provision for optional conciliation.

7 See the ICE Conciliation Procedure in Chapter 12 which provides a useful model.

panel then retires to make a recommendation. The parties are not obliged to accept the recommendation.[8]

Interim binding tribunal[9]

An independent individual (frequently called an adjudicator)[10] or panel (frequently called a disputes review board) is convened to hear representations and to make an interim decision. The decision is binding unless and until reviewed by an arbitrator or court. The procedure most frequently adopted is for each party to present his case either in writing or with brief oral submissions within a strictly limited time and for the adjudicator/board to make a decision within a relatively short time.

Binding tribunal

The parties may agree that a person or panel is to be empowered to determine any dispute without any appeal.[11] This is commonly know as an "expert determination". It has a long history; in the nineteenth century engineers under civil engineering contracts frequently issued unappealable certificates.[12] Where the parties agree to be bound by the expert's decision, it may only be upset by fraud or a manifest departure from the terms of his appointment.[13]

8 This procedure is frequently termed a "mini-trial". See Henderson D. A., Mini-trial of construction disputes, ICLR (1994) 442.

9 An Engineer under the ICE Conditions of Contract falls within this category. ADR advocates, however, assume that the tribunal is not connected with the design or management of the project. Such a tribunal is normally styled an adjudicator or disputes review panel. There is a wide literature on adjudication and dispute review boards. See e.g. McGaw M. C., Adjudicators, experts and keeping out of court (1992) 8 Constr. L.J. 332; Pike, A., Disputes review boards and adjudicators, ICLR (1993) 157; Jaynes G. J., Disputes review boards—yes!, ICLR (1993) 452.

10 Note that the Housing Grants, Construction and Regeneration Act 1996, Section 108, when in force, will require or imply adjudication provisions into construction contracts.

11 In *Lee* v. *Showmen's Guild* [1952] 2 QB 329 (CA) at 342 Lord Denning suggested that a tribunal from which there was no appeal was bound to conform with the rules of natural justice. It is submitted that Lord Denning's comments are to be viewed against the unusual background facts of that case.

12 For example *Sharpe* v. *San Paulo Railway Co.* (1873) LR 8 Ch App. 597.

13 *Jones* v. *Sherwood Computer Services* [1989] EGCS 172; *Campbell* v. *Edwards* [1976] 1 All ER 785. See also McGaw M. C., Adjudicators, experts and keeping out of court (1992) 8 Constr. L.J. 332.

Procedure in ADR

The procedures available are infinitely adaptable and may be tailored to suit the requirements of the parties.[14] If the process is commenced pursuant to a prior agreement, the terms of the agreement must be complied with, unless a later supervening agreement is made. Where an attempt at ADR process is a precondition to commencing arbitration proceedings, this is not compulsory ADR[15] as all that can be enforced is that the parties must go through the motions.

3. Arbitration

Arbitration is frequently used for the determination of disputes which arise under civil engineering contracts. The practice of arbitration is dealt with in Chapter 12. A number of general issues relating to arbitration are discussed here.

The nature of arbitration

Arbitration and other forms of dispute resolution The parties to a contract may specify the procedure by which disputes are to be resolved. They may, for example, agree that an expert has power to make a binding and final decision as to their rights and liabilities. Providing that the expert acts within the scope of the agreement and in good faith, the court will not interfere with his decision. Arbitration is different. An arbitrator's authority derives principally from the agreement of the parties; but his authority is supplemented and, to some degree, constrained[16] by his status as arbitrator under the Arbitration Acts. Furthermore, the court is given a limited, though

---

14 See e.g. Lewis D., Dispute resolution in the new Hong Kong International Airport core programme projects, ICLR (1993) 76. Here the dispute process contains four tiers: (1) a decision of the engineer; (2) a mediation; (3) an adjudication; and (4) an arbitration.

15 Which would be an "agreement to agree" and unenforceable: *May & Butcher* v. *R.* [1934] 2 KB 17 (HL).

16 An arbitrator must, for example, apply the law and allow each party a fair opportunity to present his case and to meet the case put by the other party.

important, role in supervising arbitrations. An agreement to refer an existing or future dispute to a tribunal will be construed as an arbitration agreement where that agreement is in writing and expressly, or by very clear implication, indicates that the parties intend the process to be an arbitration.[17]

Agreement and multi-party arbitrations[18] Arbitration relies upon agreement between the parties to a contract. The complexity of contractual relationships in civil engineering projects means that any one set of facts may lead to contractual disputes between more than two parties. It may be convenient for these to be dealt with in a single arbitration, or in parallel or serial arbitrations. A multi-party arbitration requires the agreement of all those to be involved. A number of devices attempt to achieve this result. A sub-contract arbitration clause may provide that if the main contractor and sub-contractor have a dispute which is substantially the same as a dispute between the main contractor and employer, then the main contractor can require that the sub-contract dispute be referred to the main contract arbitrator.[19] This is reasonably workable in practice but the arbitrator cannot be compelled to take on the new reference and the employer may have objections.[20] Furthermore, since the right to proceed to arbitration depends on the existence of a dispute,[21] the disputes for both arbitrations may not crystallise at the same time.[22] A further device often used in back-to-back contracts is to allow the sub-contractor to claim against the employer using the main contractor's name. Questions often arise here concerning the extent to which the main contractor must cooperate with the sub-contractor, the liability of

17 Normally the word "arbitration" or some derivative is used. This, however, does not appear to be decisive. See *Fordgate Bingley Ltd* v. *Argyll Stores Ltd* [1994] 39 EG 135 (2 EGLR 84).

18 See Fox-Andrews J., Construction industry disputes: Official Referee or technical arbitrator—the pros and cons, (1992), 8 Constr. L.J. 2.

19 FCEC Form of Sub-Contract, Clause 18. See the commentary in Chapter 9 and the cases cited there.

20 *Higgs & Hill Building Ltd* v. *Campbell Davis Ltd* (1982) 28 BLR 47.

21 *Re Evans, Davis and Caddick* (1870) 22 LT 507.

22 *Multi-Construction (Southern) Ltd* v. *Stent Foundations Ltd* (1988) 41 BLR 98.

the main contractor for costs and the question of discovery of the sub-contractor's documents.[23]

Binding arbitration agreements An arbitration clause is unenforceable unless it is in writing.[24] Where the contract is made by reference to standard terms which include a printed arbitration clause, this has traditionally been thought sufficient to incorporate the arbitration clause. However, this view has been doubted[25] for the following reasons: (*a*) an arbitration clause is to be viewed as a self-contained contract ancillary to the substantive contract and hence is not to be incorporated simply because the terms of the substantive contract are incorporated;[26] (*b*) an agreement to arbitrate deprives a party of his right to have disputes litigated in court and any such deprivation may only be achieved by clear words.[27] This new view suggests that care should be taken expressly to incorporate arbitration clauses.

The jurisdiction of the arbitrator The arbitrator's jurisdiction in respect of any dispute depends upon the existence of an arbitration agreement between the parties which is wide enough to encompass that dispute.[28] Furthermore, his appointment must be as prescribed by the arbitration agreement and any dispute referred to him must comply with the contract and arbitration agreement; where the

23 *Lorne Stewart Ltd* v. *William Sindall plc and NW Thames Regional Health Authority* (1986) 35 BLR 109.

24 Section 32 of the Arbitration Act 1950; Section 5 of the Arbitration Act 1996.

25 *Aughton* v. *M. F. Kent Services Ltd* (1991) 57 BLR 1 (CA); *Ben Barrett & Son (Brickwork) Ltd* v. *Henry Boot Management Ltd* (1995) CILL 1026. These authorities were not followed in *Black Country Development Corporation* v. *Kier Construction Ltd* QBD 10th July 1996.

26 There is an irony here: the doctrine of separability of the arbitration agreement was originally designed to preserve the arbitration agreement, not to deny its existence: see *Heyman* v. *Darwins Ltd* [1942] AC 365 (HL).

27 This argument does not fully account for *Northern Regional Health Authority* v. *Crouch* [1984] QB 644 (CA) (see below). To deprive a party of the right to arbitration may well deny him a substantive remedy.

28 *Ashville Investments* v. *Elmer Contractors* [1989] QB 488 (CA).

contract requires the matter to be referred to the engineer prior to its being referred to the arbitrator[29] this will be a condition precedent.

Powers of the arbitrator which cannot be exercised by the court[30]
Civil engineering contracts frequently contain an arbitration clause entitling the arbitrator to "open up, review and revise" certificates and decisions of the engineer. This puts the arbitrator retrospectively in the stead of the engineer as contract administrator. Since (*a*) the parties have agreed to give the arbitrator this special power and (*b*) the function of a court is to enforce the parties' agreement rather than to operate it for them, it may be that a court cannot "open up, review and revise" the engineer's certificates and decisions. This was the view taken in *Northern Regional Health Authority* v. *Crouch.*[31] Thus where a party commences litigation and seeks a remedy which necessarily entails the opening up of a decision made under the contract, the court will not have power to do this.[32] The decision has been criticised by some, but it has generally been followed and was presumed to represent the general law when Section 100 of the Courts and Legal Services Act 1990 was enacted. A senior Scottish judge has said:[33] "That wording ['open up review and revise any decision ... certificate ...' etc.] appears to me to show clearly that the parties' agreement is not just that they will use the arbitration machinery but that they are giving up the right to use any other means for determining the disputes covered by it. That wording does not exclude all possible resort to the courts; but it is quite inconsistent with allowing the courts an open jurisdiction to do that which an arbitrator

---

29 Clause 66 of the ICE Conditions provides that a dispute must be referred to the engineer prior to being submitted to arbitration. Note, however, that Rule 4.2 of the Arbitration Procedure 1983 clothes the arbitrator with jurisdiction over all matters "connected with and necessary to the determination of any dispute or difference already referred to him". See the commentary in Chapter 12.

30 It is submitted that where an adjudicator, disputes review board, expert, etc. is given specific powers under a contract to make decisions which involve some degree of investigation or discretion, their powers will likewise not be available to the court on the same principle as described in this section.

31 [1984] QB 644 (CA).

32 A party in this situation is said to be "Crouched".

33 Lord McCluskey in *Costain Building and Civil Engineering Ltd* v. *Scottish Rugby Union plc* (1993) 69 BLR 85 (Scottish Court of Session, Inner House) at 107; a case on the 5th Edition of the ICE Conditions of Contract.

is empowered to do." Unless and until the House of Lords overrules *Crouch*, a party takes a major risk by assuming that it may be circumvented. There seem to be two exceptions to the general rule: (*a*) where the parties have agreed that the court may exercise the powers of an arbitrator[34] and/or (*b*) where the arbitration machinery cannot be operated.

The arbitration legislation

At the time of writing, the arbitration legislation currently in force is contained in the Arbitration Acts 1950, 1975 and 1979. The Arbitration Act 1950 is the principal piece of legislation; the 1975 and 1979 Acts relate specifically to international arbitration agreements and appeals against an arbitrator's decision.

The Arbitration Act 1996 received Royal Assent in June 1996 but is not in force at the time of writing. It is designed to supersede the current legislation and will apply to arbitrations commenced after the Act comes into force, regardless of when the arbitration agreement was made.[35] Its main effect is to consolidate the legislation and to codify rules developed by the courts, although it also introduces some new powers for the arbitrator. Many of these powers are already available where there are agreed rules of procedure such as the ICE Arbitration Procedure.[36] The new Act will not greatly affect such arbitrations.

The role of the courts in supervising arbitrations

Staying court proceedings An agreement to arbitrate does not preclude parties from litigating disputes; indeed any provision which

34 Section 100 of the Courts and Legal Services Act 1990. The Engineering and Construction Contract provides that where the court is the tribunal for dispute resolution, it shall have the powers of the arbitrator.

35 There may be transitional provisions: see Section 84. The Department of Trade and Industry has sought views on commencement, but the results of this survey may not be known until late 1996.

36 See Chapter 12 where the ICE Arbitration Procedure is discussed.

purports to exclude access to the court is void.[37] Where one party to an arbitration agreement commences litigation proceedings, the court may order a stay to arbitration on the application of the other party.[38] No application will succeed unless the following conditions are met: (*a*) there must be a dispute;[39] (*b*) the application must be made before the applicant accepts the jurisdiction of the court in respect of the dispute by taking a step in the proceedings;[40] and (*c*) the applicant must also have been ready and willing at all times to arbitrate.[41] In the case of domestic agreements,[42] the courts have a discretion to refuse a stay, though it is rarely exercised. Potential grounds for the exercise of that discretion include possible multiplicity of proceedings,[43] the conduct of the applicant[44] and where the applicant raises the allegation of fraud or serious professional misconduct.[45] Where Crouch considerations apply the applicant is likely to get his stay as a

37 However, an agreement to make the arbitrator's award a condition precedent to bringing an action is enforceable: *Scott* v. *Avery* (1856) 5 HLC 811 (HL).

38 Section 4 of the Arbitration Act 1950; Section 9 of the Arbitration Act 1996, read together with Section 86.

39 *Nova (Jersey) Knit* v. *Kammgarn Spinnerei GmbH* [1977] 1 WLR 713 (HL). Thus where the plaintiff in the action establishes that he is entitled to summary judgment, the court may refuse a stay on the ground that there is no dispute: see *Imodco* v. *Wimpey and Taylor Woodrow* (1987) 40 BLR 1 (CA). There is some support, however, for the view that the stay should be granted even where there is no dispute: *R. M. Douglas* v. *Bass* (1990) 53 BLR 119. Note that under the ICE Conditions of Contract, a dispute can arise before the reference to the Engineer. Thus a stay may be granted before the matter is referred to the Engineer: *Enco Civil Engineering Ltd* v. *Zeus International Development Ltd* (1991) 8 Constr. L.J. 164.

40 *Eagle Star Insurance* v. *Yuval Insurance* [1978] 1 Lloyd's Rep. 357; *Blue Flame Mechanical Services Ltd* v. *David Lord Engineering Ltd* (1992) CILL 760.

41 *Dew & Co. Ltd* v. *Tarmac Construction* (1978) 15 BLR 22. A stay may be granted where the contract requires procedures prior to arbitration, so that at the date of commencement of the action, the applicant is in no position to arbitrate: *Channel Tunnel Group* v. *Balfour Beatty* [1993] AC 334 (HL).

42 That is, those which are not essentially international.

43 *Taunton-Collins* v. *Cromie* [1964] 1 WLR 633 (CA).

44 As where the applicant is obviously seeking a stay to create delay. See e.g. *Croudace* v. *London Borough of Lambeth* (1986) 33 BLR 20 (CA).

45 *Turner* v. *Fenton* [1982] 1 WLR 52.

matter of course.[46] In the case of non-domestic[47] agreements, the applicant is entitled to a stay.[48]

Removal of an arbitrator Where an arbitrator is biased or has misconducted the proceedings so that a reasonable party might consider that he may not receive a fair hearing, the court may remove the arbitrator.[49]

Challenging the award The decision of the arbitrator is recorded in "the award". A party who is dissatisfied with the award may challenge it by (*a*) alleging that the arbitrator has misconducted himself or the proceedings which produced the award or that the award has been improperly procured;[50] (*b*) appealing on the grounds that the arbitrator has made a serious error of law which substantially affects the rights or one or more parties to the arbitration agreement.[51] Leave to appeal under (*b*) is rarely given unless the error of law is obvious and serious or the matter is one whose resolution will be to the benefit of the body of commercial law (e.g. if it involves a standard form contract). Parties may enter into an exclusion agreement[52] to preclude appeals. In the case of a domestic agreement, such agreements are ineffective unless made after the dispute arises; in the case of international agreements, an exclusion agreement made before the dispute arises is valid.[53]

46 *Northern Regional Health Authority* v. *Crouch* [1984] QB 644 (CA).

47 That is, international arbitration agreements.

48 Section 1 of the Arbitration Act 1975; Section 9(4) of the Arbitration Act 1996. The applicant may be entitled to a stay even where there is no dispute: see *Hayter* v. *Nelson* [1990] 2 Lloyd's Rep. 265.

49 *The Elissar* [1984] 2 Lloyd's Rep. 84 (CA); *Modern Engineering (Bristol) Ltd* v. *C. Miskin & Son Ltd* [1981] 1 Lloyd's Rep. 135 (CA); see also *Stockport Metropolitan Borough Council* v. *O'Reilly* [1983] 2 Lloyd's Rep. 70 for a case where the arbitrator's authority was revoked on more general grounds.

50 Section 23 of the Arbitration Act 1950; Section 68 of the Arbitration Act 1996 uses the expression "serious irregularity".

51 Section 1 of the Arbitration Act 1979; Section 69 of the Arbitration Act 1996. See *The Nema* [1981] 2 Lloyd's Rep. 239 and *The Antaios (No. 2)* [1984] 2 Lloyd's Rep. 235.

52 Section 3 of the Arbitration Act 1979; Section 69 of the Arbitration Act 1996 read together with Section 87.

53 Except in three specified circumstances which are not relevant to civil engineering.

The powers and responsibilities of the arbitrator

Relevant skill and experience Where the arbitration agreement requires that the arbitrator possess stated qualifications, the appointment of a person who does not have such qualifications will be of no effect unless both parties waive their objections. In any event a prospective arbitrator should consider whether he has the appropriate skills and training before taking an appointment. An arbitrator who fails to demonstrate the elementary skill and knowledge of an arbitrator may misconduct himself.[54]

Independence An arbitrator should not have any commercial or other similar connection with either party which may give rise to a suspicion in the mind of a reasonable person that he may be biased.[55]

General powers Subject to any express agreement made by the parties, the arbitrator has wide powers to control the reference.[56] It has been said, indeed, that subject to the agreement of the parties and the rules of natural justice he is "master of the procedure to be followed".[57] He must, however, allow each party a fair and reasonable opportunity to put his case and to answer the case put by the other party and he must act openly. It is suggested in one case that, unless the parties agree otherwise, the arbitration must be conducted along broadly adversarial lines.[58]

Civil engineering professionals and the use of specialist knowledge Technically skilled and experienced arbitrators are generally entitled

54 *Pratt* v. *Swanmore Builders Ltd and Baker* [1980] 2 Lloyd's Rep. 504.

55 Specialist barristers frequently practise from the same chambers. Occasionally one barrister may appear as advocate, while another is arbitrator. Likewise civil engineering arbitrators and advocates frequently engage in joint activities, such as writing books and learned institution committee membership. This should not generally cause any suspicion of lack of independence: see e.g. the Paris Court of Appeal's decision in *Kuwait Foreign Trading* v. *Icori Estero SpA*, reported in [1993] ADRLJ 167 supporting this view.

56 Section 12(1) of the Arbitration Act 1950; Sections 34 to 41 of the Arbitration Act 1996.

57 *Bremer Vulcan* v. *South India Shipping* [1981] AC 909, per Lord Diplock at 985 (HL).

58 *Town & City Properties (Developments) Ltd* v. *Wiltshier Southern Ltd and Gilbert Powell* (1988) 44 BLR 109. But see also Section 34(1)(g) of the Arbitration Act 1996.

to use their own specialist knowledge in making directions as to procedure[59] and reviewing the evidence.[60] However, two types of specialist knowledge—general and particular—must be distinguished.[61] An arbitrator may use his general knowledge freely; particular knowledge acquired by the particular arbitrator and peculiar to him must, however, be disclosed so that the parties may deal with it.[62] Although an arbitrator may use his knowledge to understand the evidence presented to him, he must not gather evidence.[63] Whenever the arbitrator intends to decide against the apparent effect of the evidence, he must inform the parties so that they can make legal submissions and if appropriate provide counter-evidence to deal with the view which the arbitrator has formed.[64]

4. Litigation

Litigation is the term used to describe dispute resolution in the courts. Civil engineering cases are usually resolved by arbitration rather than by the court. Insofar as the court is involved it is usually in a supervisory or supplementary capacity; thus where questions as to the jurisdiction of an arbitrator or conduct of an arbitration are involved, these questions may be referred to the court.

Aspects of civil engineering cases are normally heard in the Queen's Bench Division of the High Court by judges known as "Official Referees".[65] High Court disputes are normally heard by a single judge,

---

59 *Carlisle Place Investments Ltd* v. *Wimpey Construction (UK) Ltd* (1980) 15 BLR 109.

60 *Jordernson & Co.* v. *Stara, Kopperbergs Bergslag Atkiebolag* [1931] 41 Lloyd's Rep. 201; *Mediterranean & Eastern Export Co. Ltd* v. *Fortress Fabrics (Manchester) Ltd* [1948] 2 All ER 186.

61 *Fox* v. *Wellfair* [1981] 2 Lloyd's Rep. 514 (CA); *F. R. Waring (UK) Ltd* v. *Administracao Geral do Acucar e do Alcool* [1983] 1 Lloyd's Rep. 45.

62 *Top Shop Estates* v. *Danino* (1984) 273 EG 197 (1 EGLR 9).

63 *Owen* v. *Nicholl* [1948] 1 All ER 707 (CA); *Fox* v. *Wellfair* [1981] 2 Lloyd's Rep. 514 (CA); *Top Shop Estates* v. *Danino* (1984) 273 EG 197 (1 EGLR 9); *Mount Charlotte Investments plc* v. *Prudential Assurance* [1995] 10 EG 129 (1 EGLR 15).

64 *Fox* v. *Wellfair* [1981] 2 Lloyd's Rep. 514 (CA) (which involved an undefended arbitration); *The Vimeira* [1984] 2 Lloyd's Rep. 66.

65 For a useful introduction see: "The preparation and presentation of cases in the Official Referees' courts", His Honour Judge Newey Q.C. 6 Constr. L.J. 216.

though they may sit with "assessors" who are technical experts who advise the judge. Where either party wishes to challenge a judgment of the High Court, they may appeal to the Court of Appeal. If not satisfied with that judgment, they may seek leave to appeal to the House of Lords; but this is not available as of right.

Aspects of the procedures of the court

The rules of court[66] Procedural rules relating to High Court litigation are set out in the "Rules of the Supreme Court". These are published with frequently updated notes in the white-bound "Supreme Court Practice", commonly referred to as the "White Book".

Remedies A plaintiff may seek a range of remedies. In addition to damages, he may seek a declaration or injunction or relief ancillary to an arbitration.

Originating process[67] Litigation may be commenced by a variety of procedures, including by writ or by originating summons. A number of specified matters may only be commenced by originating motion.[68]

Action for damages begun by writ—brief outline The plaintiff attends an office in the High Court, which issues a formal document, a writ, requiring the defendant to answer the allegations made against him. The plaintiff serves the writ on the defendant. A Statement of Claim, which sets out the plaintiff's case, is prepared by the plaintiff and served on the defendant. The defendant may then decide to capitulate, wholly or in part, or to defend the action. If he decides to defend he must answer the allegations by serving a Defence. The defendant may also serve a Counterclaim in which he will register a claim against the plaintiff. The plaintiff, on receipt of the Defence, may

---

66 In July 1996 Lord Woolf completed his review of the procedure of the civil courts. His final report "Access to Justice" recommends that a single simple set of rules be introduced governing both the Supreme Court and the County Court.

67 Order 5 of the Rules of the Supreme Court.

68 Including a number of matters ancillary to arbitration: see Order 73 of the Rules of the Supreme Court.

serve a Reply (the third and, usually, the last pleading) if he wishes, and must supply a Defence to Counterclaim if a Counterclaim is made by the defendant. When the pleadings have been served, the plaintiff will arrange for a Summons for Directions to be heard by a judge.[69] The judge will hear representations from solicitors or counsel representing both parties and will give directions concerning a number of matters relating to the conduct of the proceedings, including discovery, expert witness statements and arrangements for meetings prior to the hearing, hearing date and venue, etc. The hearing takes place before the trial judge. Parties are generally represented by advocates, who are barristers[70] or solicitors with rights of audience. Advocates will open and present their clients' cases and will examine their own witnesses and cross-examine the opponent's witnesses. Proceedings are adversarial in nature; the judge listens to the arguments put and evidence adduced by the parties. After the hearing, the judgment is given. This may be spoken from a typed draft, or a written judgment may simply be handed down.

Short forms of procedure

In order to expedite the process of the court for parties with an obviously meritorious case, a number of procedures have been devised. They are known by the names of the orders of the Rules of the Supreme Court which establish them—Orders 14, 14A and 29.

Summary judgment Order 14 summary judgment is available to a plaintiff where there is no real defence. Having issued proceedings and his statement of claim, the plaintiff applies for summary judgment upon summons supported by affidavit. Judgment will not be given if the defendant shows that "there is an issue or question in dispute which ought to be tried or that there ought for some other reason to be a trial of the claim ...". Instead, the defendant will be given leave to defend. This leave may be unconditional or conditional; typical

69 The normal practice in the Official Referee's court is to allow the parties to make further applications to the official referee following the first hearing of the application. This is normally achieved by adjourning the first hearing and restoring it at a later date, often at a "pre-trial review".

70 Also referred to as "counsel".

conditions are that the defendant makes a payment of a sum into court as security which ensures that his belief in the defence is bona fide. The court may entertain an application for summary judgment even when there is an arbitration agreement where it is satisfied that there is no real defence and hence no real dispute.[71]

Disposal of a case on a point of law Order 14A provides that where the court is satisfied that a case or any issue turns on a question of law, the court may determine the point and thereupon make such order or judgment as it thinks just.

Interim payment Order 29 provides that the court may order an interim payment if it is satisfied that should the action proceed to trial, the plaintiff would obtain judgment for a substantial amount. The test may not be identical,[72] but is similar to that required to obtain summary judgment.[73]

Proposals for reform of the civil court procedure

Lord Woolf was commissioned in 1994 to undertake a comprehensive review of the procedures of the civil courts. He has consulted widely and published his final report "Access to Justice" in July 1996. He identified a number of major problems in the system, mainly associated with costs, delay, procedural complexity and evidential matters. He has recommended streamlining the entire court system so that all cases, whether in the High Court or County Court, are governed by a new simple set of procedural rules and are closely managed by the judges. Cases will be assigned to a "track"; small cases will be assigned to the "fast track" with short durations and limited costs, while large or complex cases will be dealt with in a manner more akin to

---

71 *Croudace* v. *London Borough of Lambeth* (1986) 33 BLR 20 (CA); *Chatbrown* v. *Alfred McAlpine Construction (Southern)* (1987) 35 BLR 44 (CA); *Imodco* v. *Wimpey and Taylor Woodrow* (1987) 40 BLR 1 (CA). Where section 1 of the Arbitration Act 1975 applies, summary judgment may be refused: see *Hayter* v. *Nelson* [1990] 2 Lloyd's Rep. 265.

72 *British and Commonwealth Holdings* v. *Quadrex Holdings* [1989] QB 842 (CA).

73 *Shearson Lehman Brothers Inc.* v. *Maclaine Watson & Co. Ltd* [1987] WLR 480; *Smallman Construction* v. *Redpath Dorman Long* (1989) 47 BLR 41 (CA).

traditional litigation. Even here, however, the court will manage the case closely for time, cost and such matters as expert evidence.

A draft set of rules has been published and is available for consultation. Lord Woolf's recommendations have been widely welcomed and are being supported by the Government. It is expected that the recommendations will be put into effect from 1997 and will produce significant improvement in the provision of civil justice.

8

The ICE Conditions of Contract (6th Edition)

The ICE Conditions of Contract (6th Edition) were published in 1991, with minor corrigenda and amendments published in 1993. The Contract is drawn up by the Conditions of Contract Standing Joint Committee (the CCSJC); this is a committee set up by and with the approval and sponsorship of the Institution of Civil Engineers, the Federation of Civil Engineering Contractors and the Association of Consulting Engineers. The CCSJC is currently considering whether an updated version is required.[1] If this is to be done, it is not clear at this stage whether this will be published as a Seventh Edition or a revision to the Sixth Edition.[2] The issue of revisions does not, of course, affect contracts already let on existing versions of the ICE Conditions of Contract.

This chapter contains a clause-by-clause commentary on the Conditions of Contract. The text used here is the 1991 text as amended by the 1993 corrigenda. The Construction (Design and Management) Regulations 1994 are now in effect for new construction; a new Clause 71 has been published to provide for these.

---

1 The need for a revision is prompted by several factors: (1) the desire to make the Conditions of Contract operate more readily for a range of payment schemes; (2) to account for some of the recommendations of the Latham Report (Sir Michael Latham, *Constructing the team*, HMSO 1994); (3) to improve some aspects of the wording; and (4) to account for the provisions of the Housing Grants, Construction and Regeneration Act 1996.

2 This question remains unresolved at the time of writing.

1. The scheme of the ICE Conditions of Contract

The scheme of the Contract is broadly as follows.

Personnel and administration

The parties to the Contract are the Employer and the Contractor. Other persons described by the Contract include:

(*a*) the Engineer
(*b*) the Engineer's Representative
(*c*) the Engineer's Representative's assistants
(*d*) the Contractor's agent.

The Contract is administered by the Engineer. The Employer's role is limited to matters such as nominating the Engineer, consenting to assignments, making payment upon certificates and giving notice to determine the Contractor's employment.

The Contract

The Contract is defined as comprising:

(*a*) Conditions of Contract
(*b*) Specification
(*c*) Drawings
(*d*) Bill of Quantities
(*e*) The Tender and written acceptance thereof
(*f*) Contract Agreement (if completed).

All parts of the Contract carry equal weight and any inconsistencies are to be explained and adjusted by the Engineer.

Disputes

In the event of a dispute arising, either party may serve a Notice of Dispute on the Engineer stating the nature of the dispute; the Engineer gives his decision. Where he fails to do so or where either party is dissatisfied, the dissatisfied party may refer the matter to a Conciliator or Arbitrator.

Time

The time scheme of the Contract is as follows.

(*a*) Works Commencement Date: the date on which the works physically commence or should commence.
(*b*) The Time for Completion is the period stated in the Contract as being the time in which the Contractor has substantially to complete the Works. This time may be extended if circumstances arise entitling the Contractor to an extension of time.
(*c*) Substantial completion: the Contractor must achieve substantial completion by the end of the Time for Completion (or extended time if appropriate). Upon its attainment the Engineer issues a Certificate of Substantial Completion. There is no definition of substantial completion in the Contract.
(*d*) Defects Correction Period: upon the issue of the Certificate of Substantial Completion, the "Defects Correction Period" begins to run. This is a period specified in the Contract during which the Contractor is obliged and entitled to return to correct any problems which become apparent in the Works. Upon the expiry of the Defects Correction Period and the making good of any work which is to be made good, the Engineer issues a Defects Correction Certificate.

Payment

The ICE Conditions of Contract have the following important features.

(*a*) They create a measure and value contract. The quantities in the Bill of Quantities are estimated. The prices for elements of work are given as rates per unit of work. The Engineer determines the value by "admeasurement". The units which are measured are specified in a Standard Method of measurement (see Clause 57) and the unit rates are set out in the Bill of Quantities.
(*b*) The Contractor is paid monthly, approximately two months in arrears. The amount paid is the value of work done to date less a proportion which is retained by the Employer; the proportion is stated in the Appendix. The first half of this retained sum is released upon the issue of the Certificate of Substantial Completion; the second half is released upon the issue of the Defects Correction Certificate. Where any sum is due from the Contractor to the Employer (e.g. as liquidated damages), the Employer may deduct it by bona fide set-off against sums certified.
(*c*) The mechanism of payment is as follows. (1) The Contractor submits an account setting out the value (in his opinion) of the total work which has been performed to date. The units of work

in each class defined in the Bill of Quantities are multiplied by the unit rates for that work and the total is given. From this is deducted the retention money and the amounts paid by the Employer to date. The remainder is the sum for which the Contractor makes application. (2) The Engineer considers this application and modifies any quantities, rates or items which to him appear inaccurate or inapplicable. He prepares an account based on his approved items and rates and certifies an amount which in his opinion is due from the Employer to the Contractor. (3) Upon issue of the certificate, the Employer has 28 days from the date of the application in which to pay the certified amount. (4) In the event the sum certified is less than should have been certified, or where the Employer fails to pay it in full, the Contractor is entitled to recover interest at the rate specified in Clause 60(7).

Planning of operations

The Contractor is obliged to supply a method statement and a programme. He must advise the Engineer of any proposed changes to the method and update the programme where the Engineer requires this. He is entitled to have possession of the site so as to execute the Works in accordance with the programme. There are provisions whereby the Contractor may give notice to the Engineer of the need for further information.

Unforeseen conditions

Where unforeseen ground or other physical conditions (except weather conditions) or other obstructions are encountered, the Contractor is entitled to recover the additional cost of dealing with these, plus reasonable profit. Where other employees of the Employer hinder the Contractor, the Contractor may recover for any consequent disruption.

2. Commentary on the Conditions of Contract

DEFINITIONS AND INTERPRETATION

1 Definitions

(1) In the Contract (as hereinafter defined) the following words and expressions shall have the meanings hereby assigned to them except where the context otherwise requires.

(a) "Employer" means the person or persons firm company or other body named in the Appendix to the Form of Tender and includes the Employer's personal representatives successors and permitted assigns.

(b) "Contractor" means the person or persons firm or company to whom the Contract has been awarded by the Employer and includes the Contractor's personal representatives successors and permitted assigns.

(c) "Engineer" means the person firm or company appointed by the Employer to act as Engineer for the purposes of the Contract and named in the Appendix to the Form of Tender or such other person firm or company so appointed from time to time by the Employer and notified in writing as such to the Contractor.

(d) "Engineer's Representative" means a person notified as such from time to time by the Engineer under Clause 2(3)(a).

(e) "Contract" means the Conditions of Contract Specification Drawings Bill of Quantities the Tender the written acceptance thereof and the Contract Agreement (if completed).

(f) "Specification" means the specification referred to in the Tender and any modification thereof or addition thereto as may from time to time be furnished or approved in writing by the Engineer.

(g) "Drawings" means the drawings referred to in the Specification and any modification of such drawings approved in writing by the Engineer and such other drawings as may from time to time be furnished or approved in writing by the Engineer.

(h) "Bill of Quantities" means the priced and completed Bill of Quantities.

(i) "Tender Total: means the total of the Bill of Quantities at the date of award of the Contract or in the absence of a Bill of Quantities the agreed estimated total value of the Works at that date.

(j) "Contract Price" means the sum to be ascertained and paid in accordance with the provisions hereinafter contained for the construction and completion of the Works in accordance with the Contract.

(k) "Prime Cost (PC) Item" means an item in the Contract which contains (either wholly or in part) a sum referred to as Prime Cost (PC) which will be used for the execution of work or the supply of goods materials or services for the Works.

(l) "Provisional Sum" means a sum included and so designated in the Contract as a specific contingency for the execution of work or the supply of goods materials or services which may be used in whole or in part or not at all at the direction and discretion of the Engineer.

(m) "Nominated Sub-contractor" means any merchant tradesman specialist or other person firm or company nominated in accordance with the Contract to be employed by the Contractor for the execution of work or supply of goods materials or services for which a Prime Cost has been inserted in the Contract or ordered by the Engineer to be employed by the Contractor to execute work or supply goods materials or services under a Provisional Sum.

(n) "Permanent Works" means the permanent works to be constructed and completed in accordance with the Contract.

(o) "Temporary Works" means all temporary works of every kind required in or about the construction and completion of the Works.

(p) "Works" means the Permanent Works together with the Temporary Works.

(q) "Works Commencement Date"—as defined in Clause 41(1).

(r) "Certificate of Substantial Completion" means a certificate issued under Clause 48(2)(a) 48(3) or 48(4).

(s) "Defects Correction Period" means that period stated in the Appendix to the Form of Tender calculated from the date on which the Contractor becomes entitled to a Certificate of Substantial Completion for the Works or any Section or part thereof.

(t) "Defects Correction Certificate"—as defined in Clause 61(1).

(u) "Section" means a part of the Works separately identified in the Appendix to the Form of Tender.

(v) "Site" means the lands and other places on under in or through which the Works are to be executed and any other lands or places provided by the Employer for the purposes of the Contract together with such other places as may be designated in the Contract or subsequently agreed by the Engineer as forming part of the Site.

(w) "Contractor's Equipment" means all appliances or things of whatsoever nature required in or about the construction and completion of the Works but does not include materials or other things intended to form or forming part of the Permanent Works.

Singular and plural

(2) Words importing the singular also include the plural and vice-versa where the context requires.

Headings and marginal notes

(3) The headings and marginal notes in the Conditions of Contract shall not be deemed to be part thereof or be taken into consideration in the interpretation or construction thereof or of the Contract.

Clause references

(4) All references herein to clauses are references to clauses numbered in the Conditions of Contract and not to those in any other document forming part of the Contract.

Cost

(5) The word "cost" when used in the Conditions of Contract means all expenditure properly incurred or to be incurred whether on or off the Site including overhead finance and other charges properly allocatable thereto but does not include any allowance for profit.

Communications in writing

(6) Communications which under the Contract are required to be "in writing" may be hand-written typewritten or printed and sent by hand post telex cable or facsimile.

General

Clause 1(1) sets out the defined terms used, which are indicated by upper case initial letters. However, Clause 1(5) defines the term "cost" which appears with a lower case initial letter. Clauses 1(2), (3) and (4) deal with matters of interpretation.

Clauses 1(1)(e), (f), (g), (h)—the Contract

The Contract is defined at Clause 1(1)(e). The subsidiary definitions of Specification, Drawings and Bill of Quantities are given in Clauses 1(1)(f), (g) and (h). Read together with Clause 5 ("The several documents forming the Contract are to be taken as mutually explanatory of one another ...") these clauses produce a scheme in which all documents carry equal weight and importance. Where additional documents are to be incorporated into the Contract, this should be done expressly and clearly so as to override the presumption created by Clause 1(1)(e) that other documents have no direct contractual effect.

Clauses 1(1)(n), (o)—the Works

These clauses define the Permanent and Temporary Works respectively. These definitions may not be mutually exclusive since elements of structure used as Temporary Works may eventually be incorporated into the Permanent Works. Clause 8(2) provides, inter alia, that "the Contractor shall not be responsible for the design or specification of the Permanent Works ... or of any Temporary Works designed by the Engineer". Thus where Permanent Works are used in a temporary mode, and the Contractor is required or specifically entitled to use them in this mode, any damage they sustain in this mode will be at the Employer's risk. This is notwithstanding Clause 8(3) ("The Contractor shall take full responsibility for the adequacy stability and safety of all site operations and methods of construction"), Clause 14(9) ("... consent of the Engineer to the Contractor's proposed methods of construction ... shall not relieve the Contractor of any of his duties or responsibilities under the Contract") or Clause 20 ("(1) The Contractor shall ... take full responsibility for the care of the Works ... [save for] ... (2)(b) any fault defect error or omission in the design of the Works ...").

Clauses 1(1)(s), (t)—Defects Correction

The term "Defects Correction" is new to the Sixth Edition. Previous editions used the term "maintenance". The new expression is to be preferred as it describes more accurately the function of the period and avoids confusion where maintenance and servicing of the works is a contract obligation.

Clause 1(1)(v)—the Site

The term "Site" appears in Clauses 1(5), 11(2), 19, 22(2)(a), 32, 42, 53, 54 and 60(1)(b). The extent of the Site may be uncertain, for instance where land outside the immediate area of construction is used for purposes such as storage, materials processing, etc. Where land is designated as part of the Site, the Engineer may require the Contractor to provide at the Contractor's expense lights, guards, fencing, warning signs and watching: Clause 19(1).[3] Clause 53 deems property in the Contractor's equipment to be vested in the Employer "when on Site". And Clause 60(1)(b) envisages that goods or materials "delivered to the Site" form a special category for payment. It should be noted that where the Engineer requires the Contractor to remove any person or sub-contractor employed on the Works, that person is not just to be removed from the Site, but from the Works—Clause 16.

Clause 1(5)—cost

The term "cost" or "costs" appears in relation to claims in Clauses 7(4), 12(3), 13(3), 14(6), 17, 27(6), 31(2), 36(2), 36(3), 38(2), 40(1), 42(1) and 50. For any item of cost to become recoverable it must be "properly incurred or to be incurred". It means not just direct costs but also indirect costs such as overhead charges, finance charges[4] and any other such charges. It does not include any element of profit; however, in several claims clauses profit is expressly allowed in addition to the cost as defined: see, for example, Clauses 12(6) and 13(3).

ENGINEER AND ENGINEER'S REPRESENTATIVE

2 Duties and authority of Engineer

(1) (a) The Engineer shall carry out the duties specified in or necessarily to be implied from the Contract.

---

3 See also the Occupiers Liability Acts 1957 and 1984. These place obligations upon occupiers in respect of invitees and trespassers. The definition of Site may be important in defining the Contractor's responsibility. Generally the question turns on considerations of actual control rather than the definition of terms such as "Site".

4 See e.g. *F. G. Minter Ltd* v. *WHTSO* (1980) 13 BLR 1 (CA); *Rees & Kirby* v. *Swansea City Council* (1985) 30 BLR 1 (CA); *Ogilvie Builders Ltd* v. *Glasgow District Council* (1994) 68 BLR 122 (Scottish Court of Session). These relate to the "direct loss and/or expense" provisions of the JCT Contract.

(b) The Engineer may exercise the authority specified in or necessarily to be implied from the Contract. If the Engineer is required under the terms of his appointment by the Employer to obtain the specific approval of the Employer before exercising any such authority particulars of such requirements shall be those set out in the Appendix to the Form of Tender. Any requisite approval shall be deemed to have been given by the Employer for any such authority exercised by the Engineer.

(c) Except as expressly stated in the Contract the Engineer shall have no authority to amend the Terms and Conditions of the Contract nor to relieve the Contractor of any of his obligations under the Contract.

Named individual

(2) (a) Where the Engineer as defined in Clause 1(1)(c) is not a single named Chartered Engineer the Engineer shall within 7 days of the award of the Contract and in any event before the Works Commencement Date notify to the Contractor in writing the name of the Chartered Engineer who will act on his behalf and assume the full responsibilities of the Engineer under the Contract.

(b) The Engineer shall thereafter in like manner notify the Contractor of any replacement of the named Chartered Engineer.

Engineer's Representative

(3) (a) The Engineer's Representative shall be responsible to the Engineer who shall notify his appointment to the Contractor in writing.

(b) The Engineer's Representative shall watch and supervise the construction and completion of the Works. He shall have no authority

(i) to relieve the Contractor of any of his duties or obligations under the Contract

nor except as expressly provided hereunder

(ii) to order any work involving delay or any extra payment by the Employer or

(iii) to make any variation of or in the Works.

Delegation by Engineer

(4) The Engineer may from time to time delegate to the Engineer's Representative or any other person responsible to the Engineer

any of the duties and authorities vested in the Engineer and he may at any time revoke such delegation. Any such delegation

(a) shall be in writing and shall not take effect until such time as a copy thereof has been delivered to the Contractor or his agent appointed under Clause 15(2)

(b) shall continue in force until such time as the Engineer shall notify the Contractor in writing that the same has been revoked

(c) shall not be given in respect of any decision to be taken or certificate to be issued under Clauses 12(6) 44 46(3) 48 60(4) 61 63 or 66.

Assistants

(5) (a) The Engineer or the Engineer's Representative may appoint any number of persons to assist the Engineer's Representative in the carrying out of his duties under sub-clause (3)(b) or (4) of this Clause. He shall notify to the Contractor the names duties and scope of authority of such persons.

(b) Such assistants shall have no authority to issue any instructions to the Contractor save insofar as such instructions may be necessary to enable them to carry out their duties and to secure their acceptance of materials and workmanship as being in accordance with the Contract. Any instructions given by an assistant for these purposes shall where appropriate be in writing and be deemed to have been given by the Engineer's Representative.

(c) If the Contractor is dissatisfied by reason of any instruction of any assistant of the Engineer's Representative appointed under sub-clause (5)(a) of this Clause he shall be entitled to refer the matter to the Engineer's Representative who shall thereupon confirm reverse or vary such instruction.

Instructions

(6) (a) Instructions given by the Engineer or by any person exercising delegated duties and authorities under sub-clause (4) of this Clause shall be in writing. Provided that if for any reason it is considered necessary to give any such instruction orally the Contractor shall comply with such instruction.

(b) Any such oral instruction shall be confirmed in writing as soon as is possible under the circumstances. Provided that if the Contractor shall confirm in writing any such oral instruction and such conformation is not contradicted in writing by the Engineer or the Engineer's Representative forthwith it shall be deemed to be an instruction in writing by the Engineer.

(c) Upon the written request of the Contractor the Engineer or the Engineer's Representative exercising delegated duties or authorities under sub-clause (4) of this Clause shall specify in writing under which of his duties and authorities any instruction is given.

Reference on dissatisfaction

(7) If the Contractor is dissatisfied by reason of any act or instruction of the Engineer's Representative or any other person responsible to the Engineer he shall be entitled to refer the matter to the Engineer for his decision.

Impartiality

(8) The Engineer shall except in connection with matters requiring the specific approval of the Employer under sub-clause (1)(b) of this Clause act impartially within the terms of the Contract having regard to all the circumstances.

Clauses 2(1)(a) and 2(8)—the function of the Engineer

The Engineer has two distinct types of function under the Contract.[5]

He is the Employer's agent. It is thought that he acts in this capacity when he orders action under Clause 12(4), issues instructions under Clause 13(3), issues design criteria under Clause 14(5), instructs the uncovering of work under Clause 38(2), issues orders for the suspension of the Works under Clause 40(1), specifies the Commencement Date under Clause 41(1), requests accelerated completion under Clause 46(3), issues instructions to undertake searches or trials under Clause 50, issues variations under Clause 51, orders Provisional and/or Prime Cost work under Clause 58 or issues instructions regarding nominated

5 See Chapter 3, section 5; *Brodie* v. *Cardiff Corporation* [1919] AC 337 (HL); *London Borough of Merton* v. *Stanley Hugh Leach* (1985) 32 BLR 51.

sub-contractors under Clause 59. In his capacity as the Employer's authorised agent, the Employer is responsible for the Engineer's acts or omissions. Thus, for instance, where no commencement date is agreed, a failure by the Engineer to specify a Commencement Date in accordance with Clause 41(1)(b) is a breach of contract by the Employer.

In the exercise of many of his powers, however, the Engineer acts as an impartial decision-maker. The instances are legion and include wherever he exercises powers to value any work, variation or instruction; or where he issues any certificate or exercises any power of opinion or takes any decision under Clause 66. In this capacity the Employer warrants only that the Engineer will make his decision impartially.[6] Subject to this, the Engineer is not an agent of the Employer and the Employer is not responsible to the Contractor for his decisions.[7]

Clause2(1)(b), (c)—restrictions on the Engineer's authority

Clause 2(1)(b) enables the Employer to state explicitly what limitations he has placed upon the Engineer's authority under the terms of the Engineer's appointment. Without more, this would deprive the Engineer of authority in respect of those matters.[8] However, the final sentence deems the Engineer to have received the requisite approval; in the absence of further express notice of lack of authority, therefore, the Contractor may assume that the Engineer is fully authorised notwithstanding any limitation stated in the Appendix. Clause 2(1)(c) states the general rule that the Engineer is authorised neither to vary the terms of the Contract nor to relieve the Contractor of any of his obligations under it.[9]

---

6 The position is not clear where the Engineer is biased but the Employer neither instigates nor connives at the bias. Generally the Employer warrants that he will not interfere with the Engineer; he does not warrant the Engineer's impartiality. However, Clause 2(8) provides that in appropriate circumstances the Engineer "shall ... act impartially". Does this mean that where an Engineer acts in a partial manner, the Employer is in breach of Contract?

7 Nevertheless, the Engineer is responsible to the Employer for any loss due to negligent decision-making or certification: *Sutcliffe* v. *Thackrah* [1974] AC 727 (HL).

8 Where any agent has no actual authority; the Contractor will have to rely on his ostensible or apparent authority. Clearly he cannot do this where he has notice of the limitation.

9 *Toepfer* v. *Warinco* [1978] 2 Lloyd's Rep. 569.

Clause 2(2)—Engineer to be chartered engineer

It seems that where the Engineer or the person designated to act on his behalf is not a chartered engineer, this will be a breach of contract. No damage will ordinarily flow directly from this. However, the consequential effects may be: (*a*) that the Contractor is not obliged to comply with the instructions, variations and the like of the person who purports to act as Engineer; and (*b*) that decisions, certificates etc. issued are invalid.[10] Where the Contractor waives his objection to the Employer's continuing breach, he will be unable to claim that decisions etc. made during the relevant period were invalid.

Clauses 2(3), (4), (5)—delegation

These clauses establish a scheme for the delegation of powers by the Engineer or his delegates. Clause 2(3) establishes a defined appointment of an Engineer's Representative (who is the person normally known as the resident engineer). Despite the apparently mandatory wording of Clause 2(3) (the Engineer "shall notify his appointment to the Contractor in writing"), the appointment of an Engineer's Representative seems to be discretionary; where no such appointment is made it is thought that the Engineer carries out the Engineer's Representative's functions, such as to "watch and supervise the construction and completion of the Works": Clause 2(3)(b). Where appointed, the Engineer's Representative is granted a series of powers by virtue of his status, such as the authority to require safety measures in regard to the Site and the Works—Clause 19(1). Where the Engineer's Representative is to take any further authority, an express written delegation must be made under Clause 2(4), with notice to the Contractor. However, such powers as may be delegated may also be delegated to "any other person responsible to the Engineer". Certain powers may be exercised only by the Engineer: these are Clause 12(6)—the time and financial consequences of unforeseen ground conditions; Clause 44—extensions of time; Clause 46(3)—accelerated completion; Clause 48—Certificate of Substantial Completion; Clause 60(4)—certificate of sums due under the final account; Clause 61—

10 Any certificate etc. which is made by an unauthorised person is invalid: see Chapter 3, section 5 and e.g. *Croudace* v. *London Borough of Lambeth* (1986) 33 BLR 20 (CA).

Defects Correction Certificate; Clause 63—certificate prior to the determination of the Contractor's employment; Clause 66—Engineer's decision following a Notice of Dispute. In broad terms these are the duties of the Engineer as an impartial decision-maker which are of greatest significance. Nevertheless, a great number of discretionary powers remain delegable, including issuing of certificates for payment, except on the final account. Furthermore all the powers as Employer's agent are delegable, including issuing of instructions (Clause 13) and variation orders (Clause 51).

Clause 2(5) provides that either the Engineer or Engineer's Representative may appoint assistants. The person delegating shall "notify to the Contractor the names duties and scope of authority of such persons"; Clause 68 requires that any such notification must be in writing at the Contractor's principal place of business.

Clause 2(6)—instructions in writing

The proviso to Clause 2(6)(a) suggests that, in some circumstances, an oral instruction shall be complied with. The relevant circumstances are "if for any reason it is considered necessary". The person who is to decide upon this consideration is presumably the person authorised to make the decision. This may mean that any oral instruction issued to the Contractor, in circumstances where it is clearly intended that it should be acted upon, is valid. The procedures in Clauses 2(6)(b) and (c) seem to be directed at attaining regularity, clarity and a degree of evidential certainty. When an oral instruction is issued, the Contractor is entitled to a written confirmation immediately.

Clause 2(5)(c), 2(7)—reference

Clause 2(5)(c) provides for a reference from the Engineer's Representative's assistant to the Engineer's Representative. Clause 2(7) provides for a reference from an assistant or the Engineer's Representative (including his decision in relation to a reference under Clause 2(5)(c)) to the Engineer. No procedure is set for these references; it is thought that the proper time is a reasonable time in all the circumstances. Note that there is no provision for the Employer to seek a reference under these provisions.

ASSIGNMENT AND SUBCONTRACTING

3 Assignment

Neither the Employer nor the Contractor shall assign the Contract or any part thereof or any benefit or interest therein or thereunder without the prior written consent of the other party which consent shall not unreasonably be withheld.

Prohibition on assignment

The prohibition is on "the Contract or any part thereof or any benefit or interest therein or thereunder ...". This is comprehensive and extends to any monies due under the Contract, including, for instance, retention monies.[11] It should be noted that the Employer and Contractor are defined in Clauses 1(1)(a),(b) as including their "personal representatives [and] successors". It is thought that where the name, status or identity of the Employer changes by operation of law (e.g. by an enactment transferring authority to a new body) the new body will automatically assume the role of Employer. Where, however, the Employer or Contractor wishes to reorganise his corporate structure, an automatic assignment between corporate persons will not take effect.

Consent not to be unreasonably withheld

This provision clearly envisages that some assignments are possible. The use of the term "assignment" presumably precludes the assignments of burdens, including the Employer's obligation of cooperation and the Contractor's obligation of construction.[12] The term "unreasonably" is not defined. It may mean, for example, that consent must be given save where the other party is likely to be materially disadvantaged; or it may mean that consent may be withheld if there is any conceivable disadvantage to the party whose consent is required—which amounts to a veto. Where a dispute arises as to whether or not any consent is reasonably withheld, it seems it

11 *Helstan Securities Ltd* v. *Hertfordshire County Council* [1978] 3 All ER 262; *Linden Gardens Trust* v. *Linesta Sludge Disposals Ltd and others* [1994] 1 AC 85 (HL).

12 See generally Chapter 3, section 6.

must be referred to the Engineer under Clause 66 for his decision;[13] the parties are then required to give effect to his decision, unless and until it might be revised by an arbitrator.

4 Sub-contracting

(1) The Contractor shall not sub-contract the whole of the Works without the prior written consent of the Employer.

(2) Except where otherwise provided the Contractor may sub-contract any part of the Works or their design. The extent of the work to be sub-contracted and the name and address of the sub-contractor must be notified in writing to the Engineer prior to the sub-contractor's entry on to the Site or in the case of design on appointment.

(3) The employment of labour-only sub-contractors does not require notification to the Engineer under sub-clause (2) of this Clause.

(4) The Contractor shall be and remain liable under the Contract for all work sub-contracted under this Clause and for acts defaults or neglects of any sub-contractor his agents servants or workpeople.

(5) The Engineer shall be at liberty after due warning in writing to require the Contractor to remove from the Works any sub-contractor who mis-conducts himself or is incompetent or negligent in the performance of his duties or fails to conform with any particular provisions with regard to safety which may be set out in the Contract or persists in any conduct which is prejudicial to safety or health and such sub-contractor shall not be again employed upon the Works without the permission of the Engineer.

13 On such a technical issue, it may be that a court will entertain an application for a declaration without staying the matter to arbitration: see dicta in *Bristol Corporation* v. *Aird* [1913] AC 241 (HL); *Lakers Mechanical Services* v. *Boskalis Westminster* (1989) 5 Constr. L.J. 139. But where the Engineer makes a decision, the court may not be able to revise that decision: see *Northern Regional Health Authority* v. *Crouch* [1984] QB 644 (CA).

Clause 4(1), (2)—Contractor may sub-contract, but not the whole of the Works

Read literally, this provision seems to entitle the Contractor to sub-contract the Works providing some minor element is retained. All work, including design work may be sub-contracted; this provision defeats the general presumption that design work may not be vicariously performed.[14]

Clause 4(4)—Contractor to remain liable for neglects

The Contractor is generally liable for the breaches of contract caused by his sub-contractors; he is not, however, liable for their torts,[15] unless those torts are authorised by him[16] or the work is of a particularly hazardous nature.[17] The term "neglect", however, seems to suggest the tort of negligence and it may be that this provision renders the Contractor liable to the Employer for the torts of all his sub-contractors, even where such torts are not breaches of contract.

CONTRACT DOCUMENTS

5 Documents mutually explanatory

The several documents forming the Contract are to be taken as mutually explanatory of one another and in case of ambiguities or discrepancies the same shall be explained and adjusted by the Engineer who shall thereupon issue to the Contractor appropriate instructions in writing which shall be regarded as instructions issued in accordance with Clause 13.

---

14 *Moresk Cleaners* v. *Hicks* [1966] 2 Lloyd's Rep. 338.

15 *Honeywill and Stein Ltd* v. *Larkin (London's Commercial Photographers) Ltd* [1933] All ER 77 (CA). *D & F Estates* v. *Church Commissioners* [1989] AC 177 (HL).

16 *Ellis* v. *Sheffield Gas Consumers Co.* (1853) 2 E & B 767.

17 *Salisbury* v. *Woodland* [1970] 1 QB 324 (CA). *Honeywill and Stein Ltd* v. *Larkin (London's Commercial Photographers) Ltd* [1933] All ER 77 (CA). *Dalton* v. *Angus* (1881) 6 App Cas 740 (HL).

Documents to be mutually explanatory

This provision appears to override the canon of construction that specially prepared documents are to be given precedence.[18]

Engineer to explain and adjust ambiguities and discrepancies

This appears to require the Engineer to generate certainty in the case where the technical requirements are unclear or where two or more reasonably clear requirements conflict. Where the Engineer is of the opinion that an ambiguity or discrepancy exists, he is obliged to issue a written instruction. Any explanation or adjustment may have financial consequences, in particular where the content of the written instruction is "beyond that reasonably to have been foreseen by an experienced contractor at the time of tender"—Clause 13(3). Where appropriate, the Contractor is entitled to be paid the extra cost, profit ("in respect of any additional permanent or temporary work") and to be granted an extension of time.

6 Supply of documents

(1) Upon award of the Contract the following shall be furnished to the Contractor free of charge

(a) four copies of the Conditions of Contract Specification and (unpriced) bill of quantities and

(b) the number and type of copies as entered in the Appendix to the Form of Tender of all Drawings listed in the Specification.

(2) Upon approval by the Engineer in accordance with Clause 7(6) the Contractor shall supply to the Engineer four copies of all Drawings Specifications and other documents submitted by the Contractor. In addition the Contractor shall supply at the Employer's expense such further copies of such Drawings Specifications and other documents as the Engineer may request in writing for his use.

(3) Copyright of all Drawings Specifications and the Bill of Quantities (except the pricing thereof) supplied by the Em-

18 *Glynn* v. *Margetson* [1893] AC 351 (HL); *English Industrial Estates* v. *Wimpey* [1973] 1 Lloyd's Rep. 118 (CA).

ployer or the Engineer shall not pass to the Contractor but the Contractor may obtain or make at his own expense any further copies required by him for the purposes of the Contract. Similarly copyright of all documents supplied by the Contractor under Clause 7(6) shall remain in the Contractor but the Employer and the Engineer shall have full power to reproduce and use the same for the purpose of completing operating maintaining and adjusting the Works.

Copyright

Copyright does not pass, merely a licence to use documents and designs for the purpose of the Works.

7 Further Drawings Specifications and instructions

(1) The Engineer shall from time to time during the progress of the Works supply to the Contractor such modified or further Drawings Specifications and instructions as shall in the Engineer's opinion be necessary for the purpose of the proper and adequate construction and completion of the Works and the Contractor shall carry out and be bound by the same.

If such Drawings Specifications or instructions require any variation to any part of the Works the same shall be deemed to have been issued pursuant to Clause 51.

Contractor to provide further documents

(2) Where sub-clause (6) of this Clause applies the Engineer may require the Contractor to supply such further documents as shall in the Engineer's opinion be necessary for the purpose of the proper and adequate construction completion and maintenance of the Works and when approved by the Engineer the Contractor shall carry out and be bound by the same.

Notice by Contractor

(3) The Contractor shall give adequate notice in writing to the Engineer of any further Drawing or Specification that the Contractor may require for the construction and completion of the Works or otherwise under the Contract.

Delay in issue

(4) (a) If by reason of any failure or inability of the Engineer to issue at a time reasonable in all the circumstances Drawings Specifications or instructions requested by the Contractor

and considered necessary by the Engineer in accordance with sub-clause (1) of this Clause the Contractor suffers delay or incurs cost then the Engineer shall take such delay into account in determining any extension of time to which the Contractor is entitled under Clause 44 and the Contractor shall subject to Clause 52(4) be paid in accordance with Clause 60 the amount of such cost as may be reasonable.

(b) If the failure of the Engineer to issue any Drawing Specification or instruction is caused in whole or in part by the failure of the Contractor after due notice in writing to submit drawings specifications or other documents which he is required to submit under the Contract the Engineer shall take into account such failure by the Contractor in taking any action under sub-clause (4)(a) of this Clause.

One copy of documents to be kept on Site

(5) One copy of the Drawings and Specification furnished to the Contractor as aforesaid and of all Drawings Specifications and other documents required to be provided by the Contractor under sub-clause (6) of this Clause shall at all reasonable times be available on the Site for inspection and use by the Engineer and the Engineer's Representative and by any other person authorized by the Engineer in writing.

Permanent Works designed by Contractor

(6) Where the Contract expressly provides that part of the Permanent Works shall be designed by the Contractor he shall submit to the Engineer for approval

(a) such drawings specifications calculations and other information as shall be necessary to satisfy the Engineer as to the suitability and adequacy of the design and

(b) operation and maintenance manuals together with as completed drawings of that part of the Permanent Works in sufficient detail to enable the Employer to operate maintain dismantle reassemble and adjust the Permanent Works incorporating that design. No certificate under Clause 48 covering any part of the Permanent Works designed by the Contractor shall be issued until manuals and drawings in such detail have been submitted to and approved by the Engineer.

Responsibility unaffected by approval

(7) Approval by the Engineer in accordance with sub-clause (6) of this Clause shall not relieve the Contractor of any of his responsibilities under the Contract. The Engineer shall be responsible for the integration and co-ordination of the Contractor's design with the rest of the Works.

Supply of additional documents

It is common in civil engineering projects for many detailed drawings and some specifications to be provided or re-issued after the Contract is made. Not only is this convenient because of the interaction of structures with the detailed geology etc., but re-issues are frequently necessitated by variations to the Works. Where documents are to be issued, they must be issued in a reasonable time given all the circumstances, including the constraints under which the Engineer and his designers are working[19] and the agreed Time for Completion.[20] Any approved programme (see Clause 14) will suggest the time at which documents need to be supplied. Where the approved Clause 14 programme provides for completion in a shorter time than the Contract Time for Completion, it is thought that the Contractor is entitled to receive information in line with this shortened programme until the Engineer gives reasonable notice that information will only be released to meet completion in a longer period (not exceeding the full Time for Completion).

Delay in issue

The Engineer is not obliged to issue Drawings and/or Specifications save as is necessary for the "proper and adequate construction and completion of the Works". By Clause 7(3) the Contractor is to give adequate notice in writing of his need for these. It is unclear whether an approved Clause 14 programme is of itself sufficient, where it is reasonably clear from this that information must be provided for the

19 See *Neodox* v. *Swinton & Pendlebury Borough Council* (1958) 5 BLR 34; *London Borough of Merton* v. *Stanley Hugh Leach* (1985) 32 BLR 51.

20 *Glenlion Construction* v. *Guinness Trust* (1988) 39 BLR 89.

works to be progressed. It is always preferable for the Contractor to give specific notification.

Clauses 7(2), (6), (7)—Contractor's design

These clauses apply where the Contractor is required by express terms to design any element of the Permanent Works. The Contractor is not obliged to provide full working drawings during the currency of the Works, merely such drawings etc. as shall be necessary to satisfy the Engineer as to the suitability and adequacy of the design. The question is one of professional judgment. Excessive demands by the Engineer must, it is thought, be complied with: Clause 13(1) "the Contractor ... shall comply with and adhere strictly to the Engineer's instructions ...". However, such demands will amount to an instruction entitling the Contractor to the additional cost incurred: Clauses 13(3) and 51. Clause 7(6) requires that the drawings etc. be submitted to the Engineer for his approval. Where the Engineer is not satisfied, he may call for additional documents under Clause 7(3). Approval by the Engineer will not relieve the Contractor of any of his obligations under the Contract. Nevertheless, written comments which amount in effect to an instruction in regard to designs may rank as an instruction pursuant to Clause 13. No time is stipulated for the approval of designs (compare this with Clause 14, approval of programmes) and it seems, therefore, that a reasonable time is allowed. The reasonable time will be determined by factors such as the importance, scale and complexity of the design as well as the urgency for compliance with the programme. Failure by the Engineer to act within a reasonable time will be a breach of contract by the Employer. Before any Certificate of Substantial Completion is issued, the Contractor must supply documents required by Clause 7(6)(b).

GENERAL OBLIGATIONS

8 Contractor's general responsibilities

(1) The Contractor shall subject to the provisions of the Contract

(a) construct and complete the Works and

(b) provide all labour materials Contractor's Equipment Temporary Works transport to and from and in or about the Site and everything whether of a temporary or permanent nature required in and for such construction and completion so far as the necessity for providing the same is specified in or reasonably to be inferred from the Contract.

Design responsibility

(2) The Contractor shall not be responsible for the design or specification of the Permanent Works or any part thereof (except as may be expressly provided in the Contract) or of any Temporary Works designed by the Engineer. The Contractor shall exercise all reasonable skill care and diligence in designing any part of the Permanent Works for which he is responsible.

Contractor responsible for safety of site operations

(3) The Contractor shall take full responsibility for the adequacy stability and safety of all site operations and methods of construction.

General responsibilities

Clause 8(1) requires the Contractor to construct all the Works in accordance with the Contract and do everything needed to achieve that aim. It provides a clear restatement of the general principle that a contractor is required to do all things reasonably to be inferred, whether mentioned in the Contract or not.[21] Where the Contract is operated unamended (i.e. as a measure and value contract with a Bill of Quantities), any items of work which are not mentioned in the Bill of Quantities are omissions and the Engineer must add them in, which entitles the Contractor to be paid for them—Clause 55. Where, however, the contract is amended to a lump sum contract without a Bill of Quantities, the effect of Clause 8(1) is to require the Contractor to complete the Works for the agreed price.

Contractor's design and site operation

Clause 8(2) is new and reflects the growing practice of passing design responsibility onto the Contractor. Together with Clause 8(3) it states the general position under the Contract—the Employer is responsible for the Permanent Works and the Contractor is responsible for the Temporary Works. However, both of these assumptions may be expressly overridden. The Contractor may take on design

21 *Williams* v. *Fitzmaurice* (1858) 3 H & N 844.

responsibility where expressly stated in the Contract; thus a voluntary performance of some design will not apparently cause the Contractor to be liable for it.[22] The Engineer may however design elements of the Temporary Works, in which case the Employer is responsible for them.

Standard of design care

The expression "all reasonable skill and care and diligence ..." should be contrasted with the words "reasonable skill and care", generally used to characterise a professional's standard of care.[23] It is thought that the use of "all" and "diligence" does not create any higher standard of care. The word "diligence" may simply oblige the Contractor to ensure that his designers have all the relevant and up-to-date information. This closes down the slight (but possible) argument that the Contractor is not responsible since he delegated the design to a competent designer and he is not liable under Clause 4(4) because his designer is not in "default" or "neglect".

Exceptions to the Contractor's responsibility for site operations

Clause 8(3) purports to place the full responsibility for site operations and methods of construction on the Contractor. There are, however, a number of exceptions. Where the interaction between the method of construction and reasonably unforeseeable physical conditions (as defined in Clause 12(1)) means that the method of construction is more costly than could have been foreseen, the Contractor may be entitled to recover pursuant to Clause 12.[24] Also, where specified methods of construction become legally or physically impossible, the Contractor is entitled to an instruction under Clause 13 or a variation order or damages for breach of contract.[25] Furthermore, the responsibility of the Employer under, inter alia, the Health and Safety at Work etc. Act 1974, the Construction (Design and Management) Regulations 1994 and the Occupier's Liability Acts 1957 and 1984 will not be excluded.

---

22 In contract at any rate. The Contractor may owe a duty in the tort of negligence to third parties.

23 See Chapter 5, section 4.

24 *Humber Oil Terminals Trustee* v. *Harbour and General* (1991) 59 BLR 1 (CA).

25 See Clause 13. *Yorkshire Water Authority* v. *McAlpine* (1985) 32 BLR 114; *Holland Dredging* v. *Dredging and Construction Co.* (1987) 37 BLR 1 (CA).

9 Contract Agreement

The Contractor shall if called upon so to do enter into and execute a Contract Agreement to be prepared at the cost of the Employer in the form annexed to these Conditions.

Form annexed to this agreement

This form includes a provision for the agreement to be made under seal. The execution of a sealed contract is frequently in the Employer's interest as it extends the limitation period from 6 years to 12 years.[26]

10 Performance security

(1) If the Contract requires the Contractor to provide security for the proper performance of the Contract he shall obtain and provide to the Employer such security in a sum not exceeding 10% of the Tender Total within 28 days of the award of the Contract. The security shall be provided by a body approved by the Employer and be in the Form of Bond annexed to these Conditions. The Contractor shall pay the cost of such security unless the Contract provides otherwise.

Arbitration upon security

(2) For the purposes of the arbitration provisions in such security

(a) the Employer shall be deemed a party to the said security for the purpose of doing everything necessary to give effect to such provisions and

(b) any agreement decision award or other determination touching or concerning the relevant date for the discharge of such security shall be wholly without prejudice to the resolution or determination of any dispute or difference between the Employer and the Contractor under Clause 66.

Effect of non-compliance

Where the Contractor fails to provide the bond within 28 days, it is thought that this does not amount immediately to a breach which is

26 See Chapter 6, section 9.

sufficiently serious to enable the Employer to terminate the Contract. Obligations as to time are generally treated as "not of the essence"; in other words, their breach does not entitle the Employer to terminate. Where, however, the Employer serves notice on the Contractor that he will insist on his bond, and the Contractor fails to comply within a reasonable period, the Employer will be entitled to terminate the Contract and claim damages.[27]

The Form of Bond

This bond is a performance bond.[28] Any call by the Employer requires proof of breach and damage.[29] The bond contains its own internal arbitration provisions, hence the reference in Clause 10(2). A new form of bond is being drawn up on behalf of the CCSJC and is expected to be published shortly.[30]

11 Provision and interpretation of information

(1) The Employer shall be deemed to have made available to the Contractor before the submission of his Tender all information on

(a) the nature of the ground and subsoil including hydrological conditions and

(b) pipes and cables in on or over the ground obtained by or on behalf of the Employer from investigations undertaken relevant to the Works.

The Contractor shall be responsible for the interpretation of all such information for the purposes of constructing the Works and for any design which is the Contractor's responsibility under the Contract.

27 As to service of notice making time of the essence and its effects see *United Scientific Holdings* v. *Burnley Council* [1978] AC 904 (HL).

28 See Chapter 3, section 9.

29 See *Trafalgar House Construction (Regions) Ltd* v. *General Surety & Guarantee Co. Ltd* [1995] 3 WLR 204 (HL).

30 At the time of writing, this document was still the subject of consultation.

Inspection of Site

(2) The Contractor shall be deemed to have inspected and examined the Site and its surroundings and information available in connection therewith and to have satisfied himself so far as is practicable and reasonable before submitting his Tender as to

(a) the form and nature thereof including the ground and sub-soil

(b) the extent and nature of work and materials necessary for constructing and completing the Works and

(c) the means of communication with and access to the Site and the accommodation he may require

and in general to have obtained for himself all necessary information as to risks contingencies and all other circumstances which may influence or affect his Tender.

Basis and sufficiency of Tender

(3) The Contractor shall be deemed to have

(a) based his Tender on the information made available by the Employer and on his own inspection and examination all as aforementioned and

(b) satisfied himself before submitting his Tender as to the correctness and sufficiency of the rates and prices stated by him in the Bill of Quantities which shall (unless otherwise provided in the Contract) cover all his obligations under the Contract.

Clause 11(1)—the provision of information by the Employer[31]

The effect of the provision extends to "all information ... obtained by or on behalf of the Employer from investigations undertaken relevant to the Works", including, it seems, information obtained in the past on previous phases of the same or associated projects. The phrase "The Employer shall be deemed to have made available ..." is difficult. It is

31 See *William Sindall plc* v. *Cambridge County Council* [1994] 1 WLR 1016 (CA) where the plaintiff purchaser claimed that the vendor was obliged to disclose information.

submitted that it does not oblige the Employer to disclose information;[32] rather it establishes an evidential presumption that any data withheld were not perceived by the Employer and his advisers as relevant at the time of tender (unless the withheld information merely duplicates other data supplied by the Employer). This presumption may assist a Contractor advancing a claim in which foreseeability is at issue (e.g. a Clause 12 claim). Note, however, that such claims are generally founded on the foresight of an experienced contractor, not of the Employer and his advisers, and so any such presumption cannot be determinative of e.g. Clause 12 issues.

Clause 11(1)—Contractor's design: responsibility and interpretation

By Clause 11(1) the Employer does not warrant that the information is sufficient to design any of the Works. Interpretation for the purpose of design is to be undertaken with all reasonable skill and care and diligence (see Clause 8(2)) and with proper caution.[33] Where information is withheld which would have alerted a competent designer to exercise more care or to undertake supplementary investigation or analysis, this may afford a partial defence to a claim against the Contractor for negligent design.

Clause 11(2)—inspection of the site and its surroundings

The extent to which it is "practicable and reasonable" to inspect and examine the site etc. is a question of fact. It is submitted that, in the ordinary case, the location of the Contractor has little bearing on the matter; otherwise local contractors would take on a higher obligation. Where the Works are of major value additional effort can reasonably be expected. Likewise, where an experienced contractor would reasonably foresee that practical difficulties may be encountered, there

---

32 There seems to be no duty of general disclosure in English law and contractual obligations of this kind require clear wording. Note, however, that in Commonwealth jurisdictions employers bear a heavier responsibility for pre-contractual information than in England. See Marston D. L., The impact of subsurface ground conditions on project participants—a Canadian perspective (1992) 8 Constr. L.J. 107; *Dillingham* v. *Downs* [1972] 2 NSWLR 49 (Supreme Court of New South Wales); *Morrison-Knudsen International* v. *Commonwealth of Australia* (1972) 13 BLR 114 (High Court of Australia).

33 See Chapter 5. *Moneypenny* v. *Hartland* (1824) 2 C & P 378; *Sealand of the Pacific* v. *Robert C. McHaffie Ltd* (1974) 51 DLR (3d) 702 (British Columbia Court of Appeal).

is a greater obligation of inspection. In practice, however, there is often little opportunity to do more than examine the ground surface, open water courses, vegetation and areas surrounding the Site which are accessible to the public.

Clause 11(3)(b)—sufficiency of the tender

This appears to add little to the Contractor's obligations. Where, however, the tender rates are weighted for any reason (e.g. where "front-end-loaded" in order to promote early positive cash flow) the Contractor will be held to these prices even where they turn out to work to his disadvantage.

12 Adverse physical conditions and artificial obstructions

(1) If during the execution of the Works the Contractor shall encounter physical conditions (other than weather conditions or conditions due to weather conditions) or artificial obstructions which conditions or obstructions could not in his opinion reasonably have been foreseen by an experienced contractor the Contractor shall as early as practicable give written notice thereof to the Engineer.

Intention to claim

(2) If in addition the Contractor intends to make any claim for additional payment or extension of time arising from such condition or obstruction he shall at the same time or as soon thereafter as may be reasonable inform the Engineer in writing pursuant to Clause 52(4) and/or Clause 44(1) as may be appropriate specifying the condition or obstruction to which the claim relates.

Measures being taken

(3) When giving notification in accordance with sub-clauses (1) and (2) of this Clause or as soon as practicable thereafter the Contractor shall give details of any anticipated effects of the condition or obstruction the measures he has taken is taking or is proposing to take their estimated cost and the extent of the anticipated delay in or interference with the execution of the Works.

Action by Engineer

(4) Following receipt of any notification under sub-clauses (1) or (2) or receipt of details in accordance with sub-clause (3) of this Clause the Engineer may if he thinks fit inter alia

(a) require the Contractor to investigate and report upon the practicality cost and timing of alternative measures which may be available

(b) give written consent to measures notified under sub-clause (3) of this Clause with or without modification

(c) give written instructions as to how the physical conditions or artificial obstructions are to be dealt with

(d) order a suspension under Clause 40 or a variation under Clause 51.

Conditions reasonably foreseeable

(5) If the Engineer shall decide that the physical conditions or artificial obstructions could in whole or in part have been reasonably foreseen by an experienced contractor he shall so inform the Contractor in writing as soon as he shall have reached that decision but the value of any variation previously ordered by him pursuant to sub-clause (4)(d) of this Clause shall be ascertained in accordance with Clause 52 and included in the Contract Price.

Delay and extra cost

(6) Where an extension of time or additional payment is claimed pursuant to sub-clause (2) of this Clause the Engineer shall if in his opinion such conditions or obstructions could not reasonably have been foreseen by an experienced contractor determine the amount of any costs which may reasonably have been incurred by the Contractor by reason of such conditions or obstructions together with a reasonable percentage addition thereto in respect of profit and any extension of time to which the Contractor may be entitled and shall notify the Contractor accordingly with a copy to the Employer.

General

Clauses 11 and 12 should be read together. Clause 11 sets out the position in regard to information about the Site and its surroundings. Clause 12 divides the risk of unforeseen conditions between the Contractor and Employer and establishes a right for the Contractor to claim should unforeseeable conditions occur. The right to recover under Clause 12 is in addition to any claims for (*a*) misrepresentation,[34] (*b*) implied warranties in borehole or other survey data,[35] or (*c*) breach of contract where survey information which is incorporated into the Contract is materially wrong.

Clause 12(1)—general

The test applied in Clause 12 is based on the reasoning that it is efficient for the Employer to receive tenders which do not include speculative contingencies for unforeseen conditions. Accordingly, the Clause places the risk of adverse conditions substantially on the Employer.

"physical conditions", "artificial obstructions"

"Physical conditions" are not limited to pre-existing conditions and may include a transient condition or a combination of physical conditions and applied loading.[36] The term "artificial obstructions" clearly includes such things as buried services etc. It is not clear whether or not intangible obstructions brought about artificially (i.e. by human agency) such as local authority or pressure-group interventions fall within the scope of Clause 12.

"other than weather conditions or conditions due to weather conditions"

Rain, snow, temperature, winds, etc. clearly fall within the ambit of this phrase; so do directly consequential flooding, snow loading, immediate effects of high or low temperatures and wind loading, etc. It is not clear, however, whether conditions which are contributed to by weather, such as rising water tables, are included. It is submitted that the test is whether or not the weather was the dominant cause of the condition.

---

34 See Chapter 4, section 1.

35 *Bacal* v. *Northampton Development Corporation* (1975) 8 BLR 88 (CA).

36 *Humber Oil Terminals Trustee* v. *Harbour and General* (1991) 59 BLR 1 (CA).

"could not ... reasonably have been foreseen by an experienced contractor"

This test is given in objective terms. It is submitted that the test is to be viewed from the perspective of a contractor with reasonable skill and average experience of the work envisaged at the time of tender; the skill and experience of the particular contractor is irrelevant. It is submitted that the foreseeability test does not relate to what an experienced contractor could have visualised as a possibility if he allowed himself to speculate about it; the question is whether or not he would have considered it likely that the conditions encountered would have been encountered given the information available to him at the time of tender. The views of expert geotechnical engineers and other such witnesses as to whether the information provided at tender stage is consistent with what was actually encountered are often peripheral to the test.

Illustrations of the test

Where a contractor undertakes to excavate under an old sewer, the existence of which was known at the time of tender, the court found that the poor condition of the sewer was reasonably foreseeable.[37] Where, however, the conditions encountered or their impact on the work is surprising, this is indicative that those conditions were not reasonably foreseeable in the sense of Clause 12.[38] In a South African case[39] the wording was "sub-surface conditions which in the opinion of the engineer could not reasonably have been foreseen". The contract was for a railway tunnel. Rock was classified into six classes, ranging from Very Good (I) to Very Poor (VI). A report was provided at the time of tender which suggested that the rock condition was reasonably good. During construction the average rock encountered was $1\frac{1}{2}$ classes worse than suggested by the report. Corbett CJ. said:[40] "The differences between the geomechanical classes of rock predicted in the report and

37 *C. J. Pearce & Co. Ltd* v. *Hereford Corporation* (1968) 66 LGR 647.

38 Award No. 5 (1989) CLY 1994, 98 (Mr Uff QC, arbitrator). On appeal to the Court of Appeal: *Humber Oil Terminals Trustee* v. *Harbour and General* (1991) 59 BLR 1 (CA).

39 *Companie Interafricaine de Travaux (Comiat)* v. *South African Transport Services* (1991) CLY 1994, 149 (South African Supreme Court).

40 At p. 169. The entire court agreed with this analysis.

those alleged to have been encountered are so substantial that if the report predictions represent approximately what was reasonably foreseeable, the actual conditions were clearly not reasonably foreseeable ... It may be that in evaluating the report, Comiat should have made some allowance for predictions being overly optimistic and thus built a safety margin into its tender (I make no finding in this regard), but it seems to me to be unlikely that any such allowance would have come anywhere near to bridging the gap between the report predictions and actuality."

Partly foreseeable conditions

One interpretation of Clause 12(1) is that the conditions encountered are either reasonably foreseeable or they are not. In the latter case, Clause 12(6) entitles the Contractor to recover, inter alia, "any costs which may reasonably have been incurred by the Contractor by reason of such conditions ...". It is thought, however, that conditions encountered may be reasonably foreseeable up to a point and unforeseeable thereafter, so that only the additional costs etc. are recoverable under Clause 12(6). Thus where water inflow into an excavation is greater than could reasonably have been foreseen, this does not entitle the Contractor to all the costs of pumping it dry, merely the costs in excess of what was reasonably foreseeable.

Notifications

Clause 12(1) provides that the Contractor shall give notice of any condition etc. which is in his opinion not reasonably foreseeable. Clause 12(2) provides for a notice of intention to claim. These notices are not necessary conditions to recovery, but failure to serve them in accordance with the Contract may reduce the amount recoverable. Clause 52(4)(e) provides that the Contractor shall be entitled to recover "only to the extent that the Engineer has not been prevented from or substantially prejudiced by such failure in investigating the said claim".

Measures and action

Clause 12(3) provides for the Contractor to supply details of anticipated effects or measures being taken; there is no express requirement for him to supply written details. Clause 12(4) provides for the Engineer to take action to control the situation. Engineers are reluctant to take proactive decisions under Clause 12(4); such action, they believe, may be perceived as evidence that the conditions are not reasonably foreseeable, or may amount to a variation entitling the Contractor to recover without having to demonstrate that conditions are unforeseeable.[41] These risks may be overstated; in any event Clause 12(5) expressly contemplates the Engineer altering his view after issuing a variation order.

Clause 12(6)—recovery by the contractor

Where the Engineer decides that the conditions etc. could not reasonably have been foreseen, he is to determine the "costs which may reasonably have been incurred by the Contractor by reason of such conditions or obstructions together with a reasonable percentage addition thereto in respect of profit and any extension of time ...". It is submitted that this means: (*a*) to the extent that any conditions are not reasonably foreseeable the Contractor is entitled to reasonable cost plus profit. (*b*) This cost plus profit may be in addition to or, in substitution for (in part or in whole) the billed rates. To the extent that the original billed work is still required, despite the unforeseen conditions, cost plus profit will be paid only for the new/additional work; the original work will be paid at billed rates. (*c*) The Contractor must demonstrate that the costs were actually incurred by showing proper cost breakdowns for items of work. The fact that an item costs £*x* does not mean an automatic entitlement to £*x*; the Contractor must show that all costs were reasonable and neither extravagant nor wasteful and that he took all reasonable steps to mitigate the cost. (*d*) Any costs claimed and the need for them are to be viewed at the time they were incurred and not in hindsight. Thus, upon encountering adverse conditions, the Contractor may reasonably decide to employ a construction technique X, which may prove inappropriate as circumstances unfold, requiring a revised technique Y. Here the cost of X will be recoverable in addition to the cost of Y. This is not unreasonable as the Engineer has the option of controlling the measures being taken by instruction or variation order in accordance with Clause 12(4).

41 See *Simplex Piling v. St Pancras Borough Council* (1958) 14 BLR 80: here the Contractor was entitled to be paid for the variation despite requesting it for his own benefit; the court held that since the contractual machinery was operated, the right to payment automatically followed.

13 Work to be to satisfaction of Engineer

(1) Save insofar as it is legally or physically impossible the Contractor shall construct and complete the Works in strict accordance with the Contract to the satisfaction of the Engineer and shall comply with and adhere strictly to the Engineer's instructions on any matter connected therewith (whether mentioned in the Contract or not). The Contractor shall take instructions only from the Engineer or (subject to the limitations referred to in Clause 2) from the Engineer's Representative.

Mode and manner of construction

(2) The whole of the materials Contractor's Equipment and labour to be provided by the Contractor under Clause 8 and the mode manner and speed of construction of the Works are to be of a kind and conducted in a manner acceptable to the Engineer.

Delay and extra cost

(3) If in pursuance of Clause 5 or sub-clause (1) of this Clause the Engineer shall issue instructions which involve the Contractor in delay or disrupt his arrangements or methods of construction so as to cause him to incur cost beyond that reasonably to have been foreseen by an experienced contractor at the time of tender then the Engineer shall take such delay into account in determining any extension of time to which the Contractor is entitled under Clause 44 and the Contractor shall subject to Clause 52(4) be paid in accordance with Clause 60 the amount of such cost as may be reasonable except to the extent that such delay and extra cost result from the Contractor's default. Profit shall be added thereto in respect of any additional permanent or temporary work. If such instructions require any variation to any part of the Works the same shall be deemed to have been given pursuant to Clause 51.

General

This clause contains a number of distinct powers and provisions, including a repetition of the Contractor's primary obligation to construct and complete the Works.

Legal or physical impossibility[42]

A legal impossibility arises where any specified method of construction or necessary structure contravenes the law (e.g. the Health and Safety at Work etc. Act 1974 or the Construction (Design and Management) Regulations 1994). A physical impossibility arises where any specified method of construction cannot physically be accomplished because of any restriction on access etc.[43] Where an impossibility arises, the Contractor is not obliged to perform the relevant work as originally specified. In such a case, it seems that Clause 13(1) requires the Engineer to issue an instruction as to how progress is to be achieved, though this is not expressly stated. Where such an instruction is not forthcoming, this may well amount to a breach of contract; the quantum of damages will be the value of the instruction or variation order to which the contractor is entitled, valued in accordance with Clause 52.

Contractor to comply with the Engineer's instructions

The Contractor is required to comply with any Engineer's instruction on any matter "whether mentioned in the Contract or not"—Clause 13(1). The instruction is to be in writing save where "for any reason it is considered necessary to give any such instruction orally"—Clause 2(6)(a). Where the instruction requires a variation to the Works it shall be deemed to have been given as a variation order under Clause 51—Clause 13(3).

"to the satisfaction of the Engineer", "of a kind and conducted in a manner acceptable to the Engineer"

These references are not thought to add anything of substance to the Contractor's general obligation. They make it clear, however, that during the currency of the project, any instructions by the Engineer are

42 The expression "impossible" must not be interpreted too rigidly; it does not mean absolutely impossible: *Turriff Ltd* v. *Welsh National Water Development Authority* (1979) CLY 1994, 122.

43 As where the Works cannot be constructed in accordance with a programme or method statement which forms part of the contract terms: Award No. 2 (1985), CLY 1994, 58 (Mr Hawker, arbitrator); *Yorkshire Water Authority* v. *McAlpine* (1985) 32 BLR 114; *Holland Dredging* v. *Dredging and Construction Co.* (1987) 37 BLR 1 (CA).

to be given effect, whether or not justified by an objective reading of the specification. Where the Contractor complies with the objective meaning of the Contract but the Engineer instructs him to perform to a more onerous specification, this will be an instruction amounting to a variation and the Contractor will be entitled to claim for it.

14 Programme to be furnished

(1) (a) Within 21 days after the award of the Contract the Contractor shall submit to the Engineer for his acceptance a programme showing the order in which he proposes to carry out the Works having regard to the provisions of Clause 42(1).

(b) At the same time the Contractor shall also provide in writing for the information of the Engineer a general description of the arrangements and methods of construction which the Contractor proposes to adopt for the carrying out of the Works.

(c) Should the Engineer reject any programme under sub-clause (2)(b) of this Clause the Contractor shall within 21 days of such rejection submit a revised programme.

Action by Engineer

(2) The Engineer shall within 21 days after receipt of the Contractor's programme

(a) accept the programme in writing or

(b) reject the programme in writing with reasons or

(c) request the Contractor to supply further information to clarify or substantiate the programme or to satisfy the Engineer as to its reasonableness having regard to the Contractor's obligations under the Contract.

Provided that if none of the above actions is taken within the said period of 21 days the Engineer shall be deemed to have accepted the programme as submitted.

Provision of further information

(3) The Contractor shall within 21 days after receiving from the Engineer any request under sub-clause (2)(c) of this Clause or within such further period as the Engineer may allow provide the further information requested failing which the relevant programme shall be deemed to be rejected.

Upon receipt of such further information the Engineer shall within a further 21 days accept or reject the programme in accordance with sub-clauses (2)(a) or (2)(b) of this Clause.

Revision of programme

(4) Should it appear to the Engineer at any time that the actual progress of the work does not conform with the accepted programme referred to in sub-clause (1) of this Clause the Engineer shall be entitled to require the Contractor to produce a revised programme showing such modifications to the original programme as may be necessary to ensure completion of the Works or any Section within the time for completion as defined in Clause 43 or extended time granted pursuant to Clause 44. In such event the Contractor shall submit his revised programme within 21 days or within such further period as the Engineer may allow. Thereafter the provisions of sub-clauses (2) and (3) of this Clause shall apply.

Design criteria

(5) The Engineer shall provide to the Contractor such design criteria relevant to the Permanent Works or any Temporary Works design supplied by the Engineer as may be necessary to enable the Contractor to comply with sub-clauses (6) and (7) of this Clause.

Methods of construction

(6) If requested by the Engineer the Contractor shall submit at such times and in such further detail as the Engineer may reasonably require information pertaining to the methods of construction (including Temporary Works and the use of Contractor's Equipment) which the Contractor proposes to adopt or use and calculations of stresses strains and deflections that will arise in the Permanent Works or any parts thereof during construction so as to enable the Engineer to decide whether if these methods are adhered to the Works can be constructed and completed in accordance with the Contract and without detriment to the Permanent Works when completed.

Engineer's consent

(7) The Engineer shall inform the Contractor in writing within 21 days after receipt of the information submitted in accordance with sub-clauses (1)(b) and (6) of this Clause either

(a) that the Contractor's proposed methods have the consent of the Engineer or

(b) in what respects in the opinion of the Engineer they fail to meet the requirements of the Contract or will be detrimental to the Permanent Works.

In the latter event the Contractor shall take such steps or make such changes in the said methods as may be necessary to meet the Engineer's requirements and to obtain his consent. The Contractor shall not change the methods which have received the Engineer's consent without the further consent in writing of the Engineer which shall not be unreasonably withheld.

Delay and cost

(8) If the Contractor unavoidably incurs delay or cost because

(a) the Engineer's consent to the proposed methods of construction is unreasonably delayed or

(b) the Engineer's requirements pursuant to sub-clause (7) of this Clause or any limitations imposed by any of the design criteria supplied by the Engineer pursuant to sub-clause (5) of this Clause could not reasonably have been foreseen by an experienced contractor at the time of tender

the Engineer shall take such delay into account in determining any extension of time to which the Contractor is entitled under Clause 44 and the Contractor shall subject to Clause 52(4) be paid in accordance with Clause 60 such sum in respect of the cost incurred as the Engineer considers fair in all the circumstances. Profit shall be added thereto in respect of any additional permanent or temporary work.

Responsibility unaffected by acceptance or consent

(9) Acceptance by the Engineer of the Contractor's programme in accordance with sub-clauses (2) (3) or (4) of this Clause and the consent of the Engineer to the Contractor's proposed methods of construction in accordance with sub-clause (7) of this Clause shall not relieve the Contractor of any of his duties or responsibilities under the Contract.

What the Contractor must provide

Clause 14 calls for two distinct documents to be provided, as follows.

(*a*) *A programme.* The format of the programme is not specified, save that it must show the order in which the Contractor proposes to carry out the Works.[44] Many contractors supply detailed programmes including bar charts and networks; it seems, however, that a simple list of activities which shows the proposed order may suffice. Upon receipt of the programme, however, the Engineer may request the Contractor to supply further information to "clarify or substantiate the programme or to satisfy the Engineer as to its reasonableness"—Clause 14(2)(c).

(*b*) *A "method statement".* The expression "method statement" does not appear in Clause 14; it is used here as a convenient shorthand. What is required is a "general description of the arrangements and methods of construction which the Contractor proposes to adopt for the carrying out of the Works". The expression "general description" is thought to indicate that the statement needs to set out the broad proposals, rather than a detailed statement of them. Note, however, that Clause 14(6) entitles the Engineer to call for further reasonable information as to methods of construction. While Clause 14(6) addresses principally the impact of temporary working on the Permanent Works, its effect appears wider than that.

The importance of the programme etc. to the Engineer

The programme and method statement assist the Engineer in safeguarding the Employer's interest in having the work done expeditiously, safely and efficiently. They also enable him to plan the flow of design information and possession of the Site to suit the programme.

The Construction (Design and Management) Regulations 1994

Construction projects are now subject to these regulations—see Clause 71. These regulations provide for a Health and Safety Plan which will contain details in common with the Clause 14 programme. The plan is prepared in the first instance by the Planning Supervisor (who may be the Engineer—see Clause 71). The principal contractor (who may be

44 Having regard to Clause 42 which deals with possession, access, etc.

the Contractor—see Clause 71) then takes over responsibility for maintaining its accuracy "until the end of the construction phase"—Reg. 15(4).

The nature of the Clause 14 programme

The Clause 14 programme and method statement are not terms of the Contract. They do not create rights and obligations for the parties in the ordinary sense. Once accepted, however, the Contractor is entitled to assume that the Employer/Engineer will provide such reasonable cooperation as will enable him to complete in accordance with the programme. There is no breach of the Contract if the Contractor decides later to reschedule his work,[45] nor is it a breach of the Contract if he fails to meet any of the dates shown on the programme or to execute the work in a manner other than as shown in the original programme.[46] Likewise, the Employer will not be in breach if it is not possible to execute the Works in accordance with the programme or method statement.[47]

Revision of the programme

Once the programme has been accepted by the Engineer, it seems that it may only be formally revised where it appears to the Engineer "that the actual progress of the work does not conform with the accepted programme"—Clause 14(4). In this event, the "Engineer shall be entitled to require the Contractor to produce a revised programme" showing how the works may be completed within the Contract time for completion.[48] Where the Engineer or arbitrator later grants an

45 The general principle is that a Contractor may, subject to express terms, conduct his operations as he wishes: e.g. *Greater London Council v. Cleveland Bridge Engineering Co. Ltd* (1984) 34 BLR 50.

46 The Contractor does however have an obligation under 41(2) to "proceed with the Works with due expedition and without delay in accordance with the Contract". Furthermore, failure to proceed with "due diligence" is one of the grounds upon which the Contractor's employment may be terminated under Clause 63(1).

47 The Contractor may, however, be entitled under the Contract, e.g. under Clause 51 if the inability to perform the work amounts to a variation or under Clause 42(2) where the Employer fails to provide the Site in accordance with the accepted programme.

48 Where the Contractor is so late that he cannot possibly complete within the Contract time, it is thought that he must show a completion date as early as is practicable.

extension of time, the Contractor may be entitled to claim a re-rate for acceleration.[49] The Engineer may, it is submitted, unilaterally vary the programme under Clause 51 provided that this is for the benefit of the construction and/or functioning of the Works;[50] in this event he must value the variation. Clause 14(7) provides: "The Contractor shall not change the methods which have received the Engineer's consent without further consent in writing of the Engineer which shall not be unreasonably withheld". It is not clear whether the "methods" include the programme, although a method probably includes a sequence of work and hence entails the programme.

Timing of programmes and information

The timing in respect of the provision and approval of the programme is as follows.

(*a*) Within 21 days of the award of the Contract, the Contractor shall furnish a programme—Clause 14(1)(a).
(*b*) Within 21 days of the receipt of the programme, the Engineer shall accept or reject it or request further information. If none of these actions is taken, the Engineer is deemed to have accepted the programme—Clause 14(2).
(*c*) If the Engineer rejects the programme, the Contractor shall submit a revised programme within 21 days of the rejection—Clause 14(1)(c).
(*d*) If the Engineer perceives that the actual progress does not accord with the accepted programme he is entitled to require the Contractor to provide a revised programme within 21 days—Clause 14(4).

The timing in respect of the method statement is slightly different.

(*a*) The method statement is to be provided "at the same time" as the programme—Clause 14(1)(b).
(*b*) The Engineer shall inform the Contractor within 21 days whether the method statement has the consent of the Engineer or in what

49 That is, under Clause 52.

50 See Clause 51 and the commentary there. Variations of the programme for other reasons (e.g. to match the Employer's cash flow with the expenditure plan or to accommodate other contractors) are not permissible.

respects it fails to meet the Contract. In the latter case, the Contract shall make appropriate changes, but no time is set for this—Clause 14(7).

(*c*) Further details called for under Clause 14(6) are to be submitted "at such times ... as the Engineer may reasonably require".

Consent and responsibility

The Engineer is obliged to indicate his view on the programme; otherwise he is deemed to have accepted it—proviso to Clause 14(2). Likewise, he must indicate his view on the method statement, though there is no equivalent provision by which the method statement becomes automatically accepted—Clause 14(7). It is not clear why the terminology in each case differs: a programme is to be "accepted", while a method statement receives "consent". In either event, acceptance or consent does not relieve the Contractor of his responsibility under the Contract—Clause 14(9).

Design obligations

Where the Contractor has express design obligations, Clause 14(5) provides for the supply of relevant information to and by the Contractor.

15 Contractor's superintendence

(1) The Contractor shall give or provide all necessary superintendence during the construction and completion of the Works and as long thereafter as the Engineer may consider necessary. Such superintendence shall be given by sufficient persons having adequate knowledge of the operations to be carried out (including the methods and techniques required the hazards likely to be encountered and methods of preventing accidents) as may be requisite for the satisfactory and safe construction of the Works.

Contractor's agent

(2) The Contractor or a competent and authorized agent or representative approved of in writing by the Engineer (which approval may at any time be withdrawn) is to be constantly on the Works and shall give his whole time to the superintendence of the same. Such authorized agent or representative shall be in full charge of the Works and shall receive on behalf of the Contractor directions and instructions from the Engineer or

(subject to the limitations of Clause 2) the Engineer's Representative. The Contractor or such authorized agent or representative shall be responsible for the safety of all operations.

Where the Contractor fails to provide proper superintendence

It is clearly in the Employer's interest that the Contractor should provide proper supervision, and that there be a person on the site who is authorised to act on the Contractor's behalf. Where the Contractor fails to maintain an agent on Site in accordance with Clause 15, this will be a breach of contract. No damages will flow automatically from this breach. However, the Engineer will be able to suspend the Work until proper supervision is provided—Clause 40(1). Where the Contractor persistently fails to provide a proper agent, he will be liable to have the Contract determined pursuant to Clause 63(1)(b)(iv).

16 Removal of Contractor's employees

The Contractor shall employ or cause to be employed in and about the construction and completion of the Works and in the superintendence thereof only such persons as are careful skilled and experienced in their several trades and callings.

The Engineer shall be at liberty to object to and require the Contractor to remove or cause to be removed from the Works any person employed thereon who in the opinion of the Engineer misconducts himself or is incompetent or negligent in the performance of his duties or fails to conform with any particular provisions with regard to safety which may be set out in the Contract or persists in any conduct which is prejudicial to safety or health and such persons shall not be again employed upon the Works without the permission of the Engineer.

It is thought that the Engineer need not act reasonably in requiring the Contractor to remove a person.[51] However, it will be a precondition that the person must "in the opinion of the Engineer" (*a*) misconduct

51 See *Leedsford Ltd* v. *Bradford City Council* (1956) 24 BLR 45 (CA); there was no need for the architect to act reasonably in approving suppliers.

himself or (*b*) be incompetent or (*c*) fail to conform with safety procedures or (*d*) persist in conduct prejudicial to safety. Hence, where the Engineer cites any other reason or indicates that he is acting at the behest of the Employer without having formed the requisite opinion, any removal will be a breach of contract.

17 Setting-out

(1) The Contractor shall be responsible for the true and proper setting-out of the Works and for the correctness of the position levels dimensions and alignment of all parts of the Works and for the provision of all necessary instruments appliances and labour in connection therewith.

(2) If at any time during the progress of the Works any error shall appear or arise in the position levels dimensions or alignment of any part of the Works the Contractor on being required so to do by the Engineer shall at his own cost rectify such error to the satisfaction of the Engineer unless such error is based on incorrect data supplied in writing by the Engineer or the Engineer's Representative in which case the cost of rectifying the same shall be borne by the Employer.

(3) The checking of any setting-out or of any line or level by the Engineer or the Engineer's Representative shall not in any way relieve the Contractor of his responsibility for the correctness thereof and the Contractor shall carefully protect and preserve all bench-marks sight rails pegs and other things used in setting out the Works.

"Setting out" is the transfer of design dimensions etc. from plans and drawings to the actual work in progress by way of marks, lines, levels, etc. Incorrect setting out may lead to the Works being incorrectly located. This clause indicates clearly that the Contractor is to be responsible for his own setting out except where the error is introduced by "incorrect data supplied in writing by the Engineer or the Engineer's Representative".[52]

52 Note that, in this instance, the Engineer's Representative needs no specific authorisation under Clause 2(4) to fix the Employer with liability.

18 Boreholes and exploratory excavation

If at any time during the construction of the Works the Engineer shall require the Contractor to make boreholes or to carry out exploratory excavation such requirement shall be ordered in writing and shall be deemed to be a variation under Clause 51 unless a Provisional Sum or Prime Cost Item in respect of such anticipated work shall have been included in the Bill of Quantities.

Where difficult ground is encountered or suspected, it is frequently useful for the Engineer to order boreholes to be drilled or other exploratory excavations to be made. It is thought that the Engineer is not entitled to require the Contractor to provide an interpretive report on the boreholes.

19 Safety and security

(1) The Contractor shall throughout the progress of the Works have full regard for the safety of all persons entitled to be upon the Site and shall keep the Site (so far as the same is under his control) and the Works (so far as the same are not completed or occupied by the Employer) in an orderly state appropriate to the avoidance of danger to such persons and shall inter alia in connection with the Works provide and maintain at his own cost all lights guards fencing warning signs and watching when and where necessary or required by the Engineer or the Engineer's Representative or by any competent statutory or other authority for the protection of the Works or for the safety and convenience of the public or others.

Employer's responsibilities

(2) If under Clause 31 the Employer shall carry out work on the Site with his own workmen he shall in respect of such work

(a) have full regard to the safety of all persons entitled to be upon the Site and

(b) keep the Site in an orderly state appropriate to the avoidance of danger to such persons.

If under Clause 31 the Employer shall employ other contractors on the Site he shall require them to have the same regard for safety and avoidance of danger.

This clause sets out the general principle that the Contractor is responsible for safety on the Site, but that persons employed by the Employer shall keep their work safe and take proper care for the safety of others. This provision must be read together with the requirements of the health and safety legislation[53] and the legislation dealing with occupation of land.[54] The fact that the provisions of Clause 19 extend only to "all persons entitled to be on the Site" will not prevent the Contractor (or indeed the Employer) from being liable in appropriate cases to trespassers under the Occupier's Liability Act 1984.[55]

20 Care of the Works

(1) (a) The Contractor shall save as in paragraph (b) hereof and subject to sub-clause (2) of this Clause take full responsibility for the care of the Works and materials plant and equipment for incorporation therein from the Works Commencement Date until the date of issue of a Certificate of Substantial Completion for the whole of the Works when the responsibility for the said care shall pass to the Employer.

(b) If the Engineer issues a Certificate of Substantial Completion for any Section or part of the Permanent Works the Contractor shall cease to be responsible for the care of that Section or part from the date of issue of such Certificate of Substantial Completion when the responsibility for the care of that Section or part shall pass to the Employer.

(c) The Contractor shall take full responsibility for the care of any outstanding work and materials plant and equipment for incorporation therein which he undertakes to finish during the Defects Correction Period until such outstanding work has been completed.

Excepted Risks

(2) The Excepted Risks for which the Contractor is not liable are loss or damage to the extent that it is due to

53 Especially the Health and Safety at Work etc. Act 1974 and the Construction (Design and Management) Regulations 1994; see also Clause 71.

54 Occupier's Liability Acts 1957 and 1984.

55 See also Reg. 16(1)(c) of the Construction (Design and Management) Regulations 1994. This provides that "The principal contractor [the Contractor may be the principal contractor—see Clause 71] appointed for any project shall—... (c) take reasonable steps to ensure that only authorised persons are allowed into any premises or part of premises where construction work is being carried out". It is possible that this creates a civil liability—see Reg. 21.

(a) the use or occupation by the Employer his agents servants or other contractors (not being employed by the Contractor) of any part of the Permanent Works

(b) any fault defect error or omission in the design of the Works (other than a design provided by the Contractor pursuant to his obligations under the Contract)

(c) riot war invasion act of foreign enemies or hostilities (whether war be declared or not)

(d) civil war rebellion revolution insurrection or military or usurped power

(e) ionizing radiations or contamination by radioactivity from any nuclear fuel or from any nuclear waste from the combustion of nuclear fuel radioactive toxic explosive or other hazardous properties of any explosive nuclear assembly or nuclear component thereof and

(f) pressure waves caused by aircraft or other aerial devices travelling at sonic or supersonic speeds.

Rectification of loss or damage

(3) (a) In the event of any loss or damage to

(i) the Works or any Section or part thereof or
(ii) materials plant or equipment for incorporation therein

while the Contractor is responsible for the care thereof (except as provided in sub-clause (2) of this Clause) the Contractor shall at his own cost rectify such loss or damage so that the Permanent Works conform in every respect with the provisions of the Contract and the Engineer's instructions. The Contractor shall also be liable for any loss or damage to the Works occasioned by him in the course of any operations carried out by him for the purpose of complying with his obligations under Clauses 49 and 50.

(b) Should any such loss or damage arise from any of the Excepted Risks defined in sub-clause (2) of this Clause the Contractor shall if and to the extent required by the Engineer rectify the loss or damage at the expense of the Employer.

(c) In the event of loss or damage arising from an Excepted Risk and a risk for which the Contractor is responsible under sub-clause (1)(a) of this Clause then the Engineer shall when determining the expense to be borne by the Employer under the Contract apportion the cost of rectification into that part caused by the Excepted Risk and that part which is the responsibility of the Contractor.

21 Insurance of Works etc.

(1) The Contractor shall without limiting his or the Employer's obligations and responsibilities under Clause 20 insure in the joint names of the Contractor and the Employer the Works together with materials plant and equipment for incorporation therein to the full replacement cost plus an additional 10% to cover any additional costs that may arise incidental to the rectification of any loss or damage including professional fees cost of demolition and removal of debris.

Extent of cover

(2) (a) The insurance required under sub-clause (1) of this Clause shall cover the Employer and the Contractor against all loss or damage from whatsoever cause arising other than the Excepted Risks defined in Clause 20(2) from the Works Commencement Date until the date of issue of the relevant Certificate of Substantial Completion.

(b) The insurance shall extend to cover any loss or damage arising during the Defects Correction Period from a cause occurring prior to the issue of any Certificate of Substantial Completion and any loss or damage occasioned by the Contractor in the course of any operation carried out by him for the purpose of complying with his obligations under Clauses 49 and 50.

(c) Nothing in this Clause shall render the Contractor liable to insure against the necessity for the repair or reconstruction of any work constructed with materials or workmanship not in accordance with the requirements of the Contract unless the Bill of Quantities shall provide a special item for this insurance.

(d) Any amounts not insured or not recovered from insurers whether as excesses carried under the policy or otherwise shall be borne by the Contractor or the Employer in accordance with their respective responsibilities under Clause 20.

22 Damage to persons and property

(1) The Contractor shall except if and so far as the Contract provides otherwise and subject to the exceptions set out in sub-

clause (2) of this Clause indemnify and keep indemnified the Employer against all losses and claims in respect of

(a) death of or injury to any person or

(b) loss of or damage to any property (other than the Works)

which may arise out of or in consequence of the execution of the Works and the remedying of any defects therein and against all claims demands proceedings damages costs charges and expenses whatsoever in respect thereof or in relation thereto.

Exceptions

(2) The exceptions referred to in sub-clause (1) of this Clause which are the responsibility of the Employer are

(a) damage to crops being on the Site (save in so far as possession has not been given to the Contractor)

(b) the use or occupation of land (provided by the Employer) by the Works or any part thereof or for the purpose of executing and maintaining the Works (including consequent losses of crops) or interference whether temporary or permanent with any right of way light air or water or other easement or quasi-easement which are the unavoidable result of the construction of the Works in accordance with the Contract

(c) the right of the Employer to construct the Works or any part thereof on over under in or through any land

(d) damage which is the unavoidable result of the construction of the Works in accordance with the Contract and

(e) death of or injury to persons or loss of or damage to property resulting from any act neglect or breach of statutory duty done or committed by the Employer or his agents servants or other contractors (not being employed by the Contractor) or for or in respect of any claims demands proceedings damages costs charges and expenses in respect thereof or in relation thereto.

Indemnity by Employer

(3) The Employer shall subject to sub-clause (4) of this Clause indemnify the Contractor against all claims demands proceedings damages costs charges and expenses in respect of the matters referred to in the exceptions defined in sub-clause (2) of this Clause.

Shared responsibility

(4) (a) The Contractor's liability to indemnify the Employer under sub-clause (1) of this Clause shall be reduced in proportion to the extent that the act or neglect of the Employer his agents servants or other contractors (not being employed by the Contractor) may have contributed to the said death injury loss or damage.

(b) The Employer's liability to indemnify the Contractor under sub-clause (3) of this Clause in respect of matters referred to in sub-clause (2)(e) of this Clause shall be reduced in proportion to the extent that the act or neglect of the Contractor or his sub-contractors servants or agents may have contributed to the said death injury loss or damage.

23 Third party insurance

(1) The Contractor shall without limiting his or the Employer's obligations and responsibilities under Clause 22 insure in the joint names of the Contractor and the Employer against liabilities for death of or injury to any person (other than any operative or other person in the employment of the Contractor or any of his sub-contractors) or loss of or damage to any property (other than the Works) arising out of the execution of the Contract other than the exceptions defined in Clause 22(2)(a) (b) (c) and (d).

Cross liability clause

(2) The insurance policy shall include a cross liability clause such that the insurance shall apply to the Contractor and to the Employer as separate insured.

Amount of insurance

(3) Such insurance shall be for at least the amount stated in the Appendix to the Form of Tender.

24 Accident or injury to workpeople

The Employer shall not be liable for or in respect of any damages or compensation payable at law in respect or in consequence of any accident or injury to any operative or other person in the employment of the Contractor or any of his sub-contractors save and except to the extent that such accident or injury results from or is contributed to by any act or default of the Employer his agents or servants and the Contractor shall indemnify and keep indemnified the Employer against all such damages and compensation (save and except as aforesaid) and against all claims demands proceedings costs charges and expenses whatsoever in respect thereof or in relation thereto.

25 Evidence and terms of insurance

(1) The Contractor shall provide satisfactory evidence to the Employer prior to the Works Commencement Date that the insurances required under the Contract have been effected and shall if so required produce the insurance policies for inspection. The terms of all such insurances shall be subject to the approval of the Employer (which approval shall not unreasonably be withheld). The Contractor shall upon request produce to the Employer receipts for the payment of current insurance premiums.

Excesses

(2) Any excesses on the policies of insurance effected under Clauses 21 and 23 shall be as stated by the Contractor in the Appendix to the Form of Tender.

Remedy on Contractor's failure to insure

(3) If the Contractor shall fail upon request to produce to the Employer satisfactory evidence that there is in force any of the insurances required under the Contract then and in any such case the Employer may effect and keep in force any such insurance and pay such premium or premiums as may be necessary for that purpose and from time to time deduct the amount so paid from any monies due or which may become due to the Contractor or recover the same as a debt due from the Contractor.

Compliance with policy conditions

(4) Both the Employer and the Contractor shall comply with all conditions laid down in the insurance policies. In the event that the Contractor or the Employer fails to comply with any condition imposed by the insurance policies effected pursuant to the Contract each shall indemnify the other against all losses and claims arising from such failure.

The care of the Works and insurance

Clauses 20 to 25 cover all aspects of the care and insurance of:

(*a*) the Works
(*b*) plant, materials etc.

(*c*) third parties and their property
(*d*) workpeople.

The clauses allocate risk between the Contractor and Employer.

Synopsis of risk allocation

Clauses 20, 22 and 24 provide the general principles of risk, namely that the Contractor is responsible for damage to the Works (Clause 20), persons and property (Clause 22) and workpeople (Clause 24). There are exceptions given in Clauses 20(2) and 22(2).

Synopsis of insurance provisions

Clauses 21, 23 and 25 deal with insurance. The basic principle is that the Contractor shall insure against damage to the Works (Clause 21), persons and property (Clause 23). The insurance shall be in the joint names of the Contractor and the Employer. The terms shall be subject to the approval of the Employer (Clause 25(1)). Evidence of insurance shall be provided to the Employer and if this is not done, the Employer may effect the insurance and countercharge the Contractor.

26 Giving of notices and payment of fees

(1) The Contractor shall give all notices and pay all fees required to be given or paid by any Act of Parliament or any Regulation or Bye-law of any local or other statutory authority in relation to the construction and completion of the Works and by the rules and regulations of all public bodies and companies whose property or rights are or may be affected in any way by the Works.

Repayment by Employer

(2) The Employer shall repay or allow to the Contractor all such sums as the Engineer shall certify to have been properly payable and paid by the Contractor in respect of such fees and also all rates and taxes paid by the Contractor in respect of the Site or any part thereof or anything constructed or erected thereon or on any part thereof or any temporary structures situated elsewhere but used exclusively for the purposes of the Works or any structures used temporarily and exclusively for the purposes of the Works.

Contractor to conform with Statutes etc.

(3) The Contractor shall ascertain and conform in all respects with the provisions of any general or local Act of Parliament and the Regulations and Bye-laws of any local or other statutory authority which may be applicable to the Works and with such rules and regulations of public bodies and companies as aforesaid and shall keep the Employer indemnified against all penalties and liability of every kind for breach of any such Act Regulation or Bye-law. Provided always that

(a) the Contractor shall not be required to indemnify the Employer against the consequences of any such breach which is the unavoidable result of complying with the Contract or instructions of the Engineer

(b) if the Contract or instructions of the Engineer shall at any time be found not to be in conformity with any such Act Regulation or Bye-law the Engineer shall issue such instructions including the ordering of a variation under Clause 51 as may be necessary to ensure conformity with such Act Regulation or Bye-law and

(c) the Contractor shall not be responsible for obtaining any planning permission which may be necessary in respect of the Permanent Works or any Temporary Works design supplied by the Engineer and the Employer hereby warrants that all the said permissions have been or will in due time be obtained.

Statutory obligations

Many statutory and similar provisions apply to construction work. Clause 26(3) requires that the Contractor shall "ascertain and conform ... with" all statutes, regulations, etc. The Contractor must indemnify the Employer against any consequences of any breach of statutory requirements, save for the exceptions given in Clause 26(3). Clauses 26(1) and 26(2) deal with two common matters which arise in relation to such controls, namely notices[56] and payment of fees, taxes, etc.

56 For example notices under the Building Regulations, the New Roads and Street Works Act 1991, etc.

27 New Roads and Street Works Act 1991—Definitions

(1) (a) In this Clause "the Act" shall mean the New Roads and Street Works Act 1991 and any statutory modification or re-enactment thereof for the time being in force.

(b) For the purpose of obtaining any licence under the Act required for the Permanent Works the undertaker shall be the Employer who for the purposes of the Act will be the licensee.

(c) For all other purposes the undertaker under the licence shall be the Contractor.

(d) All other expressions common to the Act and to this Clause shall have the same meaning as those assigned to them by the Act.

Licences

(2) (a) The Employer shall obtain any street works licence and any other consent licence or permission that may be required for the carrying out of the Permanent Works and shall supply the Contractor with copies thereof including details of any conditions or limitations imposed.

(b) Any condition or limitation in any licence obtained after the award of the Contract shall be deemed to be an instruction under Clause 13.

Notices

(3) The Contractor shall be responsible for giving to any relevant authority any required notice (or advance notice where prescribed) of his proposal to commence any work. A copy of each such notice shall be given to the Employer.

Clause 27 was completely revised following the coming into force of the New Roads and Street Works Act 1991. The Act provides that an "undertaker" shall obtain a "streetworks licence" from the "street authority" before carrying out any "streetworks". The Act requires the undertaker to obtain a licence before placing, retaining and inspecting "apparatus" in the street. Apparatus includes a sewer, drain or tunnel. The Clause provides that both the Employer and the Contractor are to act as undertaker in different cases. The Employer is to obtain the licence and the Contractor shall be responsible for giving notices.

28 Patent rights

(1) The Contractor shall save harmless and indemnify the Employer from and against all claims and proceedings for or on

account of infringement of any patent right design trademark or name or other protected right in respect of any

(a) Contractor's Equipment used for or in connection with the Works

(b) materials plant and equipment for incorporation in the Works

and from and against all claims demands proceedings damages costs charges and expenses whatsoever in respect thereof or in relation thereto except where such infringement results from compliance with the design or Specification provided other than by the Contractor. In the latter event the Employer shall indemnify the Contractor from and against all claims and proceedings for or on account of infringement of any patent right design trademark or name or other protected right aforesaid.

Royalties

(2) Except where otherwise stated the Contractor shall pay all tonnage and other royalties rent and other payments or compensation (if any) for getting stone sand clay or other materials required for the Works.

Intellectual property

This clause deals with the protection of property in ideas and designs.

29 Interference with traffic and adjoining properties

(1) All operations necessary for the construction and completion of the Works shall so far as compliance with the requirements of the Contract permits be carried on so as not to interfere unnecessarily or improperly with

(a) the convenience of the public or

(b) the access to public or private roads footpaths or properties whether in the possession of the Employer or of any other person and with the use or occupation thereof.

The Contractor shall save harmless and indemnify the Employer in respect of all claims demands proceedings damages costs charges and expenses whatsoever arising out of or in relation to any such matters.

Noise disturbance and pollution

(2) All work shall be carried out without unreasonable noise or disturbance or other pollution.

Indemnity by Contractor

(3) To the extent that noise disturbance or other pollution is not the unavoidable consequence of constructing and completing the Works or performing the Contract the Contractor shall indemnify the Employer from and against any liability for damages on that account and against all claims demands proceedings damages costs charges and expenses whatsoever in regard or in relation to such liability.

Indemnity by Employer

(4) The Employer shall indemnify the Contractor from and against any liability for damages on account of noise disturbance or other pollution which is the unavoidable consequence of carrying out the Works and from and against all claims demands proceedings damages costs charges and expenses whatsoever in regard or in relation to such liability.

30 Avoidance of damage to highways etc.

(1) The Contractor shall use every reasonable means to prevent any of the highways or bridges communicating with or on the routes to the Site from being subjected to extraordinary traffic within the meaning of the Highways Act 1980 or in Scotland the Roads (Scotland) Act 1984 or any statutory modification or re-enactment thereof by any traffic of the Contractor or any of his sub-contractors and in particular shall select routes and use vehicles and restrict and distribute loads so that any such extraordinary traffic as will inevitably arise from the moving of Contractor's Equipment and materials or manufactured or fabricated articles from and to the Site shall be limited as far as reasonably possible and so that no unnecessary damage or injury may be occasioned to such highways and bridges.

Transport of Contractor's Equipment

(2) Save insofar as the Contract otherwise provides the Contractor shall be responsible for and shall pay the cost of strengthening any bridges or altering or improving any highway communicating with the Site to facilitate the movement of Contractor's Equipment or Temporary Works required in the execution of the Works and the Contractor shall indemnify and keep indemnified the Employer against all claims for damage to any highway or bridge communicating with the Site caused by such movement including such claims as may be made by any competent authority directly against the Employer pursuant to any Act of Parliament or other Statutory Instrument and shall negotiate and pay all claims arising solely out of such damage.

Transport of materials

(3) If notwithstanding sub-clause (1) of this Clause any damage shall occur to any bridge or highway communicating with the Site arising from the transport of materials or manufactured or fabricated articles in the execution of the Works the Contractor shall notify the Engineer as soon as he becomes aware of such damage or as soon as he receives any claim from the authority entitled to make such claim.

Where under any Act of Parliament or other Statutory Instrument the haulier of such materials or manufactured or fabricated articles is required to indemnify the highway authority against damage the Employer shall not be liable for any costs charges or expenses in respect thereof or in relation thereto.

In other cases the Employer shall negotiate the settlement of and pay all sums due in respect of such claim and shall indemnify the Contractor in respect thereof and in respect of all claims demands proceedings damages costs charges and expenses in relation thereto. Provided always that if and so far as any such claim or part thereof shall in the opinion of the Engineer be due to any failure on the part of the Contractor to observe and perform his obligations under sub-clause (1) of this Clause then the amount certified by the Engineer to be due to such failure shall be paid by the Contractor to the Employer or deducted from any sum due or which may become due to the Contractor.

These clauses deal with the impact of the Contractor's operations on traffic and adjoining properties and highway structures.

Nuisance etc. claims

Claims in respect of traffic and adjoining properties are dealt with in Clause 29. A system of cross-indemnities is established. The Contractor indemnifies the Employer against any claim, save where the claim is the "unavoidable consequence of carrying out the Works". In the latter case, it is the Employer who indemnifies the Contractor.

Highways

A system of cross-indemnities is also established by Clause 30. Clause 30(1) requires the Contractor to "use every reasonable means" to prevent overloading and "unnecessary damage". Where damage occurs despite the Contractor's complying with these provisions, the Employer negotiates and settles and pays any claims—Clause 30(3). Where, however, the damage occurs as a result of a breach of Clause 30(1), the Contractor indemnifies the Employer against claims—Clause 30(2).

31 Facilities for other contractors

(1) The Contractor shall in accordance with the requirements of the Engineer or Engineer's Representative afford all reasonable facilities for any other contractors employed by the Employer and their workmen and for the workmen of the Employer and of any other properly authorised authorities or statutory bodies who may be employed in the execution on or near the Site of any work not in the Contract or of any contract which the Employer may enter into in connection with or ancillary to the Works.

Delay and extra cost

(2) If compliance with sub-clause (1) of this Clause shall involve the Contractor in delay or cost beyond that reasonably to be foreseen by an experienced contractor at the time of tender then the Engineer shall take such delay into account in determining any extension of time to which the Contractor is entitled under Clause 44 and the Contractor shall subject to Clause 52(4) be paid in accordance with Clause 60 the amount of such cost as may be reasonable. Profit shall be added thereto in respect of any additional permanent or temporary work.

The extent of the Contractor's licence

This clause saves for the Employer an important right, namely to allow other contractors onto the Site. Accordingly, the Contractor is not to be regarded as having an exclusive licence. It should be noted, however, that Clause 42(2)(a), (b) requires the Employer to give the Contractor possession as will enable him to commence and carry out the work in accordance with the Clause 14 accepted programme and such "access as is necessary to enable the Contractor to proceed with the construction of the Works with due despatch".

Clause 31(2) notice

Where the Contractor intends to claim additional cost, Clause 52(4) requires him to give notice in writing. Where he fails to do so, he will be entitled only to the extent that the Engineer has been substantially prejudiced in investigating the claim.

32 Fossils etc.

All fossils coins articles of value or antiquity and structures or other remains or things of geological or archaeological interest discovered on the Site shall as between the Employer and the Contractor be deemed to be the absolute property of the Employer and the Contractor shall take reasonable precautions to prevent his workmen or any other persons from removing or damaging any such article or thing and shall immediately upon discovery thereof and before removal acquaint the Engineer of such discovery and carry out at the expense of the Employer the Engineer's orders as to the disposal of the same.

A find of fossils etc. which requires a delay or extra costs may found a claim under Clause 12.

33 Clearance of Site on completion

On the completion of the Works the Contractor shall clear away and remove from the Site all Contractor's Equipment surplus material rubbish and Temporary Works of every kind and leave the whole of the Site and Permanent Works clean and in a workmanlike condition to the satisfaction of the Engineer.

This clause states what the Contract would otherwise imply.

34 (Not used)

This clause formerly dealt with the fair wages resolution, which is no longer in effect.

35 Returns of labour and Contractor's Equipment

The Contractor shall if required by the Engineer deliver to the Engineer or the Engineer's Representative a return in such form and at such intervals as the Engineer may prescribe showing in detail the numbers of the several classes of labour from time to time employed by the Contractor on the Site and such information respecting Contractor's Equipment as the Engineer may require. The Contractor shall require his sub-contractors to observe the provisions of this Clause.

The Engineer's right to call upon the Contractor to give returns of labour and equipment is a valuable one. It enables the Employer to keep such records as may be required e.g. for assessing and valuing claims.

WORKMANSHIP AND MATERIALS

36 Quality of materials and workmanship and tests

(1) All materials and workmanship shall be of the respective kinds described in the Contract and in accordance with the Engineer's instructions and shall be subjected from time to time to such tests as the Engineer may direct at the place of manufacture or fabrication or on the Site or such other place or places as may be specified in the Contract. The Contractor shall provide such assistance instruments machines labour and materials as are normally required for examining measuring and testing any work and are the quality weight or quantity of any materials used and shall supply samples of materials before incorporation in the Works for testing as may be selected and required by the Engineer.

Cost of samples

(2) All samples shall be supplied by the Contractor at his own cost if the supply thereof is clearly intended by or provided for in the Contract but if not then at the cost of the Employer.

Cost of tests

(3) The cost of making any test shall be borne by the Contractor if such test is clearly intended by or provided for in the Contract and (in the cases only of a test under load or of a test to ascertain whether the design of any finished or partially finished work is appropriate for the purposes which it was intended to fulfil) is particularized in the Specification or Bill of Quantities in sufficient detail to enable the Contractor to have priced or allowed for the same in his Tender. If any test is ordered by the Engineer which is either

(a) not so intended by or provided for or

(b) (in the cases above mentioned) is not so particularized

then the cost of such test shall be borne by the Contractor if the test shows the workmanship or materials not to be in accordance with the provisions of the Contract or the Engineer's instructions but otherwise by the Employer.

37 Access to Site

The Engineer and any person authorized by him shall at all times have access to the Works and to the Site and to all workshops and places where work is being prepared or whence materials manufactured articles and machinery are being obtained for the Works and the Contractor shall afford every facility for and every assistance in obtaining such access or the right to such access.

38 Examination of work before covering up

(1) No work shall be covered up or put out of view without the consent of the Engineer and the Contractor shall afford full opportunity for the Engineer to examine and measure any work which is about to be covered up or put out of view and to examine foundations before permanent work is placed thereon. The Contractor shall give due notice to the Engineer whenever any such work or foundations is or are ready or about to be ready for examination and the Engineer shall without unreasonable delay unless he considers it unnecessary and advises the Contractor accordingly attend for the purpose of examining and measuring such work or of examining such foundations.

Uncovering and making openings

(2) The Contractor shall uncover any part or parts of the Works or make openings in or through the same as the Engineer may from time to time direct and shall reinstate and make good such part or parts to the satisfaction of the Engineer. If any such part or parts have been covered up or put out of view after compliance with the requirements of sub-clause (1) of this Clause and are found to be executed in accordance with the Contract the cost of uncovering making openings in or through reinstating and making good the same shall be borne by the Employer but in any other case all such cost shall be borne by the Contractor.

39 Removal of unsatisfactory work and materials

(1) The Engineer shall during the progress of the Works have power to instruct in writing the

(a) removal from the Site within such time or times specified in the instruction of any materials which in the opinion of the Engineer are not in accordance with the Contract

(b) substitution with materials in accordance with the Contract and

(c) removal and proper re-execution notwithstanding any previous test thereof or interim payment therefor of any work which in respect of

(i) material or workmanship or

(ii) design by the Contractor or for which he is responsible

is not in the opinion of the Engineer in accordance with the Contract.

Default of Contractor in compliance

(2) In case of default on the part of the Contractor in carrying out such instruction the Employer shall be entitled to employ and pay other persons to carry out the same and all costs consequent thereon or incidental thereto as determined by the Engineer shall be recoverable from the Contractor by the Employer and may be deducted by the Employer from any monies due or to become due to him and the Engineer shall notify the Contractor accordingly with a copy to the Employer.

Failure to disapprove

(3) Failure of the Engineer or any person acting under him pursuant to Clause 2 to disapprove any work or materials shall not prejudice the power of the Engineer or any such person subsequently to take action under this Clause.

Synopsis

Clauses 36 to 39 deal with workmanship and materials generally. Clause 36(1) states the general principle that materials and workmanship shall be as described in the Contract. Clause 36 also deals with testing and sampling and states who is to pay for tests and samples. Clause 37 requires the Contractor to allow the Engineer access for the purpose of inspecting work. Clause 38 requires that the Contractor shall allow the Engineer to examine work before it is covered up, and Clause 38(2) provides for uncovering work which has been covered. Clause 39 requires that the Contractor shall remove unsatisfactory work and materials.

Clause 36—quality

The Contractor's obligation is to provide work and materials in accordance with the Contract. Where the Contract is not specific, the obligation implied is that materials will be sound and fit for their intended purpose and the work will be done in a good and workmanlike manner.[57] The obligation to supply and perform the work "in accordance with the Engineer's instructions" does not, it is thought, alter the Contractor's obligations in any way. It merely emphasises that the Contractor is obliged to give effect to the instructions of the Engineer. Thus if the Engineer requires materials to be removed, the Contractor must remove them notwithstanding that they comply with the Contract. The Contractor will be entitled to recover his loss, for example, as a variation.[58]

Testing

The Works may be subject to sampling or tests as required by the Engineer. The Contractor shall provide any equipment required to carry out the tests. Where the Contract "clearly" intends or provides for such tests or samples, the Contractor shall bear the cost. Otherwise, the cost is to be borne by the Employer. The one exception is where the test is directed at determining whether the materials or workmanship correspond to the Contract—where the tests show that the materials or workmanship do not correspond, the cost is to be borne by the Contractor.

Examination

The Contractor is required to give the Employer a "full opportunity" to "examine and measure" any work which is about to be covered up.

40 Suspension of work

(1) The Contractor shall on the written order of the Engineer suspend the progress of the Works or any part thereof for such time or times and in such manner as the Engineer may consider necessary and shall during such suspension properly protect and secure the work so far as is necessary in the opinion of the Engineer. Subject to Clause 52(4) the Contractor shall be paid in accordance with Clause 60 the extra cost (if any) incurred in

57 *Young & Marten* v. *McManus Childs* [1969] 1 AC 454 (HL).

58 See Clauses 13(3) and 51.

giving effect to the Engineer's instructions under this Clause except to the extent that such suspension is

(a) otherwise provided for in the Contract or

(b) necessary by reason of weather conditions or by some default on the part of the Contractor or

(c) necessary for the proper execution or for the safety of the Works or any part thereof in as much as such necessity does not arise from any act or default of the Engineer or the Employer or from any of the Excepted Risks defined in Clause 20(2).

Profit shall be added thereto in respect of any additional permanent or temporary work.

The Engineer shall take any delay occasioned by a suspension ordered under this Clause (including that arising from any act or default of the Engineer or the Employer) into account in determining any extension of time to which the Contractor is entitled under Clause 44 except when such suspension is otherwise provided for in the Contract or is necessary by reason of some default on the part of the Contractor.

Suspension lasting more than three months

(2) If the progress of the Works or any part thereof is suspended on the written order of the Engineer and if permission to resume work is not given by the Engineer within a period of 3 months from the date of suspension then the Contractor may unless such suspension is otherwise provided for in the Contract or continues to be necessary by reason of some default on the part of the Contractor serve a written notice on the Engineer requiring permission within 28 days from the receipt of such notice to proceed with the Works or that part thereof in regard to which progress is suspended. If within the said 28 days the Engineer does not grant such permission the Contractor by a further written notice so served may (but is not bound to) elect to treat the suspension where it affects part only of the Works as an omission of such part under Clause 51 or where it affects the whole Works as an abandonment of the Contract by the Employer.

Suspension

Without this provision, suspension may amount to a breach, and may indeed amount to a fundamental breach entitling the Contractor to terminate the contract and to sue for damages. Here, the Contractor is entitled under Clause 40(1) to costs and profit plus an extension of time except to the extent that the suspension is due to one or more of the factors in Clause 40(1)(a), (b), (c). Where the work has been suspended for three months, the Contractor may (unless the suspension is due to the Contractor's default or is otherwise provided for) serve a 28 day notice upon the Engineer—Clause 40(2). Where the Engineer does not grant permission for the Contractor to proceed with the Works, the Contractor may by serving a further notice elect to treat the relevant Works as abandoned or omitted. It is thought that the Employer will not be in breach where this happens; however, where part of the Works is omitted this will amount to a variation which is to be valued in accordance with Clause 52. Where the entire works are to be treated as an "abandonment", it is thought that there will be an entitlement under the Contract to such sums as would be payable by way of damages for repudiation even though technically this will not amount to a breach.

COMMENCEMENT TIME AND DELAYS

41 Works Commencement Date

(1) The Works Commencement Date shall be

(a) the date specified in the Appendix to the Form of Tender or if no date is specified

(b) a date within 28 days of the award of the Contract to be notified by the Engineer in writing or

(c) such other date as may be agreed between the parties.

Start of Works

(2) The Contractor shall start the Works on or as soon as is reasonably practicable after the Works Commencement Date. Thereafter the Contractor shall proceed with the Works with due expedition and without delay in accordance with the Contract.

The obligation to start on time and to "proceed ... with due expedition and without delay"

It is in the Employer's interests that the Contractor should control the site until the project is complete and should complete expeditiously. Clause 41 requires the Contractor to start as soon as is reasonably practicable after the Works Commencement Date and thereafter to proceed with the Works with due expedition. While the heading for Clause 42(1) suggests that this obligation is related specifically to starting the Works, the wording of the clause is not so limited. Where the Contractor fails to start or to maintain progress, the Employer may determine the Contract.[59] Clause 41 should generally be read in conjunction with Clause 46(1) which deals with the related issue of rate of progress that is too slow to ensure substantial completion within the time for completion.

42 Possession of Site and access

(1) The Contract may prescribe

(a) the extent of portions of the Site of which the Contractor is to be given possession from time to time

(b) the order in which such portions of the Site shall be made available to the Contractor

(c) the availability and the nature of the access which is to be provided by the Employer

(d) the order in which the Works shall be constructed.

(2) (a) Subject to sub-clause (1) of this Clause the Employer shall give to the Contractor on the Works Commencement Date possession of so much of the Site and access thereto as may be required to enable the Contractor to commence and proceed with the construction of the Works.

(b) Thereafter the Employer shall during the course of the Works give to the Contractor possession of such further portions of the Site as may be required in accordance with the programme which the Engineer has accepted under Clause 14 and such further access as is necessary to enable the Contractor to proceed with the construction of the Works with due despatch.

59 See Clause 63(1)(b). Note that in some instances the Engineer is required to serve written warnings before this can happen.

Failure to give possession

(3) If the Contractor suffers delay and/or incurs additional cost from failure on the part of the Employer to give possession in accordance with the terms of this Clause the Engineer shall determine

(a) any extension of time to which the Contractor is entitled under Clause 44 and

(b) subject to Clause 52(4) the amount of any additional cost to which the Contractor may be entitled. Profit shall be added thereto in respect of any additional permanent or temporary work.

The Engineer shall notify the Contractor accordingly with a copy to the Employer.

Access and facilities provided by the Contractor

(4) The Contractor shall bear all costs and charges for any access required by him additional to those provided by the Employer. The Contractor shall also provide at his own cost and additional facilities outside the Site required by him for the purposes of the Works.

Possession of the Site

The Contractor requires access to and possession of such parts of the Site as are necessary for him to undertake and complete the Works. The Contract may specify the extent of possession to be given. Where the Contract is silent, the obligation during the commencement period is for such possession to be given as is necessary for the Contractor to commence and proceed. Thereafter the accepted Clause 14 programme sets out the extent of possession required.[60] Clause 42(2)(b) also requires the Employer to give such "access" as is necessary to enable the Contractor to proceed with due despatch. Where such possession is not given, the Contractor is entitled to cost, profit and an extension of time—Clause 42(3).

60 It should be remembered that the Clause 14 programme may not yet be accepted at the Works Commencement Date.

43 Time for completion

The whole of the Works and any Section required to be completed within a particular time as stated in the Appendix to the Form of Tender shall be substantially completed within the time so stated (or such extended time as may be allowed under Clause 44 or revised time agreed under Clause 46(3)) calculated from the Works Commencement Date.

The Contractor is obliged to complete by the time given in the Appendix, as extended or agreed under Clause 46(3). The Employer is entitled[61] to deduct liquidated damages at the rate specified in the Appendix where the Contractor completes late. Where no time for completion is included in the Appendix, a reasonable time for completion will be implied and unliquidated damages may be claimed.

44 Extension of time for completion

(1) Should the Contractor consider that

(a) any variation ordered under Clause 51(1) or

(b) increased quantities referred to in Clause 51(4) or

(c) any cause of delay referred to in these Conditions or

(d) exceptional adverse weather conditions or

(e) other special circumstances of any kind whatsoever which may occur

be such as to entitle him to an extension of time for the substantial completion of the Works or any Section thereof he shall within 28 days after the cause of any delay has arisen or as soon thereafter as is reasonable deliver to the Engineer full and detailed particulars in justification of the period of extension claimed in order that the claim may be investigated at the time.

61 It seems that the Engineer's role is limited to granting extensions of time and issuing the Certificate of Substantial Completion. He does not deduct liquidated damages in the Clause 60 certificate. It is for the Employer to decide whether or not he will deduct the damages.

Assessment of delay

(2) (a) The Engineer shall upon receipt of such particulars consider all the circumstances known to him at that time and make an assessment of the delay (if any) that has been suffered by the Contractor as a result of the alleged cause and shall so notify the Contractor in writing.

(b) The Engineer may in the absence of any claim make an assessment of the delay that he considers has been suffered by the Contractor as a result of any of the circumstances listed in sub-clause (1) of this Clause and shall so notify the Contractor in writing.

Interim grant of extension of time

(3) Should the Engineer consider that the delay suffered fairly entitles the Contractor to an extension of the time for the substantial completion of the Works or any Section thereof such interim extension shall be granted forthwith and be notified to the Contractor in writing. In the event that the Contractor has made a claim for an extension of time but the Engineer does not consider the Contractor entitled to an extension of time he shall so inform the Contractor without delay.

Assessment at due date for completion

(4) The Engineer shall not later than 14 days after the due date or extended date for completion of the Works or any Section thereof (and whether or not the Contractor shall have made any claim for an extension of time) consider all the circumstances known to him at that time and take action similar to that provided for in sub-clause (3) of this Clause. Should the Engineer consider that the Contractor is not entitled to an extension of time he shall so notify the Employer and the Contractor.

Final determination of extension

(5) The Engineer shall within 14 days of the issue of the Certificate of Substantial Completion for the Works or for any Section thereof review all the circumstances of the kind referred to in sub-clause (1) of this Clause and shall finally determine and certify to the Contractor with a copy to the Employer the overall extension of time (if any) to which he considers the Contractor entitled in respect of the Works or the relevant Section. No such final review of the circumstances shall result in a decrease in any extension of time already granted by the Engineer pursuant to sub-clauses (3) or (4) of this Clause.

The mechanism for granting extensions of time

Where the Contractor claims an extension of time he should, in the first instance, deliver particulars to the Engineer within 28 days "or as soon thereafter as is reasonable"—Clause 44(1). The scheme of the clause suggests that this is not a condition precedent; for instance Clause 44(2)(b) entitles and Clause 44(4) requires the Engineer to make an assessment of extensions whether or not a claim for an extension has been received.

The grounds for granting an extension

The Engineer may extend the time for completion in a number of given situations. The specific reasons given in Clauses 44(1)(a), (b) and (d) are supplemented by grounds which are mentioned in a number of clauses as provided for by Clause 44(1)(c).[62] The provision in Clause 44(1)(e) as to "special circumstance of any kind whatsoever" is designed to protect the clause against any matter arising for which the Employer may be liable and which would cause the liquidated damages provisions to become ineffective.[63] There is a view that a "sweeping up" provision such as this is ineffective.[64] This view suggests that acts of prevention by the Employer (however minor) do not come within Clause 44(1)(e). Accordingly, time is put at large[65] and the liquidated damages provisions fall, leaving the Employer to recover such general damages as he may prove for any late completion. It is submitted that this view is incorrect; it derives partly from the assumption that the liquidated damages provisions of a contract are for the Employer's benefit and hence are to be construed strictly against the Employer.[66] In fact liquidated damages provisions are valuable to the Contractor; they provide a valuable limit to his exposure for

62 Clauses 7(4), 12(2), 13(3), 14(8), 31(2), 40(1), 42(1) and 59(4)(f).

63 *Peak Construction (Liverpool)* v. *McKinney Foundations* (1970) 1 BLR 111 (CA).

64 See *Fernbrook Trading Co. Ltd* v. *Taggart* [1979] 1 NZLR 556 (New Zealand Supreme Court); *Perini Pacific Ltd* v. *Greater Vancouver Sewerage and Drainage District* (1966) 57 DLR (2d) 307 (British Columbia Court of Appeal). In these cases the "sweeping up" provisions were similar to those found in Clause 44. In each case the court found that they had to be construed narrowly and in each case breaches of the employer were held to fall outside their scope; time was put at large.

65 See Chapter 3, section 2.

66 *Peak Construction (Liverpool)* v. *McKinney Foundations* (1970) 1 BLR 111 (CA).

damages.[67] It is submitted, therefore, that the provision should be given a commercial meaning and that all events, including breaches by the Employer, entitle and require the Engineer to grant an extension, where appropriate.

Assessment

Clauses 44(2) and (3) imply a two-stage approach:

(*a*) an assessment by the Engineer of the delay suffered by the Contractor—Clause 44(2)

(*b*) a consideration whether any delay suffered fairly entitles the Contractor to an extension of the time for substantial completion—Clause 44(3). The mere fact that the Contractor has been delayed in some elements of the Works does not mean that the overall completion date is delayed.

Interim, due date and final determinations

The Contract entitles and/or requires the Engineer to make an assessment of the extension of time due in three different circumstances.

(*a*) The Engineer shall assess any claim made by the Contractor during the currency of the Works; in addition he may of his own volition make such an assessment—Clause 44(2). As a result of this determination he may grant an "interim extension"—Clause 44(3).

(*b*) At the date which is the current due date for completion of the Works, the Engineer is obliged to consider whether an extension of time is due. This obligation arises whether or not the Contractor has claimed an extension of time—Clause 44(4).

(*c*) Within 14 days of the issue of the Certificate of Substantial Completion the Engineer certifies his final determination of the overall extension of time to which he considers the Contractor entitled—Clause 44(5). He may not reduce any extension already granted.

67 *Temloc* v. *Errill Properties Ltd* (1988) 39 BLR 30 (CA).

45 Night and Sunday work

Subject to any provision to the contrary contained in the Contract none of the Works shall be executed during the night or on Sundays without the permission in writing of the Engineer save when the work is unavoidable or absolutely necessary for the saving of life or property or for the safety of the Works in which case the Contractor shall immediately advise the Engineer or the Engineer's Representative. Provided always that this Clause shall not be applicable in the case of any work which it is customary to carry out outside normal working hours or by rotary or double shifts.

Generally speaking, the Engineer may unreasonably withhold his permission for night and/or Sunday work.[68] Where, however, the conditions in Clause 46(1) are met, the permission shall not be unreasonably withheld—see Clause 46(2).

46 Rate of progress

(1) If for any reason which does not entitle the Contractor to an extension of time the rate of progress of the Works or any Section is at any time in the opinion of the Engineer too slow to ensure substantial completion by the time or extended time for completion prescribed by Clause 43 and 44 as appropriate or the revised time for completion agreed under sub-clause (3) of this Clause the Engineer shall notify the Contractor in writing and the Contractor shall thereupon take such steps as are necessary and to which the Engineer may consent to expedite the progress so as substantially to complete the Works or such Section by that prescribed time or extended time. The Contractor shall not be entitled to any additional payment for taking such steps.

Permission to work at night or on Sundays

(2) If as a result of any notice given by the Engineer under sub-clause (1) of this Clause the Contractor shall seek the Engineer's permission to do any work on Site at night or on Sundays such permission shall not be unreasonably refused.

---

68 *Leedsford Ltd* v. *Bradford City Council* (1956) 24 BLR 45 (CA).

Provision for accelerated completion

(3) If the Contractor is requested by the Employer or the Engineer to complete the Works or any Section within a revised time being less than the time or extended time for completion prescribed by Clauses 43 and 44 as appropriate and the Contractor agrees so to do then any special terms and conditions of payment shall be agreed between the Contractor and the Employer before any such action is taken.

Acceleration notice under Clause 46(1)

Where the Engineer gives notice under Clause 46(1), the Contractor may incur costs associated with acceleration. Where an extension of time is later granted by the Engineer or arbitrator, so that the Clause 46(1) notice was wrongly given, the Contractor may claim that he has incurred cost as a result of complying with the Engineer's instruction and hence is entitled to recover under Clause 13(3).

Acceleration agreement—Clause 46(3)

This clause acknowledges that the parties may make an agreement to accelerate the work. It is thought that in the absence of agreement, the Engineer may in any event order a variation which has the effect of reducing the time for completion under Clause 51. Such a variation must, however, be "desirable for the completion of and/or improved functioning of the Works".

LIQUIDATED DAMAGES FOR DELAY

47 Liquidated damages for delay in substantial completion of the whole of the Works

(1) (a) Where the whole of the Works is not divided into Sections the Appendix to the Form of Tender shall include a sum which represents the Employer's genuine pre-estimate (expressed per week or per day as the case may be) of the damages likely to be suffered by him if the whole of the Works is not substantially completed within the time prescribed by Clause 43 or by any extension thereof granted under Clause 44 or by any revision thereof agreed under Clause 46(3) as the case may be.

(b) If the Contractor fails to complete the whole of the Works within the time so prescribed he shall pay to the Employer the said sum for every week or day (as the case may be) which shall elapse between the date on which the prescribed time expired and the date the whole of the Works is substantially completed.

Provided that if any part of the Works is certified as complete pursuant to Clause 48 before the completion of the whole of the Works the said sum shall be reduced by the proportion which the value of the part so completed bears to the value of the whole of the Works.

Liquidated damages for delay in substantial completion where the whole of the Works is divided into Sections

(2) (a) Where the Works is divided into Sections (together comprising the whole of the Works) which are required to be completed within particular times as stated in the Appendix to the Form of Tender sub-clause (1) of this Clause shall not apply and the said Appendix shall include a sum in respect of each Section which represents the Employer's genuine pre-estimate (expressed per week or per day as the case may be) of the damages likely to be suffered by him if that Section is not substantially completed within the time prescribed by Clause 43 or by any extension thereof granted under Clause 44 or by any revision thereof agreed under Clause 46(3) as the case may be.

(b) If the Contractor fails to complete any Section within the time so prescribed he shall pay to the Employer the appropriate stated sum for every week or day (as the case may be) which shall elapse between the date on which the prescribed time expired and the date of substantial completion of that Section.

Provided that if any part of that Section is certified as complete pursuant to Clause 48 before the completion of the whole thereof the appropriate stated sum shall be reduced by the proportion which the value of the part so completed bears to the value of the whole of that Section.

(c) Liquidated damages in respect of two or more Sections may where circumstances so dictate run concurrently.

Damages not a penalty

(3) All sums payable by the Contractor to the Employer pursuant to this Clause shall be paid as liquidated damages for delay and not as a penalty.

Limitation of liquidated damages

(4) (a) The total amount of liquidated damages in respect of the whole of the Works or any Section thereof shall be limited to the appropriate sum stated in the Appendix to the Form of Tender. If no such limit is stated therein then liquidated damages without limit shall apply.

(b) Should there be omitted from the Appendix to the Form of Tender any sum required to be inserted therein either by sub-clause (1)(a) or by sub-clause (2)(a) of this Clause as the case may be or if any such sum is stated to be "nil" then to that extent damages shall not be payable.

Recovery and reimbursement of liquidated damages

(5) The Employer may

(a) deduct and retain the amount of any liquidated damages becoming due under the provision of this Clause from any sums due or which become due to the Contractor or

(b) require the Contractor to pay such amount to the Employer forthwith.

If upon a subsequent or final review of the circumstances causing delay the Engineer grants a relevant extension or further extension of time the Employer shall no longer be entitled to liquidated damages in respect of the period of such extension.

Any sum in respect of such period which may already have been recovered under this Clause shall be reimbursed forthwith to the Contractor together with interest compounded monthly at the rate provided for in Clause 60(7) from the date on which such sums were recovered from the Contractor.

Intervention of variations etc.

(6) If after liquidated damages have become payable in respect of any part of the Works the Engineer issues a variation order under Clause 51 or adverse physical conditions or artificial obstructions within the meaning of Clause 12 are encountered or any other situation outside the Contractor's control arises any of which in the Engineer's opinion results in further delay to that part of the Works

(a) the Engineer shall so inform the Contractor and the Employer in writing and

(b) the Employer's entitlement to liquidated damages in respect of that part of the Works shall be suspended until the Engineer notifies the Contractor and the Employer in writing that the further delay has come to an end.

Such suspension shall not invalidate any entitlement to liquidated damages which accrued before the period of delay started to run and any monies deducted or paid in accordance with sub-clause (5) of this Clause may be retained by the Employer without incurring liability for interest thereon under Clause 60(7).

Synopsis

This clause sets out detailed provisions as to liquidated damages for delay. There are two basic schemes. In the standard case where there is a single completion date for all the Works, Clause 47(1) applies but Clause 47(2) does not. Where there are sectional completion provisions, Clause 47(2) applies but Clause 47(1) does not.

The general case

Clause 47(1) applies unless the Works are divided into Sections in the Appendix. In this case the Contractor "shall pay to the Employer" the agreed rate of liquidated damages when he has failed to complete in the time for completion. The proviso states that where any part of the Works has been certified as substantially complete,[69] the liquidated damages shall be reduced proportionately.

Sectional completion

Where the Works are divided into sections, this is stated in the Appendix. In this case, Clause 47(1) does not apply. Here each section is to be treated independently[70] and liquidated damages levied in respect to each section separately.

69 Clause 48(4).

70 For the purpose of assessing extensions of time, the interconnectedness between sections may, of course, be taken into account.

Liquidated damages not a penalty—Clause 47(3)

Where the sum is excessive so that it could not be a genuine pre-estimate of the likely damage, this clause will not save it.[71]

Limitation

Clause 47(4)(a) allows the parties to agree upon a ceiling to liquidated damages payable.

Where no sum is inserted in the Appendix

Clause 47(4)(b) provides that where the appropriate entry is left blank or where it is given as "nil", no damages shall be payable.[72] This is an important provision, since many civil engineering contracts are drawn up informally by reference to the ICE Conditions of Contract without agreeing liquidated damages. Where liquidated damages are not agreed, the effect which ordinarily results is that the Employer may recover unliquidated damages. The effect of Clause 47(4)(b), however, seems to be that no damages at all, liquidated or unliquidated, may be recovered. It is submitted, however, that very clear wording is required before an entitlement to a damages is excluded and the efficacy of this purported exclusion must therefore be in doubt.

Delays at the Employer's risk occurring during any period in which liquidated damages are to apply

Clause 47(6) deals with the common situation where a variation is ordered (or other event at the Employer's risk occurs) during a period of the Contractor's culpable delay. The Employer's entitlement to liquidated damages is suspended during the period in which the variation is to be done or other delay is caused without invalidating the Employer's right to liquidated damages incurred before and after the event in question.[73]

---

71 As to penalties generally see Chapter 6, section 4.

72 See *Temloc* v. *Errill Properties Ltd* (1988) 39 BLR 30 (CA).

73 This may well be the general law: see *McAlpine Humberoak* v. *McDermott International* (1992) 58 BLR 1 (CA); *Balfour Beatty Building Limited* v. *Chestermount Properties Ltd* (1993) 9 Constr. L.J. 117.

CERTIFICATE OF SUBSTANTIAL COMPLETION

48 Notification of substantial completion

(1) When the Contractor considers that

(a) the whole of the Works or

(b) any Section in respect of which a separate time for completion is provided in the Appendix to the Form of Tender

has been substantially completed and has satisfactorily passed any final test that may be prescribed by the Contract he may give notice in writing to that effect to the Engineer or to the Engineer's Representative. Such notice shall be accompanied by an undertaking to finish any outstanding work in accordance with the provisions of Clause 49(1).

Certification of substantial completion

(2) The Engineer shall within 21 days of the date of delivery of such notice either

(a) issue to the Contractor (with a copy to the Employer) a Certificate of Substantial Completion stating the date on which in his opinion the Works were or the Section was substantially completed in accordance with the Contract or

(b) give instructions in writing to the Contractor specifying all the work which in the Engineer's opinion requires to be done by the Contractor before the issue of such certificate.

If the Engineer gives such instructions the Contractor shall be entitled to receive a Certificate of Substantial Completion within 21 days of completion to the satisfaction of the Engineer of the work specified in the said instructions.

Premature use by Employer

(3) If any substantial part of the Works has been occupied or used by the Employer other than as provided in the Contract the Contractor may request in writing and the Engineer shall issue a Certificate of Substantial Completion in respect thereof. Such certificate shall take effect from the date of delivery of the Contractor's request and upon the issue of such certificate the Contractor shall be deemed to have undertaken to complete any outstanding work in that part of the Works during the Defects Correction Period.

Substantial completion of other parts of the Works

(4) If the Engineer considers that any part of the Works has been substantially completed and has passed any final test that may be prescribed by the Contract he may issue a Certificate of Substantial Completion in respect of that part of the Works before completion of the whole of the Works and upon the issue of such certificate the Contractor shall be deemed to have undertaken to complete any outstanding work in that part of the Works during the Defects Correction Period.

Reinstatement of ground

(5) A Certificate of Substantial Completion given in respect of any Section or part of the Works before completion of the whole shall not be deemed to certify completion of any ground or surfaces requiring reinstatement unless such certificate shall expressly so state.

Clause 48(1) "... substantially completed"

The terms "substantial completion" and "substantially completed" are used in various clauses. The term is not defined.[74] It is submitted that it means a state in which all work is complete, save that which is latently defective or that which is unimportant for the reasonable functioning of the Works.[75] The use of the term "substantial" clearly indicates that absolute completion is not required.

The certificate

The giving of a certificate of substantial completion is important in a number of situations. Most importantly, it defines the time at which liquidated damages cease to be incurred.

74 The expression "substantial performance" is used in the general law of contracts to mean a state which entitled a contracting party to be paid under an entire contract: however, it is submitted that the two expressions have quite different meanings.

75 See *Westminster Corporation* v. *Jarvis* [1970] 1 WLR 637 (HL) which points to this conclusion even though the expression "substantial completion" was not in issue.

Certificates issued in respect of sections or parts of the Works

The standard case is where the whole Works are to be completed by a single completion date and a single Certificate of Substantial Completion is issued. Clause 48 provides for a number of other scenarios. Thus where separate Sections are specified in the Appendix, the Contractor is entitled to a certificate in respect of each: Clause 48(1)(b). Furthermore, where any substantial part of the Works is used by the Employer before the issue of a Certificate of Substantial Completion, the Engineer shall issue a certificate upon the Contractor's request: Clause 48(3). In addition, where any "part of the Works" (which does not amount to a Section) is substantially completed the Engineer may in his discretion issue a certificate in respect of that part of the Works—Clause 48(4).

OUTSTANDING WORK AND DEFECTS

49 Work outstanding

(1) The undertaking to be given under Clause 48(1) may after agreement between the Engineer and the Contractor specify a time or times within which the outstanding work shall be completed. If no such times are specified any outstanding work shall be completed as soon as practicable during the Defects Correction Period.

Execution of work of repair etc.

(2) The Contractor shall deliver up to the Employer the Works and each Section and part thereof at or as soon as practicable after the expiry of the relevant Defects Correction Period in the condition required by the Contract (fair wear and tear excepted) to the satisfaction of the Engineer. To this end the Contractor shall as soon as practicable execute all work of repair amendment reconstruction rectification and making good of defects of whatever nature as may be required of him in writing by the Engineer during the relevant Defects Correction Period or within 14 days after its expiry as a result of an inspection made by or on behalf of the Engineer prior to its expiry.

Cost of execution of work of repair etc.

(3) All work required under sub-clause (2) of this Clause shall be carried out by the Contractor at his own expense if in the Engineer's opinion it is necessary due to the use of materials or workmanship not in accordance with the Contract or to neglect

or failure by the Contractor to comply with any of his obligations under the Contract. In any other event the value of such work shall be ascertained and paid for as if it were additional work.

Remedy on Contractor's failure to carry out work required

(4) If the Contractor fails to do any such work as aforesaid the Employer shall be entitled to carry out such work by his own workpeople or by other contractors and if such work is work which the Contractor should have carried out at his own expense the Employer shall be entitled to recover the cost thereof from the Contractor and may deduct the same from any monies that are or may become due to the Contractor.

Work which is not in accordance with the Contract will render the Contractor in breach of contract. Minor latent defects in the Works may quickly come to light after substantial completion. The Contractor will not normally welcome the Employer employing another contractor to remedy these defects, the costs of which will be deducted from the retention fund which the Employer holds. Such an arrangement gives the Contractor little control and little opportunity to mitigate the loss. Clause 49 provides for a period—the Defects Correction Period—following substantial completion during which the Contractor is entitled and obliged to remedy defects. The general position is that the Contractor pays for defects which result from his breach of contract. The Employer pays for defects which result from poor specification or design, as though such work were additional work: Clause 49(3).

50 Contractor to search

The Contractor shall if required by the Engineer in writing carry out such searches tests or trials as may be necessary to determine the cause of any defect imperfection or fault under the directions of the Engineer. Unless such defect imperfection or fault shall be one for which the Contractor is liable under the Contract the cost of the work carried out by the Contractor as aforesaid shall be borne by the Employer. But if such defect imperfection or fault shall be one for which the Contractor is liable the cost of the work carried out as aforesaid shall be borne by the Contractor and he shall in such case repair rectify and make good such defect imperfection or fault at his own expense in accordance with Clause 49.

This provision clearly deals with matters analogous to those in Clauses 36(3) and 38(2).

ALTERATIONS, ADDITIONS AND OMISSIONS

51 Ordered variations

(1) The Engineer

(a) shall order any variation to any part of the Works that is in his opinion necessary for the completion of the Works and

(b) may order any variation that for any other reason shall in his opinion be desirable for the completion and/or improved functioning of the Works.

Such variations may include additions omissions substitutions alterations changes in quality form character kind position dimension level or line and changes in any specified sequence method or timing of construction required by the Contract and may be ordered during the Defects Correction Period.

Ordered variations to be in writing

(2) All variations shall be ordered in writing but the provisions of Clause 2(6) in respect of oral instructions shall apply.

Variation not to affect Contract

(3) No variation ordered in accordance with sub-clauses (1) and (2) of this Clause shall in any way vitiate or invalidate the Contract but the value (if any) of all such variations shall be taken into account in ascertaining the amount of the Contract Price except to the extent that such variation is necessitated by the Contractor's default.

Changes in quantities

(4) No order in writing shall be required for increase or decrease in the quantity of any work where such increase or decrease is not the result of an order given under this Clause but is the result of the quantities exceeding or being less than those stated in the Bill of Quantities.

General

A variations clause is desirable in civil engineering contracts where there are works of any complexity. Changes to the specification and design are very frequently required, not only because they may seem desirable as construction proceeds, but sometimes because the Works cannot be completed satisfactorily without a variation. The Engineer's right to vary the Contract Works must be expressly agreed.

Necessary variations—Clause 51(1)(a)

The Engineer is obliged to order a variation which is in his opinion necessary for the completion of the Works.

Desirable variations—Clause 51(1)(b)

Where a variation is merely "desirable for the completion and/or improved functioning of the Works", the Engineer's power to order variations is discretionary.

Permitted and non-permitted variations

The Engineer has no power to order a variation unless it is necessary or desirable for the completion or functioning of the Works. Thus, the Engineer is not entitled to vary the Works in order to suit the Employer's financial or management arrangements. For instance, an omission which will improve the performance of the Works may be ordered: but an omission which is desirable simply to save money may not.[76] A variety of types of variations are mentioned in Clause 51(1). These may be classified as follows:

(*a*) physical—a variation to the definition of the completed structure
(*b*) positional—the line, level, orientation, etc. may be varied

76 Where such an omission is in fact ordered, the Contractor will not be entitled to construct the wrongly omitted works regardless. His remedy will be as damages for breach of contract. In particular he will claim loss of profit suffered insofar as he is not able to mitigate that loss by redeploying his resources elsewhere.

(*c*) management — the specified[77] sequence method or timing may be varied.

The words in Clause 51 from which this classification is derived are clearly illustrative and do not define the expression "variation". However, it is frequently important to be able to decide whether an instruction or constraint is or is not a variation. Where it is a variation, the Contractor is entitled to be paid in accordance with Clause 52. It is thought that an instruction amounts to a variation where it imposes any new obligation or constraint.[78]

Variations to be in writing

It is clearly desirable that any variation be in writing and Clause 51(2) provides for this. Frequently, however, the Contractor is required to modify his work or his work pattern for reasons which he claims amount to a variation; for example he may have to modify the design of Works for which he bears no responsibility under the Contract.[79] Where Clause 51(1)(a) applies, the issue of a variation order is obligatory; thus where the Engineer fails to order a variation which falls within the scope of Clause 51(1)(a), the Employer may not rely on a lack of written variation order to refuse to pay such additional sums as may become due.[80]

77 The significance of the expression "changes in specified sequence ..." etc. is not wholly clear. One interpretation is that a sequence etc. cannot be varied unless it is specified in the Contract in the first place. This would produce the result that where no sequence is specified, the Engineer is powerless to specify a sequence (although he may still be entitled to do so under Clause 13(1)). It is thought that this cannot be the proper construction. While a variation clearly needs to relate to some contractual obligation or entitlement, it is thought that where no sequence etc. is specified, it may still be varied so that it becomes more closely specified.

78 See *English Industrial Estates* v. *Kier* (1991) 56 BLR 93.

79 *Shanks & McEwan* v. *Strathclyde Regional Council* (1994) CILL 916 (Scottish Court of Session).

80 *Brodie* v. *Cardiff Corporation* [1919] AC 337 (HL). Here the variation clause provided that no extras would be paid for unless there was a written order. The engineer required work to be done which he said was within the original scope of works and which the contractor claimed was an extra. The work was done. Upon completion, the contractor commenced arbitration proceedings and the arbitrator accepted that the works were extra to that originally agreed. But the employer claimed not to be required to pay for such extras because of the lack of a written order. The House of Lords held that the arbitrator's award perfected the missing variation order.

Variations necessitated by the Contractor's default

The Contractor is not entitled to claim the cost of any variation to the extent that it is made necessary by his own default. In other forms of contract, the absence of such a provision has in some circumstances caused the Employer to be liable for the cost of a variation which is instructed following the Contractor's default.[81]

52 Valuation of ordered variations

(1) The value of all variations ordered by the Engineer in accordance with Clause 51 shall be ascertained by the Engineer after consultation with the Contractor in accordance with the following principles.

(a) Where work is of similar character and executed under similar conditions to work priced in the Bill of Quantities it shall be valued at such rates and prices contained therein as may be applicable.

(b) Where work is not of a similar character or is not executed under similar conditions or is ordered during the Defects Correction Period the rates and prices in the Bill of Quantities shall be used as the basis for valuation so far as may be reasonable failing which a fair valuation shall be made.

Failing agreement between the Engineer and the Contractor as to any rate or price to be applied in the valuation of any variation the Engineer shall determine the rate or price in accordance with the foregoing principles and he shall notify the Contractor accordingly.

Engineer to fix rates

(2) If the nature or amount of any variation relative to the nature or amount of the whole of the contract work or to any part thereof shall be such that in the opinion of the Engineer or the Contractor any rate or price contained in the Contract for any item of work is by reason of such variation rendered unreasonable or inapplicable either the Engineer shall give to the

81 *Simplex Concrete Piles* v. *St Pancras Borough Council* (1958) 14 BLR 80; compare with *Howard de Walden Estates Ltd* v. *Costain Management Design Ltd* (1991) 55 BLR 124.

Contractor or the Contractor shall give to the Engineer notice before the varied work is commenced or as soon thereafter as is reasonable in all the circumstances that such rate or price should be varied and the Engineer shall fix such rate or price as in the circumstances he shall think reasonable and proper.

Daywork

(3) The Engineer may if in his opinion it is necessary or desirable order in writing that any additional or substituted work shall be executed on a daywork basis in accordance with the provisions of Clause 56(4).

Notice of claims

(4) (a) If the Contractor intends to claim a higher rate or price than one notified to him by the Engineer pursuant to sub-clauses (1) and (2) of this Clause or Clause 56(2) the Contractor shall within 28 days after such notification give notice in writing of his intention to the Engineer.

(b) If the Contractor intends to claim any additional payment pursuant to any Clause of these Conditions other than sub-clauses (1) and (2) of this Clause or Clause 56(2) he shall give notice in writing of his intention to the Engineer as soon as may be reasonable and in any event within 28 days after the happening of the events giving rise to the claim. Upon the happening of such events the Contractor shall keep such contemporary records as may reasonably be necessary to support any claim he may subsequently wish to make.

(c) Without necessarily admitting the Employer's liability the Engineer may upon receipt of a notice under this Clause instruct the Contractor to keep such contemporary records or further contemporary records as the case may be as are reasonable and may be material to the claim of which notice has been given and the Contractor shall keep such records. The Contractor shall permit the Engineer to inspect all records kept pursuant to this Clause and shall supply him with copies thereof as and when the Engineer shall so instruct.

(d) After the giving of a notice to the Engineer under this Clause the Contractor shall as soon as is reasonable in all the circumstances send to the Engineer a first interim account giving full and detailed particulars of the amount claimed to that date and of the grounds upon which the claim is based. Thereafter at such intervals as the Engineer may reasonably require the Contractor shall send to the Engineer further up to date accounts giving the accumulated total of the claim and any further grounds upon which it is based.

(e) If the Contractor fails to comply with any of the provisions of this Clause in respect of any claim which he shall seek to make then the Contractor shall be entitled to payment in respect thereof only to the extent that the Engineer has not been prevented from or substantially prejudiced by such failure in investigating the said claim.

(f) The Contractor shall be entitled to have included in any interim payment certified by the Engineer pursuant to Clause 60 such amount in respect of any claim as the Engineer may consider due to the Contractor provided that the Contractor shall have supplied sufficient particulars to enable the Engineer to determine the amount due. If such particulars are insufficient to substantiate the whole of the claim the Contractor shall be entitled to payment in respect of such part of the claim as the particulars may substantiate to the satisfaction of the Engineer.

General

This clause is headed "Valuation of ordered variations". It is wider than this. It deals first with the valuation of variations generally, even variations which are not acknowledged by the Engineer to be variations. Secondly, it deals with claims procedures, including notification of claims.

The valuation of variations

Clause 52(1) establishes a three-level approach to the valuation of variations:

(*a*) evaluate variations using the existing rates where this is both possible and reasonable
(*b*) where appropriate use the existing rates as "the basis for valuation"
(*c*) where there are no applicable or analogous rates, use a fair valuation.

Fixing rates

Clause 52(2) provides that where rates are rendered unreasonable or inapplicable by reason of any variation, the Engineer shall fix a varied rate.[82] This procedure is to be commenced by notice given either by the Contractor or Engineer. This notice seems at first sight to be a condition precedent to be given "before the varied work is commenced or as soon thereafter as is reasonable in all the circumstances ...". However, Clause 52(4)(e) provides that a failure to give notice will defeat the claim "only to the extent that the Engineer has not been prevented from or substantially prejudiced by such failure in investigating the said claim". The means or principles by which the rate is to be fixed are not set out in the clause. Normally the Engineer will request the Contractor to supply him with a breakdown of his rates into components such as labour, plant, materials, on-site and off-site overheads and profit. Using these the Engineer will determine how each of these components is or will be affected and adjust the quantities and/or proportion of the composite rate accordingly.

Dayworks—Clause 52(3)

Contractors frequently submit accounts for varied work in the form of "daywork sheets". Dayworks are valued in accordance with a schedule of rates produced by representative contractors' bodies.[83] An hourly or daily rate is specified in the schedule. This basis of valuation assumes, in essence, that the Contractor's labour, plant, etc. are being hired to the Employer, so that payment is not made on the basis of work output, but time input. Such a valuation frequently suits contractors. However, this basis is only formally applicable where the Engineer orders dayworks in writing. In other cases, where a reasonable rate is to be paid, for example in accordance with Clause 52(1)(b), or where the Contractor is entitled to his cost plus profit, daywork rates may provide a starting basis for valuation.

---

82 Clause 56(2) contains analogous provisions to those in Clause 52(2) for the case where the rates are rendered unreasonable or inapplicable by virtue of changes in quantities.

83 Clause 56(4) provides that the daywork rates are those "contained in the Schedule of Dayworks carried out incidental to Contract Work issued by the Federation of Civil Engineering Contractors current at the date of execution of the daywork".

Notices, records and particulars—Clause 52(4)

Clause 52(4) applies to claims generally under the Contract. Its approach is to set out the basic requirement for notices to be given; Clause 52(4)(e), however, provides that lack of notice will only defeat a claim to the extent that the failure to notify disadvantages the Engineer. In addition to notices, the clause also deals with records and particulars of claims in Clauses 52(4)(c), (d), (f).

PROPERTY IN MATERIALS AND CONTRACTOR'S EQUIPMENT

53 Vesting of Contractor's Equipment

(1) All Contractor's Equipment Temporary Works materials for Temporary Works or other goods or materials owned by the Contractor shall when on Site be deemed to be the property of the Employer and shall not be removed therefrom without the written consent of the Engineer which consent shall not unreasonably be withheld where the items in question are no longer immediately required for the purposes of the completion of the Works.

Liability for loss or damage to Contractor's Equipment

(2) The Employer shall not at any time be liable save as mentioned in Clauses 22 and 65 for the loss of or damage to any Contractor's Equipment Temporary Works goods or materials.

Disposal of Contractor's Equipment

(3) If the Contractor fails to remove any of the said Contractor's Equipment Temporary Works goods or materials as required by Clause 33 within such reasonable time after completion of the Works as the Engineer may allow then the Employer may sell or otherwise dispose of such items. From the proceeds of the sale of any such items the Employer shall be entitled to retain any costs or expenses incurred in connection with their sale and disposal before paying the balance (if any) to the Contractor.

54 Vesting of goods and materials not on Site

(1) With a view to securing payment under Clause 60(1)(c) the Contractor may (and shall if the Engineer so directs) transfer to the Employer the property in goods and materials listed in the Appendix to the Form of Tender before the same are delivered to the Site provided that the goods and materials

(a) have been manufactured or prepared and are substantially ready for incorporation in the Works and

(b) are the property of the Contractor or the contract for the supply of the same expressly provides that the property therein shall pass unconditionally to the Contractor upon the Contractor taking the action referred to in sub-clause (2) of this Clause.

Action by Contractor

(2) The intention of the Contractor to transfer the property in any goods or materials to the Employer in accordance with this Clause shall be evidenced by the Contractor taking or causing the supplier of those goods or materials to take the following actions.

(a) Provide to the Engineer documentary evidence that the property in the said goods or materials has vested in the Contractor.

(b) Suitably mark or otherwise plainly identify the goods and materials so as to show that their destination is the Site that they are the property of the Employer and (where they are not stored at the premises of the Contractor) to whose order they are held.

(c) Set aside and store the said goods and materials so marked or identified to the satisfaction of the Engineer.

(d) Send to the Engineer a schedule listing and giving the value of every item of the goods and materials so set aside and stored and inviting him to inspect them.

Vesting in Employer

(3) Upon the Engineer approving in writing the transfer in ownership of any goods and materials for the purposes of this Clause they shall vest in and become the absolute property of the Employer and thereafter shall be in possession of the Contractor for the sole purpose of delivering them to the Employer and incorporating them in the Works and shall not be within the ownership control or disposition of the Contractor.

Provided always that

(a) approval by the Engineer for the purposes of this Clause or any payment certified by him in respect of goods and materials pursuant to Clause 60 shall be without prejudice to the exercise of any power of the Engineer contained in this Contract to reject any goods or materials which are not in accordance with the provisions of the Contract and upon any such rejection the property in the rejected goods or materials shall immediately revest in the Contractor and

(b) the Contractor shall be responsible for any loss or damage to such goods and materials and for the cost of storing handling and transporting the same and shall effect such additional insurance as may be necessary to cover the risk of such loss or damage from any cause.

Lien on goods or materials

(4) Neither the Contractor nor a sub-contractor nor any other person shall have a lien on any goods or materials which have vested in the Employer under sub-clause (3) of this Clause for any sum due to the Contractor sub-contractor or other person and the Contractor shall take all such steps as may reasonably be necessary to ensure that the title of the Employer and the exclusion of any such lien are brought to the notice of sub-contractors and other persons dealing with any such goods or materials.

Delivery to the Employer of vested goods or materials

(5) Upon cessation of the employment of the Contractor under this Contract before the completion of the Works whether as a result of the operation of Clause 63 or otherwise the Contractor shall deliver to the Employer any goods or materials the property in which has vested in the Employer by virtue of sub-clause (3) of this Clause and if he shall fail to do so the Employer may enter any premises of the Contractor or of any sub-contractor and remove such goods and materials and recover the cost of so doing from the Contractor.

Incorporation in sub-contracts

(6) The Contractor shall incorporate provisions equivalent to those provided in this Clause in every sub-contract in which provision is to be made for payment in respect of goods or materials before the same have been delivered to the Site.

These clauses are designed to ensure: (*a*) that equipment on site is available to perform the work should the Contractor default; and (*b*) to transfer property in goods and materials to the Employer, where the Employer is to pay some element of their cost in advance. The effectiveness of both clauses, particularly against third parties with interests in plant[84] and goods,[85] is uncertain.

MEASUREMENT

55 Quantities

(1) The quantities set out in the Bill of Quantities are the estimated quantities of the work but they are not to be taken as the actual and correct quantities of the Works to be executed by the Contractor in fulfilment of his obligations under the Contract.

Correction of errors

(2) Any error in description in the Bill of Quantities or omission therefrom shall not vitiate the Contract nor release the Contractor from the execution of the whole or any part of the Works according to the Drawings and Specification or from any of his obligations or liabilities under the Contract. Any such error or omission shall be corrected by the Engineer and the value of the work actually carried out shall be ascertained in accordance with Clause 52. Provided that there shall be no rectification of any errors omissions or wrong estimates in the descriptions rates and prices inserted by the Contractor in the Bill of Quantities.

56 Measurement and valuation

(1) The Engineer shall except as otherwise stated ascertain and determine by admeasurement the value in accordance with the Contract of the work done in accordance with the Contract.

Increase or decrease of rate

(2) Should the actual quantities executed in respect of any item be greater or less than those stated in the Bill of Quantities and if

84 But see many forms of sub-contract (e.g. the FCEC Form of Sub-Contract) which provide that the sub-contractor's property in equipment is to pass through the main contractor onto the employer.

85 See the body of law on retention of title Clauses: *Aluminium Industrie Vaasen* v. *Romalpa Aluminium* [1976] 1 WLR 676 (CA). *Armour* v. *Thyssen Edalstahlwerke AG* [1991] 2 AC 339 (HL) (Scotland).

in the opinion of the Engineer such increase or decrease of itself shall so warrant the Engineer shall after consultation with the Contractor determine an appropriate increase or decrease of any rates or prices rendered unreasonable or inapplicable in consequence thereof and shall notify the Contractor accordingly.

Attending for measurement

(3) The Engineer shall when he requires any part or parts of the work to be measured give reasonable notice to the Contractor who shall attend or send a qualified agent to assist the Engineer or the Engineer's Representative in making such measurement and shall furnish all particulars required by either of them. Should the Contractor not attend or neglect or omit to send such agent then the measurement made by the Engineer or approved by him shall be taken to be the correct measurement of the work.

Daywork

(4) Where any work is carried out on a daywork basis the Contractor shall be paid for such work under the conditions and at the rates and prices set out in the daywork schedule included in the Contract or failing the inclusion of a daywork schedule he shall be paid at the rates and prices and under the conditions contained in the "Schedule of Dayworks carried out incidental to Contract Work" issued by The Federation of Civil Engineering Contractors current at the date of the execution of the daywork.

The Contractor shall furnish to the Engineer such records receipts and other documentation as may be necessary to prove amounts paid and/or costs incurred. Such returns shall be in the form and delivered at the times the Engineer shall direct and shall be agreed within a reasonable time.

Before ordering materials the Contractor shall if so required submit to the Engineer quotations for the same for his approval.

57 Method of measurement

Unless otherwise provided in the Contract or unless general or detailed description of the work in the Bill of Quantities or any other statement clearly shows to the contrary the Bill of Quantities shall be deemed to have been prepared and measurements shall be made according to the procedure set out in the "Civil Engineering Standard Method of Measurement Second Edition 1985" approved by the Institution of Civil Engineers and the Federation of Civil Engineering Contractors in association with the Association of Consulting Engineers or such later or amended edition thereof as may be stated in the Appendix to the Form of Tender to have been adopted in its preparation.

Clauses 55, 56 and 57 deal with measurement and should be read together. They establish a "measure and value" arrangement. The measurement shall be done by the Engineer (after giving notice to the Contractor) in accordance with the Method of measurement specified in Clause 57.[86]

PROVISIONAL AND PRIME COST SUMS AND NOMINATED SUB-CONTRACTS

58 Use of Provisional Sums

(1) In respect of every Provisional Sum the Engineer may order either or both of the following.

(a) Work to be executed or goods materials or services to be supplied by the Contractor the value thereof being determined in accordance with Clause 52 and included in the Contract Price.

(b) Work to be executed or goods materials or services to be supplied by a Nominated Sub-contractor in accordance with Clause 59.

Use of Prime Cost Items

(2) In respect of every Prime Cost Item the Engineer may order either or both of the following.

(a) Subject to Clause 59 that the Contractor employ a sub-contractor nominated by the Engineer for the execution of any work or the supply of any goods materials or services included therein.

86 The Method of Measurement mentioned in Clause 57 is the CESSM 2nd Edition. The 3rd Edition has now been published (1991). Clause 57 enables the parties to specify in the Appendix that the 3rd Edition or other method of measurement be used.

(b) With the consent of the Contractor that the Contractor himself execute any such work or supply any such goods materials or services in which event the Contractor shall be paid in accordance with the terms of a quotation submitted by him and accepted by the Engineer or in the absence thereof the value shall be determined in accordance with Clause 52 and included in the Contract Price.

Design requirements to be expressly stated

(3) If in connection with any Provisional Sum or Prime Cost Item the services to be provided include any matter of design or specification of any part of the Permanent Works or of any equipment or plant to be incorporated therein such requirement shall be expressly stated in the Contract and shall be included in any Nominated Sub-contract. The obligation of the Contractor in respect thereof shall be only that which has been expressly stated in accordance with this sub-clause.

Provisional Sums and Prime Cost Items—general terminology

These terms are partially defined in Clauses 1(1)(k), (l). The term "Prime Cost Item" derives from the prime cost, or actual cost, of a piece of work. A Prime Cost Item is, therefore, a piece of work for which the actual cost is determined in advance. It relates specifically to work which is to be done by a nominated supplier or sub-contractor and for which an advance quote is obtained. The Contractor then includes a mark-up on the item for attendance and profit. In practice, however, an advance quotation is frequently not obtained and so the price may have to be adjusted. A Provisional Sum relates to work which may be executed at the discretion of the Engineer; it need not be executed.

59 Nominated Sub-contractors—objection to nomination

(1) The Contractor shall not be under any obligation to enter into a sub-contract with any Nominated Sub-contractor against whom the Contractor may raise reasonable objection or who declines to enter into a sub-contract with the Contractor containing provisions

(a) that in respect of the work goods materials or services the subject of the sub-contract the Nominated Sub-contractor will undertake towards the Contractor such obligations and liabilities as will enable the Contractor to discharge his own obligations and liabilities towards the Employer under the terms of the Contract

(b) that the Nominated Sub-contractor will save harmless and indemnify the Contractor against all claims demands and proceedings damages costs charges and expenses whatsoever arising out of or in connection with any failure by the Nominated Sub-contractor to perform such obligations or fulfil such liabilities

(c) that the Nominated Sub-contractor will save harmless and indemnify the Contractor from and against any negligence by the Nominated Sub-contractor his agents workmen and servants and against any misuse by him or them of any Contractor's Equipment or Temporary Works provided by the Contractor for the purposes of the Contract and for all claims as aforesaid

(d) that the Nominated Sub-contractor will provide the Contractor with security for the proper performance of the sub-contract and

(e) equivalent to those contained in Clause 63.

Engineer's action upon objection to nomination or upon determination of Nominated Sub-contract

(2) If pursuant to sub-clause (1) of this Clause the Contractor declines to enter into a sub-contract with a sub-contractor nominated by the Engineer or if during the course of the Nominated Sub-contract the Contractor shall validly terminate the employment of the Nominated Sub-contractor as a result of his default the Engineer shall

(a) nominate an alternative sub-contractor in which case sub-clause (1) of this Clause shall apply or

(b) by order under Clause 51 vary the Works or the work goods materials or services in question or

(c) by order under Clause 51 omit any or any part of such works goods materials or services so that they may be provided by workmen contractors or suppliers employed by the Employer either

(i) concurrently with the Works (in which case Clause 31 shall apply) or
(ii) at some other date

and in either case there shall nevertheless be included in the Contract Price such sum (if any) in respect of the Contractor's charges and profit being a percentage of the estimated value of such omission as would have been payable had there been no such omission and the value thereof had been that estimated in the Bill of Quantities or inserted in the Appendix to the Form of Tender as the case may be or

(d) instruct the Contractor to secure a sub-contractor of his own choice and to submit a quotation for the work goods materials or services in question to be so performed or provided for the Engineer's consideration and action or

(e) invite the Contractor himself to execute or supply the work goods materials or services in question under Clause 58(1)(a) or Clause 58(2)(b) or on a daywork basis as the case may be.

Contractor responsible for Nominated Sub-contractors

(3) Except as otherwise provided in Clause 58 (3) the Contractor shall be as responsible for the work executed or goods materials or services supplied by a Nominated Sub-contractor employed by him as if he had himself executed such work or supplied such goods materials or services.

Nominated Sub-contractor's default

(4) (a) If any event arises which in the opinion of the Contractor justifies the exercise of his right under any forfeiture clause to terminate the sub-contract or to treat the sub-contract as repudiated by the Nominated Sub-contractor he shall at once notify the Engineer in writing.

Termination of Sub-contract

(b) With the consent in writing of the Engineer the Contractor may give notice to the Nominated Sub-contractor expelling him from the Sub-contract works pursuant to any forfeiture clause or rescinding the Sub-contract as the case may be. If however the Engineer's consent is withheld the Contractor shall be entitled to appropriate instructions under Clause 13.

Engineer's action upon termination

(c) In the event that the Nominated Sub-contractor is expelled from the Sub-contract works the Engineer shall at once take such action as is required under sub-clause (2) of this Clause.

Recovery of additional expense

(d) Having with the Engineer's consent terminated the Nominated Sub-contract the Contractor shall take all necessary steps and proceedings as are available to him to recover all additional expenses that are incurred from the Sub-contractor or under the security provided pursuant to sub-clause (1)(d) of this Clause. Such expenses shall include any additional expenses incurred by the Employer as a result of the termination.

Reimbursement of Contractor's loss

(e) If and to the extent that the Contractor fails to recover all his reasonable expenses of completing the Sub-contract works and all his proper additional expenses arising from the termination the Employer will reimburse the Contractor his unrecovered expenses.

Consequent delay

(f) The Engineer shall take any delay to the completion of the Works consequent upon the Nominated Sub-contractor's default into account in determining any extension of time to which the Contractor is entitled under Clause 44.

Provisions for payment

(5) For all work executed or goods materials or services supplied by Nominated Sub-contractors there shall be included in the Contract Price

(a) the actual price paid or due to be paid by the Contractor in accordance with the terms of the sub-contract (unless and to the extent that any such payment is the result of a default of the Contractor) net of all trade and other discounts rebates and allowances other than any discount obtainable by the Contractor for prompt payment

(b) the sum (if any) provided in the Bill of Quantities for labours in connection therewith and

(c) in respect of all other charges and profit a sum being a percentage of the actual price paid or due to be paid calculated (where provision has been made in the Bill of Quantities for a rate to be set against the relevant item of prime cost) at the rate inserted by the Contractor against that item or (where no such provision has been made) at the rate inserted by the Contractor in the Appendix to the Form of Tender as the percentage for adjustment of sums set against Prime Cost Items.

Production of vouchers etc.

(6) The Contractor shall when required by the Engineer produce all quotations invoices vouchers sub-contract documents accounts and receipts in connection with expenditure in respect of work carried out by all Nominated Sub-contractors.

Payment to Nominated Sub-contractors

(7) Before issuing any certificate under Clause 60 the Engineer shall be entitled to demand from the Contractor reasonable proof that all sums (less retentions provided for in the Sub-contract) included in previous certificates in respect of the work executed or goods or materials or services supplied by Nominated Sub-contractors have been paid to the Nominated Sub-contractors or discharged by the Contractor in default whereof unless the Contractor shall

(a) give details to the Engineer in writing of any reasonable cause he may have for withholding or refusing to make such payment and

(b) produce to the Engineer reasonable proof that he has so informed such Nominated Sub-contractor in writing

the Employer shall be entitled to pay to such Nominated Sub-contractor direct upon the certification of the Engineer all payments (less retentions provided for in the Sub-contract) which the Contractor has failed to make to such Nominated Sub-contractor and to deduct by way of set-off the amount so paid by the Employer from any sums due or which become due from the Employer to the Contractor. Provided always that where the Engineer has certified and the Employer has made direct payment to the Nominated Sub-contractor the Engineer shall in issuing any further certificate in favour of the Contractor deduct from the amount thereof the amount so paid but shall not withhold or delay the issue of the certificate itself when due to be issued under the terms of the Contract.

Nomination[87]

Clause 59 sets out a detailed scheme for the nomination of sub-contractors.

Grounds for objection to a nomination

Clause 59(1) provides that the Contractor shall not be obliged to enter into a contract with a prospective nominated sub-contractor where the Contractor raises a "reasonable objection". This expression is not defined. It is thought that the sub-contractor's reputation, experience, financial standing as well as its safety and quality procedures are relevant factors. Clause 59(1) also provides an outline of the minimum terms which the Contractor may insist on in any sub-contract; the form or amount of security in Clause 59(1)(d), which is a most material consideration is not, however, specified.

Payment for work done by nominated sub-contractor

Clause 59(5) sets out the elements of payment which may be applied for by the Contractor, namely: (*a*) the actual price to be paid to the nominated sub-contractor; (*b*) any billed sum for specific labours associated with the nominated work; (*c*) the mark-up inserted by the Contractor in the Bill of Quantities. Reference should also be made to Clauses 60(1), 60(2), 60(8)(a), (b) which provide technical rules concerning the application for payment in respect of nominated sub-contractors.

Payments to Nominated Sub-Contractors

Clause 59(7) provides a scheme whereby the Contractor is required to show that he has paid nominated sub-contractors the relevant proportion of sums certified. If the Contractor does not provide this evidence, the Employer may pay the Nominated Sub-Contractor directly. However, the Contractor may refuse to pay the Nominated Sub-Contractor where the latter is in breach of the Sub-Contract and the Engineer and Nominated Sub-Contractor have both received details of the sums deducted. It seems that the Engineer must, in this

87 See Chapter 2, section 4.

case, determine whether the amount deducted is proper; if the Engineer values the proper deduction at a smaller sum, the Engineer should inform the Contractor of the sum which may be deducted. Should the Contractor's deduction be vindicated at a later date, the Contractor may recover the difference from the Sub-Contractor.[88]

CERTIFICATES AND PAYMENT

60 Monthly statements

(1) The Contractor shall submit to the Engineer at monthly intervals a statement (in such form if any as may be prescribed in the Specification) showing

(a) the estimated contract value of the Permanent Works executed up to the end of that month

(b) a list of any goods or materials delivered to the Site for but not yet incorporated in the Permanent Works and their value

(c) a list of any of those goods or materials identified in the Appendix to the Form of Tender which have not yet been delivered to the Site but of which the property has vested in the Employer pursuant to Clause 54 and their value and

(d) the estimated amounts to which the Contractor considers himself entitled in connection with all other matters for which provision is made under the Contract including any Temporary Works or Contractor's Equipment for which separate amounts are included in the Bill of Quantities

unless in the opinion of the Contractor such values and amounts together will not justify the issue of an interim certificate.

Amounts payable in respect of Nominated Sub-contracts are to be listed separately.

Monthly payments

(2) Within 28 days of the date of delivery to the Engineer or Engineer's Representative in accordance with sub-clause (1) of

88 Or, where the Sub-Contractor cannot pay, from the Employer under Clause 59(4)(e).

this Clause of the Contractor's monthly statement the Engineer shall certify and the Employer shall pay to the Contractor (after deducting any previous payments on account)

(a) the amount which in the opinion of the Engineer on the basis of the monthly statement is due to the Contractor on account of sub-clauses (1)(a) and (1)(d) of this Clause less a retention as provided in sub-clause (5) of this Clause and

(b) such amounts (if any) as the Engineer may consider proper (but in no case exceeding the percentage of the value stated in the Appendix to the Form of Tender) in respect of sub-clauses (1)(b) and (1)(c) of this Clause.

The amounts certified in respect of Nominated Sub-contracts shall be shown separately in the certificate.

Minimum amount of certificate

(3) Until the whole of the Works has been certified as substantially complete in accordance with Clause 48 the Engineer shall not be bound to issue an interim certificate for a sum less than that stated in the Appendix to the Form of Tender but thereafter he shall be bound to do so and the certification and payment of amounts due to the Contractor shall be in accordance with the time limits contained in this Clause.

Final account

(4) Not later than 3 months after the date of the Defects Correction Certificate the Contractor shall submit to the Engineer a statement of final account and supporting documentation showing in detail the value in accordance with the Contract of the Works executed together with all further sums which the Contractor considers to be due to him under the Contract up to the date of the Defects Correction Certificate.

Within 3 months after receipt of this final account and of all information reasonably required for its verification the Engineer shall issue a certificate stating the amount which in his opinion is finally due under the Contract from the Employer to the Contractor or from the Contractor to the Employer as the case may be up to the date of the Defects Correction Certificate and after giving credit to the Employer for all amounts previously paid by the Employer and for all sums to which the Employer is entitled under the Contract.

Such amount shall subject to Clause 47 be paid to or by the Contractor as the case may require within 28 days of the date of the certificate.

Retention

(5) The retention to be made pursuant to sub-clause (2)(a) of this Clause shall be the difference between

(a) an amount calculated at the rate indicated in and up to the limit set out in the Appendix to the Form of Tender upon the amount due to the Contractor on account of sub-clauses (1)(a) and (1)(d) of this Clause and

(b) any payment which shall have become due under sub-clause (6) of this Clause.

Payment of retention

(6) (a) Upon the issue of a Certificate of Substantial Completion in respect of any Section or part of the Works there shall become due to the Contractor one half of such proportion of the retention money deductible to date under sub-clause (5)(a) of this Clause as the value of the Section or part bears to the value of the whole of the Works completed to date as certified under sub-clause (2)(a) of this Clause and such amount shall be added to the amount next certified as due to the Contractor under sub-clause (2) of this Clause.

The total of the amounts released shall in no event exceed one half of the limit of retention set out in the Appendix to the Form of Tender.

(b) Upon issue of the Certificate of Substantial Completion in respect of the whole of the Works there shall become due to the Contractor one half of the retention money calculated in accordance with sub-clause (5)(a) of this Clause. The amount so due (or the balance thereof over and above such payments already made pursuant to sub-clause (6)(a) of this Clause) shall be paid within 14 days of the issue of the said Certificate.

(c) Upon the expiry of the Defects Correction Period or if more than one the last of such periods the remainder of the retention money shall be paid to the Contractor within 14 days notwithstanding that at that time there may be outstanding claims by the Contractor against the Employer.

Provided that if at that time there remains to be executed by the Contractor any outstanding work referred to under Clause 48 or any work ordered pursuant to Clauses 49 or 50 the Employer may withhold payment until the completion of such work of so much of the said remainder as shall in the opinion of the Engineer represent the cost of the work remaining to be executed.

Interest on overdue payments

(7) In the event of

(a) failure by the Engineer to certify or the Employer to make payment in accordance with sub-clauses (2) (4) or (6) of this Clause or

(b) any finding of an arbitrator to such effect

the Employer shall pay to the Contractor interest compounded monthly for each day on which any payment is overdue or which should have been certified and paid at a rate equivalent to 2% per annum above the base lending rate of the bank specified in the Appendix to the Form of Tender.

If in an arbitration pursuant to Clause 66 the arbitrator holds that any sum or additional sum should have been certified by a particular date in accordance with the aforementioned sub-clauses but was not so certified this shall be regarded for the purposes of this sub-clause as a failure to certify such sum or additional sum. Such sum or additional sum shall be regarded as overdue for payment 28 days after the date by which the arbitrator holds that the Engineer should have certified the sum or if no such date is identified by the arbitrator shall be regarded as overdue for payment from the date of the Certificate of Substantial Completion for the whole of the Works.

Correction and withholding of certificates

(8) The Engineer shall have power to omit from any certificate the value of any work done goods or materials supplied or services rendered with which he may for the time being be dissatisfied and for that purpose or for any other reason which to him may seem proper may by any certificate delete correct or modify any sum previously certified by him. Provided that

(a) the Engineer shall not in any interim certificate delete or reduce any sum previously certified in respect of work done goods or materials supplied or services rendered by a Nominated Sub-contractor if the Contractor shall have already paid or be bound to pay that sum to the Nominated Sub-contractor and

(b) if the Engineer in the final certificate shall delete or reduce any sum previously certified in respect of work done goods or materials supplied or services rendered by a Nominated Sub-contractor which sum shall have been already paid by the Contractor to the Nominated Sub-contractor the Employer shall reimburse to the Contractor the amount of any sum overpaid by the Contractor to the Sub-contractor in accordance with the certificates issued under sub-clause (2) of this Clause which the Contractor shall be unable to recover from the Nominated Sub-contractor together with interest thereon at the rate stated in sub-clause (7) of this Clause from 28 days after the date of the final certificate issued under sub-clause (4) of this Clause until the date of such reimbursement.

Copy of certificate for Contractor

(9) Every certificate issued by the Engineer pursuant to this Clause shall be sent to the Employer and at the same time copied to the Contractor with such detailed explanation as may be necessary.

Payment advice

(10) Where a payment made in accordance with sub-clause (2) of this Clause differs in any respect from the amount certified by the Engineer the Employer shall notify the Contractor forthwith with full details showing how the amount being paid has been calculated.

Synopsis

The Contractor submits monthly statements in accordance with Clause 60(1). These statements (frequently called "interim applications"), list all the work done to date. Within 28 days of the delivery of the statement, the Engineer certifies and the Employer pays the amount which, in the opinion of the Engineer, is due: Clause 60(2). The amount due is the value of work calculated in accordance with the Contract less (*a*) any sums already paid on account and (*b*) an element for retentions, as set out in Clauses 60(5) and (6). In the event that payment is late, the Contractor is entitled to interest: Clause 60(7). Following the issue of the Defects Correction Certificate the Contractor submits a final account: Clause 60(4).

The monthly statement

Clause 60(1) sets out the headings which should be used for compiling the monthly statement. Any claims will be included under Clause 60(1)(d).

The Engineer's valuation and certificate

The Engineer is required to compute the amount due and to certify that sum in sufficient time to enable the Employer to pay it within 28 days of the delivery of the monthly statement. As to items applied for in accordance with Clause 60(1)(a), the Engineer must satisfy himself that the rates and quantities are correct. As to items under Clauses 60(1)(b), (c) the Engineer must satisfy himself as to the quantities, values and state of security of goods and materials and come to a view as to the "proper" amount to be paid. As to items claimed under Clause 60(1)(d) the Engineer must come to a view as to the Contractor's entitlement under the Contract, which may require him to evaluate claims. Generally speaking, the issue of a certificate is a condition precedent to payment;[89] in other words, the Contractor will not be entitled to be paid any sum unless that sum is certified. Where, however, there has been any interference or duplicity in relation to the issue of certificates, a certificate will not be a condition precedent.[90]

Payment by the Employer

The Employer is required to pay the certified sum within 28 days of the delivery of the monthly application: Clause 60(2). The express terms do not derogate from the Employer's right to set-off for breaches by the Contractor; accordingly, where the Contractor is in breach the Employer may evaluate any damages (including liquidated damages) which flow and deduct them from the sum certified.

Final account—Clause 60(4)

Within three months of the date of the Defects Correction Certificate, the Contractor submits "a statement of final account". This may be a statement in the same format as the monthly statements. Within three months of the receipt of the statement, the Engineer issues a certificate stating the sum due from the Employer to the Contractor or vice versa. It is thought that the Engineer should not take into account any damages for breach of contract which the Employer claims, but should confine himself to entitlements "under the Contract"; the Employer's right to set-off damages for breach when paying for certified sums remains.

---

89 *Costain Building and Civil Engineering Ltd* v. *Scottish Rugby Union plc* (1993) 69 BLR 85 (Scottish Court of Session, Inner House): a case on the 5th Edition Conditions of Contract.

90 As to certificates generally see Chapter 3, section 5.

Retentions—Clauses 60(5), (6)

Sums are retained in order that the Employer has a fund to call upon in the event that the Contractor defaults and the Employer must employ others to remedy defects or to complete work. Clause 60(5) sets out the amount of retention to be withheld. Clause 60(6) sets out the timetable for its repayment; in short, half is paid when each of the defined completion certificates (i.e. the Certificate of Substantial Completion and the Defects Correction Certificate) is issued.

Interest—Clause 60(7)

Claims[91] under civil engineering contracts may not be resolved for many years. Where a party has been kept out of his money for a long period, the interest may be very substantial. Interest clauses in previous editions of the ICE Conditions of Contract have caused many problems; issues have included whether interest is payable on a compound or simple basis and whether certification which is merely inadequate entitled the Contractor to any interest at all.[92] The present Clause 60(7) is designed to provide the Contractor with a right to payment of compound interest from the date when the principal sum to which it relates should have been paid. It is not clear whether or not it achieves this objective.[93] Assuming that it is effective, the Contractor may include for interest in his monthly statements in accordance with Clause 60(1)(d) and he may update his interest claim in each interim application; caution must be exercised, however, to ensure that the Contractor is not double recovering on interest.

---

91 Including defects claims by Employers.

92 See e.g. *Secretary of State for Transport* v. *Birse-Farr Joint Venture* (1993) 62 BLR 36.

93 The clause is not clearly drafted. Where the Employer pays all sums actually certified, the operation of the clause is alternatively conditioned upon (*a*) failure by the Engineer to certify or (*b*) a finding of the arbitrator to such effect. Where the Engineer issues a certificate, albeit for a smaller sum than is properly due, it seems that condition (*a*) is not met and there is no evidence for condition (*b*). See *Secretary of State for Transport* v. *Birse-Farr Joint Venture* (1993) 62 BLR 36 and other associated cases for the question of "failure to certify".

Correcting certificates—Clause 60(8)

A certificate under this form of contract has no significant evidential weight and the Engineer may revise and correct earlier certificates and deduct any monies overpaid at a later stage.[94]

61 Defects Correction Certificate

(1) Upon the expiry of the Defects Correction Period or where there is more than one such period upon the expiration of the last of such periods and when all outstanding work referred to under Clause 48 and all work of repair amendment reconstruction rectification and making good of defects imperfections shrinkages and other faults referred to under Clauses 49 and 50 shall have been completed the Engineer shall issue to the Employer (with a copy to the Contractor) a Defects Correction Certificate stating the date on which the Contractor shall have completed his obligations to construct and complete the Works to the Engineer's satisfaction.

Unfulfilled obligations

(2) The issue of the Defects Correction Certificate shall not be taken as relieving either the Contractor or the Employer from any liability the one towards the other arising out of or in any way connected with the performance of their respective obligations under the Contract.

The Defects Correction Certificate is an administrative step. It has no evidential status, but merely triggers payment of the second half of the retention money and sets in motion the procedure under Clause 60(4) in relation to the certificate as to the final sums due.

REMEDIES AND POWERS

62 Urgent repairs

If by reason of any accident or failure or other event occurring to in or in connection with the Works or any part thereof either during the execution of the Works or during the Defects Correction Period any

94 See *Mears Construction* v. *Samuel Williams* (1977) 16 BLR 49 (4th Edition of the ICE Conditions of Contract).

remedial or other work or repair shall in the opinion of the Engineer or the Engineer's Representative be urgently necessary and the Contractor is unable or unwilling at once to do such work or repair the Employer may by his own or other workpeople do such work or repair. If the work or repair so done by the Employer is work which in the opinion of the Engineer the Contractor was liable to do at his own expense under the Contract all costs and charges properly incurred by the Employer in so doing shall on demand be paid by the Contractor to the Employer or may be deducted by the Employer from any monies due or which may become due to the Contractor. Provided that the Engineer shall as soon after the occurrence of any such emergency as may be reasonably practicable notify the Contractor thereof in writing.

The Contract scheme is that the Contractor is generally obliged and entitled to undertake remedial and other such works during the Defects Correction Period. In emergencies this may not always be practicable. Clause 62 entitles the Employer to make other arrangements in emergencies.

63 Determination of the Contractor's employment

(1) If

(a) the Contractor shall be in default in that he

(i) becomes bankrupt or has a receiving order or administration order made against him or presents his petition in bankruptcy or makes an arrangement with or assignment in favour of his creditors or agrees to carry out the Contract under a committee of inspection of his creditors or (being a corporation) goes into liquidation (other than a voluntary liquidation for the purposes of amalgamation or reconstruction) or

(ii) assigns the Contract without the consent in writing of the Employer first obtained or

(iii) has an execution levied on his goods which is not stayed or discharged within 28 days

or

(b) the Engineer certifies in writing to the Employer with a copy to the Contractor that in his opinion the Contractor

(i) has abandoned the Contract or

(ii) without reasonable excuse has failed to commence the Works in accordance with Clause 41 or has suspended the progress of the Works for 14 days after receiving from the Engineer written notice to proceed or

(iii) has failed to remove goods or materials from the Site or to pull down and replace work for 14 days after receiving from the Engineer written notice that the said goods materials or work have been condemned and rejected by the Engineer or

(iv) despite previous warnings by the Engineer in writing is failing to proceed with the Works with due diligence or is otherwise persistently or fundamentally in breach of his obligations under Contract

then the Employer may after giving 7 days' notice in writing to the Contractor specifying the default enter upon the Site and the Works and expel the Contractor therefrom without thereby avoiding the Contract or releasing the Contractor from any of his obligations or liabilities under the Contract. Provided that the Employer may extend the period of notice to give the Contractor opportunity to remedy the default.

Where a notice of determination is given pursuant to this sub-clause it shall be given as soon as is reasonably possible after receipt of the Engineer's certificate.

Completing the Works

(2) Where the Employer has entered upon the Site and the Works as hereinbefore provided he may himself complete the Works or may employ any other contractor to complete the Works and the Employer or such other contractor may use for such completion so much of the Contractor's Equipment Temporary Works goods and materials which have been deemed to become the property of the Employer under Clauses 53 and 54 as he or they may think proper and the Employer may at any time sell any of the said Contractor's Equipment Temporary Works and unused goods and materials and apply the proceeds of sale in or towards the satisfaction of any sums due or which may become due to him from the Contractor under the Contract.

Assignment to Employer

(3) By the said notice or by further notice in writing within 7 days of the date of expiry thereof the Engineer may require the Contractor to assign to the Employer and if so required the Contractor shall forthwith assign to the Employer the benefit of any agreement for the supply of any goods or materials and/or for the execution of any work for the purposes of this Contract which the Contractor may have entered into.

Payment after determination

(4) If the Employer enters and expels the Contractor under this Clause he shall not be liable to pay to the Contractor any money on account of the Contract until the expiration of the Defects Correction Period and thereafter until the costs of completion damages for delay in completion (if any) and all other expenses incurred by the Employer have been ascertained and the amount thereof certified by the Engineer.

The Contractor shall then be entitled to receive only such sum or sums (if any) as the Engineer may certify would have been due to him upon due completion by him after deducting the said amount. But if such amount shall exceed the sum which would have been payable to the Contractor on due completion by him then the Contractor shall upon demand pay to the Employer the amount of such excess and it shall be deemed a debt due by the Contractor to the Employer and shall be recoverable accordingly.

Valuation at date of determination

(5) As soon as may be practicable after any such entry and expulsion by the Employer the Engineer shall fix and determine as at the time of such entry and expulsion

(a) the amount (if any) which had been reasonably earned by or would reasonably accrue to the Contractor in respect of work actually done by him under the Contract and

(b) the value of any unused or partially used goods and materials and any Contractor's Equipment and Temporary Works which had been deemed to become the property of the Employer under Clauses 53 and 54

and shall certify accordingly.

The said determination may be carried out ex parte or by or after reference to the parties or after such investigation or enquiry as the Engineer may think fit to make or institute.

Synopsis

Clause 63 entitles the Employer to terminate the Contract on 7 days notice if:

(*a*) specified events defined in Clause 63(1)(a) have occurred; or
(*b*) the Engineer certifies that the Contractor has committed any of the defaults defined in Clause 63(1)(b).

Upon receipt of the notice, the Contractor removes himself[95] from the site—Clause 63(1). He assigns to the Employer the benefit of any supply agreements—Clause 63(3).[96] The Employer may employ others to complete the Works—Clause 63(2). The Contractor is entitled to the value of the Works done until the date of termination less the additional cost to which the Employer has been put. In any event the Employer is not obliged to pay the Contractor anything until the expiry of the Defects Correction Period—Clause 63(4).[97]

Repudiation generally

Clause 63 provides for the determination of the Contractor's employment. It is not expressed to be exclusive and so it is thought that the rights of either party to elect to treat the obligation to perform the Contract as being at an end are preserved.[98]

The Engineer's certificate

Clause 63(1)(b) sets out a number of circumstances which might be reckoned as repudiations of contract in any event. The definition of these grounds, however, gives the Employer a good deal of security, since repudiation cannot be readily recognised in practice. Before issuing the certificate the Engineer should carefully consider the terms of Clause 63(1)(b) since each sub-clause contains conditions which must be met to avoid causing the certificate to be defective; for example in (iv) the phrase "warnings ... in writing ..." suggests that at least two written warnings have been issued. Caution should always be exercised when determining any contract, and an abundance of caution is always advisable.

---

95 Clause 63(1) in fact talks of the Employer expelling the Contractor. This is rarely required, though in at least one instance a Contractor has sought an injunction to entitle him to remain on site: *Tara Civil Engineering Ltd* v. *Moorfield Developments Ltd* (1989) 46 BLR 72. See, generally, Chapter 3, section 8 for a discussion on the contractor's licence to remain on site.

96 See also Clauses 53 and 54 which provide for the transfer of property in equipment and materials.

97 Where this is some time off, the Contractor may commence arbitration proceedings and obtain an award in his favour. Where the award is to the effect that the determination was wrong, such award will overreach the provisions of Clause 63(4) and will entitle the Contractor to be paid immediately.

98 See Chapter 2, section 7.

The timing of the notice

The clause provides that the notice must be given "as soon as is reasonably possible after receipt of the Engineer's certificate". This cannot apply to determinations which are dependent on Clause 63(1)(a) for which no Engineer's certificate is required. It is thought that the expression "Engineer's certificate" should be extended to read "... or notice or knowledge that the relevant event in Clause 63(1)(a) has occurred". Such a requirement accords with the general principle that the innocent party is entitled to elect whether or not to treat the contract as being at an end within a reasonable time; failing an election to determine the contract, the contract is affirmed.[99] Where the Contractor commits a new default, the Employer will have a renewed opportunity to terminate.

The content of the notice

The notice must: (*a*) give 7 or more days' notice; (*b*) specify the defaults. It is appropriate to set out the fact that the notice is served pursuant to Clause 63 and to specify the grounds of default together with relevant clause numbers and a succinct narrative setting out the facts which support the grounds.

Due diligence—Clause 63(1)(b)(iv)

It is thought that due diligence refers to a rate of progress and industriousness which is appropriate taking into account the currently accepted Clause 14 programme, the progress made to date and any difficulties being encountered. A number of related expressions have been considered by the courts. "Regularly and diligently" is said to "convey a sense of activity, of orderly progress".[100] "Reasonable diligence" is said to mean "not only the personal industriousness of the defendant [contractor] himself but his efficiency and that of those who work for him".[101]

99 *Vitol SA* v. *Norelf Ltd* [1995] 3 WLR 549 (CA).

100 *Hounslow Borough Council* v. *Twickenham Garden Developments Ltd* [1971] ChD 233. See also *West Faulkner Associates* v. *London Borough of Newham* (1994) 71 BLR I (CA).

101 *Hooker Constructions* v. *Chris's Engineering Contracting Company* [1970] ALR 821 (Supreme Court of the Northern Territory of Australia).

FRUSTRATION

64 Payment in event of frustration

In the event of the Contract being frustrated whether by war or by any other supervening event which may occur independently of the will of the parties the sum payable by the Employer to the Contractor in respect of the work executed shall be the same as that which would have been payable under Clause 65(5) if the Contract had been determined by the Employer under Clause 65.

This clause deals with the same subject matter as frustration at common law. The position as regards payment upon frustration is specified.

WAR CLAUSE

65 Works to continue for 28 days on outbreak of war

(1) If during the currency of the Contract there shall be an outbreak of war (whether war is declared or not) in which Great Britain shall be engaged on a scale involving general mobilization of the armed forces of the Crown the Contractor shall for a period of 28 days reckoned from midnight on the date that the order for general mobilization is given continue so far as is physically possible to execute the Works in accordance with the Contract.

Effect of completion within 28 days

(2) If at any time before the expiration of the said period of 28 days the Works shall have been completed or completed so far as to be usable all provisions of the Contract shall continue to have full force and effect save that

(a) the Contractor shall in lieu of fulfilling his obligations under Clauses 49 and 50 be entitled at his option to allow against the sum due to him under the provisions hereof the cost (calculated at the prices ruling at the beginning of the said period of 28 days) as certified by the Engineer at the expiration of the Defects Correction Period of repair rectification and making good any work for the repair rectification or making good of which the Contractor would have been liable under the said Clauses had they continued to be applicable

(b) the Employer shall not be entitled at the expiry of the Defects Correction Period to withhold payment under Clause 60(6)(c) of the second half of the retention money or any part thereof except such sum as may be allowable by the Contractor under the provisions of the last preceding paragraph which sum may (without prejudice to any other mode of recovery thereof) be deducted by the Employer from such second half.

Right of Employer to determine Contract

(3) If the Works shall not have been completed as aforesaid the Employer shall be entitled to determine the Contract (with the exception of this Clause and Clauses 66 and 68) by giving notice in writing to the Contractor at any time after the aforesaid period of 28 days has expired and upon such notice being given the Contract shall (except as above mentioned) forthwith determine but without prejudice to the claims of either party in respect of any antecedent breach thereof.

Removal of Contractor's Equipment on determination

(4) If the Contract shall be determined under the provisions of the last preceding sub-clause the Contractor shall with all reasonable despatch remove from the Site all his Contractor's Equipment and shall give facilities to his sub-contractors to remove similarly all Contractor's Equipment belonging to them and in the event of any failure so to do the Employer shall have the like powers as are contained in Clause 53(3) in regard to failure to remove Contractor's Equipment on completion of the Works but subject to the same condition as is contained in Clause 53(2).

Payment on determination

(5) If the Contract shall be determined as aforesaid the Contractor shall be paid by the Employer (insofar as such amounts or items shall not have been already covered by payment on account made to the Contractor) for all work executed prior to the date of determination at the rates and prices provided in the Contract and in addition

(a) the amounts payable in respect of any preliminary items so far as the work or service comprised therein has been carried out or performed and a proper proportion as certified by the Engineer of any such items the work or service comprised in which has been partially carried out or performed

(b) the cost of materials or goods reasonably ordered for the Works which have been delivered to the Contractor or of which the Contractor is legally liable to accept delivery (such materials or goods becoming the property of the Employer upon such payment being made by him)

(c) a sum to be certified by the Engineer being the amount of any expenditure reasonably incurred by the Contractor in the expectation of completing the whole of the Works in so far as such expenditure shall not have been covered by the payments in this sub-clause before mentioned

(d) any additional sum payable under sub-clauses (6)(b)(c) and (d) of this Clause and

(e) the reasonable cost of removal under sub-clause (4) of this Clause.

Provisions to apply as from outbreak of war

(6) Whether the Contract shall be determined under the provisions of sub-clause (3) of this Clause or not the following provisions shall apply or be deemed to have applied as from the date of the said outbreak of war notwithstanding anything expressed in or implied by the other terms of the Contract viz

(a) The Contractor shall be under no liability whatsoever by way of indemnity or otherwise for or in respect of damage to the Works or to property (other than property of the Contractor or property hired by him for the purposes of executing the Works) whether of the Employer or of third parties or for or in respect of injury or loss of life to persons which is the consequence whether direct or indirect of war hostilities (whether war has been declared or not) invasion act of the Queen's enemies civil war rebellion revolution insurrection military or usurped power and the Employer shall indemnify the Contractor against all such liabilities and against all claims demands proceedings damages costs charges and expenses whatsoever arising thereout or in connection therewith.

(b) If the Works shall sustain destruction or any damage by reason of any of the causes mentioned in the last preceding paragraph the Contractor shall nevertheless be entitled to payment for any part of the Works so destroyed or damaged and the Contractor shall be entitled to be paid by the Employer the cost of making good any such destruction or damage so far as may be required by the Engineer or as may be necessary for the completion of the Works on a cost basis plus such profit as the Engineer may certify to be reasonable.

(c) In the event that the Contract includes the Contract Price Fluctuations Clause the terms of that Clause shall continue to apply but if subsequent to the outbreak of war the index figures therein referred to shall cease to be published or in the event that the Contract shall not include a Contract Price Fluctuations Clause in that form the following paragraph shall have effect:

If under decision of the Civil Engineering Construction Conciliation Board or of any other body recognized as an appropriate body for regulating the rates of wages in any trade or industry other than the Civil Engineering Construction Industry to which Contractors undertaking works of civil engineering construction give effect by agreement or in practice or by reason of any Statute or Statutory Instrument there shall during the currency of the Contract be any increase or decrease in the wages or the rates of wages or in the allowances or rates of allowances (including allowances in respect of holidays) payable to or in respect of labour of any kind prevailing at the date of outbreak of war as then fixed by the said Board or such other body as aforesaid or by Statute or Statutory Instrument or any increase in the amount payable by the Contractor by virtue or in respect of any Scheme of State Insurance or if there shall be any increase or decrease in the cost prevailing at the date of the said outbreak of war of any materials consumable stores fuel or power (and whether for permanent or temporary works) which increase or increases decrease or decreases shall result in an increase or decrease of cost to the Contractor in carrying out the Works the net increase or decrease of cost shall form an addition or deduction as the case may be to or from the Contract Price and be paid to or allowed by the Contractor accordingly.

(d) If the cost of the Works to the Contractor shall be increased or decreased by reason of the provisions of any Statute or Statutory Instrument or other Government or Local Government Order or Regulation becoming applicable to the Works after the date of the said outbreak of war or by reason of any trade or industrial agreement entered in to after such date to which the Civil Engineering Construction Conciliation Board or any other body as aforesaid is party or gives effect or by reason of any amendment of whatsoever nature of the Working Rule Agreement of the said Board or of any other body as aforesaid or by reason of any other circumstance or thing attributable to or consequent on such outbreak of war such increase or decrease of cost as certified by the Engineer shall be reimbursed by the Employer to the Contractor or allowed by the Contractor as the case may be.

(e) Damage or injury caused by the explosion whenever occurring of any mine bomb shell grenade or other projectile missile or munition of war and whether occurring before or after the cessation of hostilities shall be deemed to be the consequence of any of the events mentioned in sub-clause (6)(a) of this Clause.

SETTLEMENT OF DISPUTES

66 Settlement of disputes

(1) Except as otherwise provided in these Conditions if a dispute of any kind whatsoever arises between the Employer and the Contractor in connection with or arising out of the Contract or the carrying out of the Works including any dispute as to any decision opinion instruction direction certificate or valuation of the Engineer (whether during the progress of the Works or after their completion and whether before or after the determination abandonment or breach of the Contract) it shall be settled in accordance with the following provisions.

Notice of Dispute

(2) For the purpose of sub-clauses (2) to (6) inclusive of this Clause a dispute shall be deemed to arise when one party serves on the Engineer a notice in writing (hereinafter called the Notice of Dispute) stating the nature of the dispute. Provided that no Notice of Dispute may be served unless the party wishing to do so has first taken any steps or invoked any procedure available elsewhere in the Contract in connection with the subject matter of such dispute and the other party or the Engineer as the case may be has

(a) taken such step as may be required or

(b) been allowed a reasonable time to take any such action.

Engineer's decision

(3) Every dispute notified under sub-clause (2) of this Clause shall be settled by the Engineer who shall state his decision in

writing and give notice of the same to the Employer and the Contractor within the time limits set out in sub-clause (6) of this Clause.

Effect on Contractor and Employer

(4) Unless the Contract has already been determined or abandoned the Contractor shall in every case continue to proceed with the Works with all due diligence and the Contractor and the Employer shall both give effect forthwith to every such decision of the Engineer. Such decisions shall be final and binding upon the Contractor and the Employer unless and until as hereinafter provided either

(a) the recommendation of a conciliator has been accepted by both parties or

(b) the decision of the Engineer is revised by an arbitrator and an award made and published.

Conciliation

(5) In relation to any dispute notified under sub-clause (2) of this Clause and in respect of which

(a) the Engineer has given his decision or

(b) the time for giving an Engineer's decision as set out in sub-clause (3) of this Clause has expired

and no Notice to Refer under sub-clause (6) of this Clause has been served either party may within one calendar month after receiving notice of such decision of within one calendar month after the expiry of the said period give notice in writing requiring the dispute to be considered under the Institution of Civil Engineers' Conciliation Procedure (1988) or any amendment or modification thereof being in force at the date of such notice and the dispute shall thereafter be referred and considered in accordance with the said Procedure. The recommendation of the conciliator shall be deemed to have been accepted in settlement of the dispute unless a written Notice to Refer under sub-clause (6) of this Clause is served within one calendar month of its receipt.

Arbitration

(6) (a) Where a Certificate of Substantial Completion of the whole of the Works has not been issued and either

(i) the Employer or the Contractor is dissatisfied with any decision of the Engineer given under sub-clause (3) of this Clause or

(ii) the Engineer fails to give such decision for a period of one calendar month after the service of the Notice of Dispute or

(iii) the Employer or the Contractor is dissatisfied with any recommendation of a conciliator appointed under sub-clause (5) of this Clause

then either the Employer or the Contractor may within 3 calendar months after receiving notice of such decision or within 3 calendar months after the expiry of the said period of one month or within one calendar month of receipt of the conciliator's recommendation (as the case may be) refer the dispute to the arbitration of a person to be agreed upon by the parties by serving on the other party a written Notice to Refer.

(b) Where a Certificate of Substantial Completion of the whole of the Works has been issued the foregoing provisions shall apply save that the said period of one calendar month referred to in (a)(ii) above shall be read as 3 calendar months.

President or Vice-President to act

(7) (a) If the parties fail to appoint an arbitrator within one calendar month of either party serving on the other party written Notice to Concur in the appointment of an arbitrator the dispute or difference shall be referred to a person to be appointed on the application of either party by the President for the time being of the Institution of Civil Engineers.

(b) If an arbitrator declines the appointment or after appointment is removed by order of a competent court or is incapable of acting or dies and the parties do not within one calendar month of the vacancy arising fill the vacancy then either party may apply to the President for the time being of the Institution of Civil Engineers to appoint another arbitrator to fill the vacancy.

(c) In any case where the President for the time being of the Institution of Civil Engineers is not able to exercise the functions conferred on him by this Clause the said functions shall be exercised on his behalf by a Vice-President for the time being of the said Institution.

Arbitration—procedure and powers

(8) (a) Any reference to arbitration under this Clause shall be deemed to be a submission to arbitration within the meaning of the Arbitration Acts 1950 to 1979 or any statutory re-enactment or amendment thereof for the time being in force. The reference shall be conducted in accordance with the Institution of Civil Engineers' Arbitration Procedure (1983) or any amendment or modification thereof being in force at the time of the appointment of the arbitrator. Such arbitrator shall have full power to open up review and revise any decision opinion instruction direction certificate or valuation of the Engineer.

(b) Neither party shall be limited in the proceedings before such arbitrator to the evidence or arguments put before the Engineer for the purpose of obtaining his decision under sub-clause (3) of this Clause.

(c) The award of the arbitrator shall be binding on all parties.

(d) Unless the parties otherwise agree in writing any reference to arbitration may proceed notwithstanding that the Works are not then complete or alleged to be complete.

Engineer as witness

(9) No decision given by the Engineer in accordance with the foregoing provisions shall disqualify him from being called as witness and giving evidence before the arbitrator on any matter whatsoever relevant to the dispute or difference so referred to the arbitrator.

Synopsis

Clause 66 sets out a detailed procedure for the resolution of disputes. It provides for three distinct dispute resolution procedures:

(*a*) an Engineer's formal decision under Clause 66(3)
(*b*) conciliation pursuant to Clause 66(5)
(*c*) arbitration pursuant to Clauses 66(6)–(9).

The clause provides that the decision in (*a*) (or the Engineer's failure to give a decision within the specified time) is a condition precedent to the right to have the matter conciliated[102] or arbitrated. This commentary should be read in conjunction with the commentaries on the associated Arbitration and Conciliation Procedures in Chapter 12.

The nature of disputes which are dealt with in Clause 66

Clause 66(1) purports to apply to "a dispute of any kind whatsoever [which] arises ... in connection with or arising out of the Contract or the carrying out of the Works ... [etc.]". This is extremely wide. It is thought that the Engineer, Conciliator and/or Arbitrator will have jurisdiction in respect of any dispute between the Employer and Contractor in relation to the Contract or project save one in which a necessary element is a determination as to whether Clause 66 is part of the agreement.[103]

A valid Engineer's decision

The question of whether or not the Engineer has made a valid decision in accordance with Clause 66(3) frequently arises. Until such a decision is made in respect of any matter (or not made within the allowed time) that matter may not be referred to conciliation or arbitration. Equally importantly, when a decision has been made in respect of any matter, a party who is dissatisfied with the decision is entitled to have the decision reviewed during a strictly limited time.[104] The key elements are as follows.

(*a*) *A dispute.* A matter may be referred to the Engineer for his formal decision when a dispute arises. A dispute is deemed to arise when a Notice of Dispute is served in accordance with Clause 66 (2). Such notice in ineffective, however, until the serving party has exhausted other procedures or has allowed the other party (or

102 The statement in the text as regards conciliation is the formal position. Conciliation is unlikely to work unless the parties are committed, so considerations of "conditions precedent" are perhaps irrelevant.

103 For example a claim for quantum meruit on the grounds that no contract exists, or a claim for rectification to exclude the arbitration clause. As to the extent of an arbitrator's jurisdiction generally see Chapter 12.

104 But see Section 27 of the Arbitration Act 1950 and Section 12 of the Arbitration Act 1996 by which the court has a discretionary power to extend such time limits: this power is exercised sparingly. See *McLaughlin & Harvey plc* v. *P & O Developments* (1991) 55 BLR 101.

the Engineer) time to take any appropriate action. It is thought that no matter may be included in a Notice of Dispute unless the Engineer has had an opportunity of considering the claim and has indicated, expressly or by a refusal to acknowledge it, that he rejects it. Where the matter is plainly the subject of a dispute, the Engineer may not attempt to maintain the fiction that there is no dispute by, for instance, persistently asking for additional substantiation under Clause 52(4)(d).

(*b*) *The format of the Notice.* The Notice of Dispute must be in writing. It must also state "the nature of the dispute". The meaning of the latter expression is not easy to discern. In order to ensure compliance with this requirement, it is suggested that the party serving the notice should state the background facts upon which he relies and outline his broad interpretation of the consequences for which he argues. In this way the extent of the dispute can be more readily ascertained. Where, however, the nature of the dispute is very briefly stated, this will not render the notice ineffective where the recipient is aware of the meaning and context of the notice.

(*c*) *The timing of the decision.*[105] Where a Notice of Dispute is served prior to the Certificate of Substantial Completion the Engineer must give his decision within one month. Where the Certificate of Substantial Completion has been issued, the decision must be made within three months—see Clauses 66(3) and (6).

(*d*) *The format of the decision.* The Engineer is not required to give his decision in any particular format, save that it be in writing. He need not afford the parties a hearing, though he is entitled to seek their views and to collect further information as he considers proper. A decision under Clause 66(3) cannot be delegated,[106] however, the Engineer may use the assistance of others providing he takes the final decision himself.[107] The consequences which flow from his having made a decision are important and for this reason it is thought that a formal decision in accordance with Clause 66(3) must plainly be such a

105 See generally *EEC Quarries Ltd* v. *Merriman Ltd* (1988) 45 BLR 90.

106 See Clause 2(4).

107 *Anglian Water Authority* v. *RDL Contracting* (1988) 43 BLR 98.

decision.[108] The majority of Engineers use the formula "Engineer's decision in accordance with Clause 66 of the Conditions of Contract" or similar words; this is to be encouraged.

(e) *Effect of the decision on the parties.* The parties are to continue with the Works and to give effect to the decision of the Engineer unless that decision is revised by agreement or an arbitrator's award—Clause 66(4).

Conciliation and arbitration

Following the Engineer's decision (or failure to make a decision within the time limit) either party may serve a notice of conciliation or arbitration in accordance with Clauses 66(5) and/or (6). It seems that the first notice in time will be effective. Thus where the Engineer issues his decision and one party immediately issues a Notice to Refer to arbitration, the possibility of conciliation is lost. A detailed review of conciliation and arbitration under the respective ICE Procedures is given in Chapter 12.

APPLICATION TO SCOTLAND AND NORTHERN IRELAND

67 Application to Scotland

(1) If the Works are situated in Scotland the Contract shall in all respects be construed and operate as a Scottish contract and shall be interpreted in accordance with Scots Law and the provisions of this Clause shall apply.

(2) In the application of these Conditions and in particular Clause 66 thereof

(a) the word "arbiter" shall be substituted for the word "arbitrator"

(b) for any reference to the "Arbitration Acts" there shall be substituted reference to the "Arbitration (Scotland) Act 1894"

108 *Monmouthshire County Council* v. *Costelloe and Kemple* (1965) 5 BLR 83 (CA).

(c) for any reference to the Institution of Civil Engineers Arbitration Procedure (1983) there shall be substituted a reference to the Institution of Civil Engineers Arbitration Procedure (Scotland) (1983) and

(d) notwithstanding any of the other provisions of these Conditions nothing therein shall be construed as excluding or otherwise affecting the right of a party to arbitration to call in terms of Section 3 of the Administration of Justice (Scotland) Act 1972 for the arbiter to state a case.

Application to Northern Ireland

(3) If the Works are situated in Northern Ireland the Contract shall in all respects be construed and operate as a Northern Irish contract and shall be interpreted in accordance with the law of Northern Ireland and the provisions of sub-clause (4) of this Clause shall apply.

(4) In the application of these Conditions and in particular Clause 66 thereof for any reference to the "Arbitration Acts" there shall be substituted reference to the "Arbitration (Northern Ireland) Act 1937".

NOTICES

68 Service of notices on Contractor

(1) Any notice to be given to the Contractor under the terms of the Contract shall be served in writing at the Contractor's principal place of business (or in the event of the Contractor being a Company to or at its registered office).

Service of notices on Employer

(2) Any notice to be given to the Employer under the terms of the Contract shall be served in writing at the Employer's last known address (or in the event of the Employer being a Company to or at its registered office).

Notices are required at many points in the Contract. Clause 68 requires that notices are to be in writing and specifies, in rather quaint terms, that they are to be served at the Contractor's "principal place of business" or the "Employer's last known address". The term "served" does not, it is thought, imply any technical legal meaning; rather the requirement is to bring the matter to the notice of the other party. Throughout the Conditions there are requirements as to notices: see e.g. Clauses 1(6) and Clause 52(4). Note also that some clauses specify slightly altered arrangements; for instance, Clause 2(4)(a) provides that notice of the Engineer's Representative may be given to the Contractor's agent. In any event, the question of waiver is likely to be important in many instances. Parties tend to have several addresses. It may be quite inconvenient to communicate with the "principal place of business". Where the parties have established alternative arrangements neither will be able to claim that a notice served in accordance with that practice was invalidly served unless they expressly indicate that they wish to revert to the strict terms of the Contract.

TAX MATTERS

69 Labour-tax fluctuations

(1) The rates and prices contained in the Bill of Quantities shall be deemed to take account of the levels and incidence at the date for return of tenders of the taxes levies contributions premiums or refunds (including national insurance contributions but excluding income tax and any levy payable under the Industrial Training Act 1964) which are by law payable by or to the Contractor and his sub-contractors in respect of their workpeople engaged on the Contract.

The rates and prices contained in the Bill of Quantities do not take account of any level or incidence of the aforesaid matters where at the date for return of tenders such level or incidence does not then have effect but although then known is to take effect at some later date.

(2) If after the date for return of tenders there shall occur any change in the level and/or incidence of any such taxes levies contributions premiums or refunds the Contractor shall so inform the Engineer and the net increase or decrease shall be taken into account in arriving at the Contract Price. The Contractor shall supply the information necessary to support any consequent adjustment to the Contract Price. All certificates for payment issued after submission of such information shall take due account of the additions or deductions to which such information relates.

70 Value Added Tax

(1) The Contractor shall be deemed not to have allowed in his tender for the tax payable by him as a taxable person to the Commissioners of Customs and Excise being tax chargeable on

any taxable supplies to the Employer which are to be made under the Contract.

Engineer's certificates net of Value Added Tax

(2) All certificates issued by the Engineer under Clause 60 shall be net of Value Added Tax.

In addition to the payments due under such certificates the Employer shall separately identify and pay to the Contractor any Value Added Tax properly chargeable by the Commissioners of Customs and Excise on the supply to the Employer of any goods and/or services by the Contractor under the Contract.

Disputes

(3) If any dispute difference or question arises between either the Employer or the Contractor and the Commissioners of Customs and Excise in relation to any tax chargeable or alleged to be chargeable in connection with the Contract or the Works each shall render to the other such support and assistance as may be necessary to resolve the dispute difference or question.

Clause 66 not applicable

(4) Clause 66 shall not apply to any dispute difference or question arising under this Clause.

THE CONSTRUCTION (DESIGN AND MANAGEMENT) REGULATIONS 1994

71 CDM Regulations 1994

(1) In this clause

(a) "the Regulations" means the Construction (Design and Management) Regulations 1994 or any statutory re-enactment or amendment thereof for the time being in force

(b) "Planning Supervisor" and "Principal Contractor" mean the persons so described in regulation 2(1) of the Regulations

(c) "Health and Safety Plan" means the plan prepared by virtue of regulation 15 of the Regulations.

(2) Where and to the extent that the Regulations apply to the Works and

(a) the Engineer is appointed Planning Supervisor and/or

(b) the Contractor is appointed Principal Contractor

then in taking any action as such they shall state in writing that the action is being taken under the Regulations.

(3) (a) Any action under the Regulations taken by either the Planning Supervisor or the Principal Contractor and in particular any alteration or amendment to the Health and Safety Plan shall be deemed to be an Engineer's instruction pursuant to Clause 13. Provided that the Contractor shall in no event be entitled to any additional payment and/or extension of time in respect of any such action to the extent that it results from any action lack of action or default on the part of the Contractor.

(b) If any such action of either the Planning Supervisor or the Principal Contractor could not in the Contractor's opinion reasonably have been foreseen by an experienced contractor the Contractor shall as early as practicable give written notice thereof to the Engineer.

These regulations are now fully in force.

The Planning Supervisor (*a*) coordinates the health and safety aspects of the design and planning phase of the project, (*b*) ensures that a Health and Safety Plan is prepared, (*c*) ensures that designers have taken safety properly into account, (*d*) advises the client (Employer) on health and safety and (*e*) prepares a safety file. The Engineer may be the Planning Supervisor and it is convenient for him to act in this role. Clause 71(2) applies where he is the Planning Supervisor; whenever he acts in this capacity he must state in writing that he does so.

Where a Contractor is appointed under the ICE Conditions of Contract, he will ordinarily be the "Principal Contractor" and Clause 71(2) makes provision for this.

The regulations affect the Employer and Contractor (the Principal Contractor) in a number of ways.[109]

---

109 See the Health and Safety Executive approved code of practice, *Managing construction for health and safety*, HSE, 1995.

(*a*) The Contractor takes over the Health and Safety Plan. This contains many of the features normally to be found in the Clause 14 programme and method statement. It includes a general description of the construction work involved, details of the time within which it is intended that the project and any intermediate stages will be completed, and a statement of risks—Regulation 15(2). The Contractor maintains and updates the plan to show such arrangements for the project which will ensure, as far as is reasonably practicable, health and safety—Regulation 15(3).

(*b*) The Contractor takes reasonable steps to ensure cooperation between all contractors and to ensure that they comply with the rules in the Health and Safety Plan. Furthermore, the Contractor is required to take reasonable steps to ensure that only authorised persons are allowed in the vicinity of construction in progress—Regulation 16(1).

(*c*) Where the Contractor has a design responsibility, his designers or any design sub-contractors are obliged to take full account of the health and safety aspects which their designs entail and to cooperate with the Planning Supervisor and other designers in this regard—Regulation 13.

Clause 71(3) provides that any alteration or amendment of the Health and Safety Plan shall be deemed to be an Engineer's instruction pursuant to Clause 13(1). At first sight this seems somewhat curious given that responsibility for the health and safety plan is transferred to the Contractor who develops the plan so as to suit not only the health and safety requirements of the project but his own construction plans. Where any change is made to the health and safety plan, the Contractor's entitlement to additional payment or extensions of time is set out in Clause 13(3). This entitles the Contractor, as a general principle, to claim where the instruction causes "him to incur cost beyond that reasonably to have been foreseen by an experienced contractor at the time of tender". However, by Clause 71(3)(a) he will not be entitled to additional payment or an extension of time to the extent that it "results from any action, lack of action or default on the part of the Contractor". The word "action" seems misleading, since any change which the Contractor makes to the plan is, in a superficial sense, necessarily a result of his action. It is thought that where the Contractor is the principal contractor he may amend the health and safety plan in any way which is necessary to comply with his health and safety obligations and he is entitled to be paid in accordance with Clause 13(3) to the extent that the test in Clause 13(3) is satisfied.

SPECIAL CONDITIONS

72 Special Conditions

The following special conditions form part of the Conditions of Contract. (Note. Any special conditions including contract price fluctuation which it is desired to incorporate in the Conditions of Contract should be numbered consecutively with the foregoing Conditions of Contract).

The Conditions of Contract are frequently supplemented with special conditions. Clause 5 gives equal weight to all terms of the Contract and hence an identical effect will be produced by including any additional terms in any other document.

SHORT DESCRIPTION OF WORKS

All Permanent and Temporary Works in connection with* ....
....
....

Form of Tender

(NOTE: The Appendix forms part of the Tender)

To ....
....
....

GENTLEMEN,

Having examined the Drawings, Conditions of Contract, Specification and Bill of Quantities for the construction of the above-mentioned Works (and the matters set out in the Appendix hereto) we offer to construct and complete the whole of the said Works in conformity with the said Drawings, Conditions of Contract, Specification and Bill of Quantities for such sum as may be ascertained in accordance with the said Conditions of Contract.

We undertake to complete and deliver the whole of the Permanent Works comprised in the Contract within the time stated in the Appendix hereto.

If our tender is accepted we will, if required, provide security for the due performance of the Contract as stipulated in the Conditions of Contract and the Appendix hereto.

Unless and until a formal Agreement is prepared and executed this Tender together with your written acceptance thereof, shall constitute a binding Contract between us.

We understand that you are not bound to accept the lowest or any tender you may receive.

We are, Gentlemen,

Yours faithfully,

Signature ....

Address ....

....

Date ....

* Complete as appropriate

FORM OF TENDER (APPENDIX)

(NOTE: Relevant Clause numbers are shown in brackets)

Appendix—Part 1 (to be completed prior to the invitation of Tenders)

1 Name of the Employer (Clause 1(1)(a)) . . . .
Address . . . .

2 Name of the Engineer (Clause 1(1)(c)) . . . .
Address . . . .

3 Defects Correction Period (Clause 1(1)(s)) . . . . .weeks

4 Number and type of copies of Drawings to be provided
(Clause 6(1)(b)) . . . .
. . . .

5 Contract Agreement (Clause 9) Required/Not required

6 Performance Bond (Clause 10(1)) Required/Not required
Amount of Bond (if required) to be . . . .% of Tender Total

7 Minimum amount of third party insurance (persons and
property) (Clause 23(3)) £ . . . .
each and every occurrence

8 Works Commencement Date (if known) (Clause 41(1)(a))
. . . .

9 Time for Completion (Clause 43)[a] . . . . .weeks
EITHER for the whole of the Works
OR for Sections of the Works (Clause 1(1)(u))[b]
Section A . . . . weeks
Section B . . . . .weeks
Section C . . . . .weeks
Section D. . . . . weeks
the Remainder of the Works . . . . weeks

10 Liquidated damages for delay (Clause 47)

	per day/week	limit of liability[c]
EITHER for the whole of the Works	£	£ . . . .
OR for Section A (as above)	£	£ . . . .
Section B (as above)	£	£ . . . .
Section C (as above)	£	£ . . . .

Section D (as above) £ . . . . £ . . . .
the Remainder of the Works £ . . . . £ . . . .
(as above)

11 Vesting of materials not on Site (Clauses 54(1) and 60(1)(c)) (if required by the Employer)[d]

1 . . . . 4 . . . .
2 . . . . 5 . . . .
3 . . . . 6 . . . .

12 Method of measurement adopted in preparation of Bills of Quantities (Clause 57)[e] . . . .

. . . .

13 Percentage of the value of goods and materials to be included in Interim Certificates (Clause 60(2)(b)) . . . .%

14 Minimum amount of Interim Certificates (Clause 60(3)) £ . . . .

15 Rate of retention (recommended not to exceed 5%) (Clause 60(5)) . . . .%

16 Limit of retention (% of Tender Total) (Clause 60(5)) (Recommended not to exceed 3%) . . . .%

17 Bank whose Base Lending Rate is to be used (Clause 60(7)) . . . .

18 Requirement for prior approval by the Employer before the Engineer can act.
DETAILS TO BE GIVEN AND CLAUSE NUMBER STATED (Clause 2(1)(b))[f]

. . . .
. . . .
. . . .

19 Name of the Planning Supervisor (Clause 71(1)(b)) . . . .
Address . . . .

20 Name of the Principal Contractor (Clause 71(1)(b)) . . . .
Address . . . .

[a] If not stated is to be completed by Contractor in Part 2 of the Appendix.
[b] To be completed if required, with brief description. Where Sectional completion applies the item for "the Remainder of the Works" must be used to cover the balance of the Works if the Sections described do not in total comprise the whole of the Works.
[c] Delete where not required.
[d] (If used) Materials to which the Clauses apply must be listed in Part 1 (Employer's option) or Part 2 (Contractor's option) .
[e] Insert here any amendment or modification adopted if different from that stated in Clause 57.
[f] If there is any requirement that the Engineer has to obtain prior approval from the Employer before he can act full particulars of such requirements must be set out above.

Appendix—Part 2

(To be completed by Contractor)

1 Insurance Policy Excesses (Clause 25(2))

Insurance of the Works (Clause 21(1)) £ . . . .

Third party (property damage) (Clause 23(1)) £ . . . .

2 Time for Completion (Clause 43) (if not completed in Part 1 of the Appendix)

EITHER for the whole of the Works . . . . .weeks

OR for Sections of the Works (Clause 1(1)(u)) (as detailed in Part 1 of the Appendix)

Section A . . . . .weeks

Section B . . . . .weeks

Section C . . . . .weeks

Section D . . . . .weeks

the Remainder of the Works . . . . .weeks

3 Vesting of materials not on site (Clauses 54(1) and 60(1)(c)) (at the option of the Contractor—see [d] in Part 1)

1 . . . . 4. . . . .

2 . . . . 5. . . . .

3 . . . . 6. . . . .

4 Percentage(s) for adjustment of PC sums (Clauses 59(2)(c) and 59(5)(c)) (with details if required)

. . . .

. . . .

FORM OF AGREEMENT

THIS AGREEMENT made the . . . . . . .day of . . . . 19 . . . . . .
BETWEEN . . . .
of . . . .
in the County of ....
(hereinafter called "the Employer")
and . . . .
of . . . .
in the County of . . . .(hereinafter called "the Contractor").
WHEREAS the Employer is desirous that certain Works should be constructed, namely the Permanent and Temporary Works in connection with . . . .
. . . .
and has accepted a Tender by the Contractor for the construction and completion of such Works.

NOW THIS AGREEMENT WITNESSETH as follows

1. In this Agreement words and expressions shall have the same meanings as are respectively assigned to them in the Conditions of Contract hereinafter referred to.

2. The following documents shall be deemed to form and be read and construed as part of this Agreement, namely

   (a) the said Tender and the written acceptance thereof
   (b) the Drawings
   (c) the Conditions of Contract
   (d) the Specification
   (e) the priced Bill of Quantities.

3. In consideration of the payments to be made by the Employer to the Contractor as hereinafter mentioned the Contractor hereby covenants with the Employer to construct and complete the Works in conformity in all respects with the provisions of the Contract.

4. The Employer hereby covenants to pay to the Contractor in consideration of the construction and completion of the Works the Contract Price at the times and in the manner prescribed by the Contract.

IN WITNESS whereof the parties hereto have caused this Agreement to be executed the day and year first above written.

SIGNED on behalf of the said . . . . Ltd/plc

Signature	Signature . . . .
Position	Position . . . .
In the presence of.	In the presence of. . . . .

or SIGNED SEALED AND DELIVERED AS A DEED
by the said . . . . Ltd/plc
in the presence of . . . .

FORM OF BOND

The text of the bond is not published here. The CCSJC has commissioned the drafting of a new form of bond and consultations are currently under way. It is expected that this bond will be published shortly.[110]

110 Probably in late 1996 or early 1997.

9

The FCEC Form of Sub-Contract (September 1991 Edition)

1. Introduction

The FCEC Form of Sub-Contract (1991) is published by the Federation of Civil Engineering Contractors and is designed for use in conjunction with the ICE Conditions of Contract 6th Edition. With minor amendments it may be used with other main contracts.

The relationship between the Main Contract and the Sub-Contract

It is assumed that there is a Main Contract and the Sub-Contractor is deemed to have full knowledge of its provisions—Clause 3(1). The terms of the Main Contract are not incorporated into the Sub-Contract. Nevertheless, the Sub-Contract is in some respects back-to-back[1] with the Main Contract. The Sub-Contractor's liability for breaches of the Sub-Contract includes providing an indemnity to the Contractor for any losses which the Contractor incurs—Clause 3. The Sub-Contractor's right to claim for adverse physical conditions etc. is through the Main Contractor on the same basis as the Main Contractor

1 The expression "back-to-back" is frequently used in the context of sub-contracts. Fully back-to-back main and sub-contracts exist where the terms in one are mirrored in the other so that whenever the main contractor is liable to the employer, the sub-contractor is correspondingly liable to him; and whenever he is liable to sub-contractor, the employer is correspondingly liable to him. In a sense, providing he is not personally in default, the main contractor acts as a mere conduit between the employer and sub-contractor, having an indemnity either way and collecting any "percentage mark-up" in his prices as the work is done.

is entitled against the Employer—Clause 10(2). Likewise, payment under the Sub-Contract may be withheld where payment under the Main Contract has either not been certified or not been paid—Clause 15(3)(b).

The relationship between the Main Works and Sub-Contract Works

The Sub-Contract is an agreement to carry out the Sub-Contract Works. These works are to "form part of the Works to be executed by the Contractor under the Main Contract"—2nd Recital. Accordingly, the Contractor is not obliged to undertake work which does not form part of the Main Works. Where such works are undertaken, they fall outside the scope of the Sub-Contract and entitle the Sub-Contractor to claim a reasonable sum.[2] Where the extra works are delayed, the Contractor may have no right to counterclaim against the Sub-Contractor in respect of them for delay and/or disruption unless the parties agree that they should be covered by the Sub-Contract.[3] By Clause 7(1) an Engineer's instruction which is not validly given under the Main Contract entitles the Sub-Contractor to recover "such costs as may be reasonable" and by Clause 7(2) the Contractor has the like powers and the Sub-Contractor has the like rights as the Engineer and Contractor have respectively under the Main Contract. Since work which falls outside the Main Works will entitle the Contractor to a *quantum meruit*, the Sub-Contractor will also be entitled to claim a *quantum meruit*.

2. Commentary on the Form of Sub-Contract

AGREEMENT AND RECITALS

AN AGREEMENT made the ..... day of .... 19 .....

BETWEEN ....of/whose

registered office is at ....

.... (hereinafter called "the Contractor") of the one

2 *Sir Lindsay Parkinson & Co. v. Commissioners of Works* [1949] 2 KB 632 (CA).

3 *British Steel Corporation v. Cleveland Bridge & Engineering Co. Ltd* [1984] 1 All ER 504.

part and .... .of/whose

registered office is at ....

...... (hereinafter called "the Sub-Contractor") of the other part.

WHEREAS the Contractor has entered into a Contract (hereinafter called "the Main Contract") particulars of which are set out in the First Schedule hereto:

AND WHEREAS the Sub-Contractor having been afforded the opportunity to read and note the provisions of the Main Contract (other than details of the Contractor's prices thereunder), has agreed to execute upon the terms hereinafter appearing the works which are described in the documents specified in the Second Schedule hereto and which form part of the Works to be executed by the Contractor under the Main Contract:

NOW IT IS HEREBY AGREED as follows:

The recitals[4] should not in general be used in construing the contract, save where it is necessary to clarify any matter.[5] Here the recitals act at most as a guide to construction. They set out the basic assumptions, namely: (*a*) there is a Main Contract; and (*b*) the Sub-Contractor agrees to execute work which forms part of the Main Works.

1 Definitions

(1) In this Sub-Contract (as hereinafter defined) all words and expressions have the same meaning as in the Main Contract unless otherwise provided or where the context otherwise requires.

(a) "the Main Contract" means the contract, particulars of which are given in the First Schedule hereto.

---

4 That is the sections beginning "Whereas ...".

5 *Leggot* v. *Barrett* (1880) 15 ChD 306 (CA) per Brett LJ at 311: "If there is any doubt about the construction of the governing words of the document, the recitals may be looked at in order to determine what is the true construction; but if there is no doubt about the construction, the rights of the parties are governed entirely by the operative part of the writing".

(b) "the Sub-Contract" means this Agreement together with such other documents as are specified in the Second Schedule hereto, but excluding any standard printed conditions that may be included in such other documents unless separately specified in the said Schedule.

(c) "the Sub-Contract Works" means the works described in the documents specified in the Second Schedule hereto.

(d) "the Main Works" means the Works as defined in the Main Contract.

(e) "the Price" means the sum specified in the Third Schedule hereto as payable to the Sub-Contractor for the Sub-Contract Works or such other sum as may become payable under the Sub-Contract.

(f) "Sub-Contractor's Equipment" means all appliances or things of whatsoever nature required by the Sub-Contractor in the fulfilment of his obligations under the Sub-Contract but does not include materials or other things intended to form or forming part of the Sub-Contract Works.

(g) "maintain" means the execution of outstanding work and the correction of defects as required by Clause 49 of the Main Contract and the words "maintenance" and "maintaining" shall be construed accordingly.

(2) Words importing the singular also include the plural and vice-versa where the context requires.

(3) The headings, marginal notes and notes for guidance in this Agreement shall not be deemed to be part thereof or taken into consideration in the interpretation or construction thereof or of the Sub-Contract.

(4) All references herein to clauses are references to clauses numbered in this Agreement and not to those in any other document forming part of the Sub-Contract unless otherwise stated.

Clause 1(1)(b)—"the Sub-Contract"

The Second Schedule, part A invites the parties to identify letters etc. which are to form part of the Sub-Contract. Any "standard printed conditions" are excluded unless they are separately specified. The expression "standard printed conditions" presumably means any list of conditions which appear upon inspection to be designed for use on a range of projects.[6]

Clause 1(1)(e)—"the Price"

The price is generally broken down into items and/or rates in a document referred to in the Third Schedule. The Sub-Contract form is equally applicable to a measure and value or a lump sum payment scheme.[7] Where the price is to be computed by measurement, a method of measurement may be agreed; otherwise, the rates given will be construed to include for all those things reasonably to be inferred from the item descriptions. Where a lump sum price is agreed, this will include for all those things which are necessary to do the works.[8]

Clause 1(1)(g)—"maintain"

The traditional expression "maintain" is no longer used in the ICE Conditions of Contract 6th Edition. The expression "correction of defects" is now used.

2 General

(1) The Sub-Contractor shall execute, complete and maintain the Sub-Contract Works in accordance with the Sub-Contract and to the reasonable satisfaction of the Contractor and of the Engineer. The Sub-Contractor shall exercise all reasonable skill care and diligence in designing any part of the Sub-Contract Works for which design he is responsible.

6 See *Chester Grosvenor Hotel* v. *Alfred McAlpine* (1991) 56 BLR 115. This case dealt with the question whether terms were standard for the purpose of the Unfair Contract Terms Act 1977. Judge Stannard said at 133 "... what is required for terms to be standard is that they should be regarded by the party which advances them as its standard terms and that it should habitually contract on those terms". It is submitted that this test is inappropriate here. It is submitted that Clause 1(1)(b) is designed to deal with lists of terms which are advanced in negotiations and which are not drawn up specifically for the instant Sub-Contract; it is submitted that the clause should be construed accordingly.

7 Note that there are expressions in the Sub-Contract which suggest a measure and value payment scheme. In Clause 9(4), for instance, it is agreed that "any difference between the quantity so billed and the actual quantity so executed shall be ascertained by measurement".

8 *Williams* v. *Fitzmaurice* (1858) 3 H & N 844.

(2) The Sub-Contractor shall provide all labour, materials, Sub-Contractor's Equipment, Temporary Works and everything whether of a permanent or temporary nature required for the execution, completion and maintenance of the Sub-Contract Works, except as otherwise agreed in accordance with Clause 4 and set out in the Fourth Schedule hereto.

(3) The Sub-Contractor shall not assign the whole or any part of the benefit of this Sub-Contract nor shall he sub-let the whole or any part of the Sub-Contract Works without the previous written consent of the Contractor.

Provided always that the Sub-Contractor may without such consent assign either absolutely or by way of charge any sum which is or may become due and payable to him under this Sub-Contract.

(4) Copyright of all Drawings, Specifications and the Bill of Quantities (except the pricing thereof) supplied by the Employer or the Engineer or the Contractor shall not pass to the Sub-Contractor but the Sub-Contractor may obtain or make at his own expense any further copies required by him for the purposes of the Sub-Contract. Similarly copyright of all documents supplied by the Sub-Contractor under the Sub-Contract shall remain in the Sub-Contractor but the Employer and the Engineer and the Contractor shall have full power to reproduce and use the same for the purpose of completing operating maintaining and adjusting the Works.

General obligation of the Sub-Contractor

Clauses 2(1) and 2(2) set out the Contractor's general obligations. Note that the standard of care for design work is identical to that required for design work under the ICE Conditions of Contract.[9]

Sub-letting—Clause 2(3)

This clause prohibits sub-letting of the Sub-Contract Works without the Contractor's consent. There is no obligation upon the Contractor to

9 ICE Conditions of Contract, Clause 8(2).

give consent where it is reasonable to do so.[10] In some cases, there may be some uncertainty about whether work to be done is "sub-letting" in the sense meant by this provision or "supply"; thus, for example, where a proprietary system is supplied together with personnel to erect it, this may constitute sub-letting. It is thought that the test is whether the supplier supplies an ingredient to an element of finished work, or the work itself; in the former case he is a supplier and no consent is required, while in the latter case he is a sub-contractor and consent is required.

Assignment—Clause 2(3)

There is no prohibition on assignment of sums due under the Sub-Contract. There is an absolute prohibition on all other assignments, but it is not easy to see what else apart from sums due, or to become due, the Sub-Contractor could properly assign.

Copyright—Clause 2(4)

No copyright is to pass unless otherwise stated.

3 Main Contract

(1) The Sub-Contractor shall be deemed to have full knowledge of the provisions of the Main Contract (other than the details of the Contractor's prices thereunder as stated in the bills of quantities or schedules of rates and prices as the case may be), and the Contractor shall, if so requested by the Sub-Contractor, provide the Sub-Contractor with a true copy of the Main Contract (less such details), at the Sub-Contractor's expense. The Main Contractor shall on request provide the Sub-Contractor with a copy of the Appendix to the Form of Tender to the Main Contract together with details of any contract conditions which apply to the Main Contract which differ from the applicable standard ICE Conditions of Contract.

10 By analogy with *Leedsford Ltd* v. *Bradford Corporation* (1956) 24 BLR 45 (CA), where it was held that a right to disapprove suppliers did not have to be exercised reasonably unless the contract required this.

(2) Save where the provisions of the Sub-Contract otherwise require, the Sub-Contractor shall so execute, complete and maintain the Sub-Contract Works that no act or omission of his in relation thereto shall constitute, cause or contribute to any breach by the Contractor of any of his obligations under the Main Contract and the Sub-Contractor shall, save as aforesaid, assume and perform hereunder all the obligations and liabilities of the Contractor under the Main Contract in relation to the Sub-Contract Works.

Nothing herein shall be construed as creating any privity of contract between the Sub-Contractor and the Employer.

(3) The Sub-Contractor shall indemnify the Contractor against every liability which the Contractor may incur to any other person whatsoever and against all claims, demands, proceedings, damages, costs and expenses made against or incurred by the Contractor by reason of any breach by the Sub-Contractor of the Sub-Contract.

(4) The Sub-Contractor hereby acknowledges that any breach by him of the Sub-Contract may result in the Contractor's committing breaches of and becoming liable in damages under the Main Contract and other contracts made by him in connection with the Main Works and may occasion further loss or expense to the Contractor in connection with the Main Works and all such damages loss and expense are hereby agreed to be within the contemplation of the parties as being probable results of any such breach by the Sub-Contractor.

General

This clause imports the technical requirements of the Main Contract into the Sub-Contract. In addition, each of four sub-clauses aims to ensure that wherever the Sub-Contractor is in breach of his Sub-Contract and thereby causes the Contractor to incur any liability, the Contractor can recover his loss in full from the Sub-Contractor. While the clause is headed "Main Contract", its scope is wider than this; Clause 3(3) is not limited to claims which are connected with the Main Contract.

Clauses 3(1) and 3(4)—notice and special damages

These two provisions are designed to fix the Sub-Contractor with notice of the terms of the Main Contract and with an express acknowledgement that any loss or liability which may be incurred by the Contractor is a foreseeable loss. The expression "within the contemplation of the parties as being probable results of any breach" is modified from the leading case on damages.[11]

Clause 3(2)—assumption of all liabilities

This requires the Sub-Contractor not only to provide Sub-Contract Works which comply with the Sub-Contract at the date at which they are completed, but to ensure that at no stage are they such as put the Contractor in breach.

Clause 3(3)—indemnity

This indemnity is wide in its effect. It covers not only damages which the Contractor incurs but also "costs and expenses" suffered by the Contractor. It is, however, limited to the losses "incurred by the Contractor by reason of any breach by the Sub-Contractor of the Sub-Contract" and so there is no indemnity for "claims" and "demands" unless the Sub-Contractor is in breach.[12] The styling of this obligation as an indemnity may be important in that the Contractor's cause of action does not arise until a claim is made against him; thus, the limitation period may be considerably longer than where the claim is based on a straightforward breach of contract terms.

Liquidated damages

It is common for main contracts to include liquidated damages provisions for delay, but less common for them to be agreed under a sub-contract. The reason is that the Contractor's losses cannot generally be evaluated in advance with any certainty, unless the sub-contractor is the principal contractor on the site. There is no specific provision for liquidated damages under the FCEC Sub-Contract. Clause 3 fixes the Sub-Contractor with knowledge of the liquidated

11 *Hadley* v. *Baxendale* (1854) 9 Ex 341. The original words are given in Chapter 6, section 4.

12 In the normal course of events, the alleged losses, costs, expenses, etc. will be deducted by the Contractor as an alleged set-off against sums owed to the Sub-Contractor. It will thus fall to the Sub-Contractor to show that they were wrongly deducted.

damages provisions of the Main Contract. Thus where the Sub-Contractor delays the Main Works so as to cause the Contractor to incur liquidated damages, the Contractor may contra-charge these; however, as between the Contractor and Sub-Contractor, these are not liquidated damages and the Contractor must prove his loss, including showing that he has reasonably mitigated it.

4 Contractor's facilities

(1) The Contractor shall permit the Sub-Contractor for the purpose of executing and completing the Sub-Contract Works to use such scaffolding as is from time to time provided by the Contractor in connection with the Main Works, but the Contractor shall not be bound to provide or retain any such scaffolding for the Sub-Contractor's use. No such permission shall imply any warranty by the Contractor as to the fitness, condition and suitability of such scaffolding, nor shall it impose any liability upon the Contractor in respect of its use by the Sub-Contractor, his servants or agents, nor relieve the Sub-Contractor of any statutory or other obligation to test or inspect the scaffolding to be used by his servants or agents or to provide suitable scaffolding for their use.

(2) The Contractor shall provide at the Site the Contractor's Equipment and/or other facilities (if any) specified in Part 1 of the Fourth Schedule hereto and shall permit the Sub-Contractor in common with such other contractors as the Contractor may allow, to have the use thereof for the purposes of executing and completing but not of maintaining the Sub-Contract Works, upon such terms and conditions (if any) as are specified in the said Schedule, but the Contractor shall have no liability to the Sub-Contractor in respect of any failure to provide such Contractor's Equipment or facilities, if such failure is due to circumstances outside the Contractor's control, nor in respect of any inadequacy or unfitness for the Sub-Contractor's purposes of any Contractor's Equipment or facilities so provided.

(3) The Contractor shall also provide for the exclusive use of the Sub-Contractor the Contractor's Equipment and/or other facilities (if any) specified in Part II of the Fourth Schedule upon such terms and conditions (if any) as are specified in the said Schedule, but the Contractor shall have no liability to the Sub-Contractor in respect of any failure to provide such Contractor's Equipment and facilities, if such failure is due to circumstances outside the Contractor's control, nor in respect of any inadequacy or unfitness for the Sub-Contractor's purposes of any such Contractor's Equipment or facilities so provided.

(4) The Sub-Contractor shall indemnify the Contractor against any damage or loss whatsoever arising from the misuse by the Sub-Contractor, his servants or agents of the Contractor's Equipment and/or other facilities provided for his use by the Contractor.

General

It is commonly the case that several sub-contractors should need access or cranage to the same areas at the same time. It would be inefficient, if not unworkable, for each to provide his own scaffolding, cranes, etc. In such a case, the Contractor may provide the general site "infrastructure", which the sub-contractors may use. Exclusive use is not, of course, guaranteed. There may also be parts of the site and equipment which is exclusively for the use of the particular Sub-Contractor. This clause divides the equipment into two categories. Contractor's Equipment for general use includes scaffolding (Clause 4(1)) and any other equipment described in Part 1 of the Fourth Schedule (Clause 4(2)), while Contractor's Equipment for the Sub-Contractor's exclusive use is set out in Part 2 of the Fourth Schedule (Clause 4(3)).

Clause 4(1)—scaffolding

The Contractor does not warrant that he will provide any scaffolding. However, when "from time to time" it may be in place, the Sub-Contractor shall be permitted to use it. If the Sub-Contractor requires scaffolding, therefore, this should be recorded in the Fourth Schedule. The Contractor also puts the Sub-Contractor on notice that he must test or inspect the scaffolding to ensure that it is safe for his own requirements.

Clause 4(2)—equipment which is to be provided (non-exclusive use)

The clause refers to Part 1 of the Fourth Schedule where equipment which is to be available for the Sub-Contractor's non-exclusive use is recorded. There are a number of limitations, as follows.

(*a*) The equipment is available only during construction, and will not be provided for "maintaining" the Sub-Contract works. It is thought that the Contractor's obligation is to provide the equipment until the Sub-Contract Works are completed; where however, this exceeds any agreed Sub-Contract period (Third

Schedule, Part C plus due extensions) and the equipment must be kept on at extra cost to the Contractor, that cost may be charged against the Sub-Contractor's account.

(*b*) Where the Contractor fails to provide the equipment for circumstances outside his control. The phrase "outside the Contractor's control" has been held in one instance to include where the contractor waits for supplies from suppliers over whom he has no direct control.[13] It is thought, however, that in the present case, the Contractor will be responsible unless it can be shown that he took all reasonable measures to ensure that the equipment was available.

(*c*) There is no liability for any inadequacy or unfitness. In one sense, the Contractor is in no position to warrant the adequacy or fitness of the equipment as he cannot know what loads etc. are to be applied. Where, however, the equipment is closely specified, and the Contractor does not provide equipment which meets that specification, or provides defective equipment this will be a breach of contract.[14]

Clause 4(3)—equipment which is to be provided (exclusive use)

This clause refers to the Fourth Schedule, Part 2 where equipment for exclusive use is set out. The reservations noted at (*b*) and (*c*) in the commentary on Clause 4(2) relate equally to this Clause. There is no restriction on the time for which the equipment is provided, it being assumed that details will be included in the Schedule. Where no time is specified, it will be implied that the time for provision will be a reasonable time and period given the function of the equipment, the known Sub-Contract period and any other relevant factors.

5 Site working and access

(1) The Sub-Contractor shall in the execution of the Sub-Contract Works on the Site observe the same hours of working as the Contractor, unless otherwise agreed, and shall comply with all reasonable rules and regulations of the Contractor governing the

13 *Scott Lithgow Limited* v. *Secretary of State for Defence* (1989) 45 BLR 6. See also *John Mowlem & Co.* v. *Eagle Star* (1995) CILL 1047.

14 To the extent that this provision is an exclusion clause it will be subject to the test of reasonableness in the Unfair Contract Terms Act 1977.

execution of the work, the arrival at and the departure from the Site of materials and Sub-Contractor's Equipment and the storage of materials and Sub-Contractor's Equipment on the Site.

(2) The Contractor shall from time to time make available to the Sub-Contractor such part or parts of the Site and such means of access thereto within the Site as shall be necessary to enable the Sub-Contractor to execute the Sub-Contract Works in accordance with the Sub-Contract, but the Contractor shall not be bound to give the Sub-Contractor exclusive possession or exclusive control of any part of the Site.

(3) The Sub-Contractor shall permit the Engineer, the Engineer's Representative and other the Engineer's servants and agents and the Contractor, his servants and agents (including any other sub-contractors engaged in the execution of the Main Works), during working hours to have reasonable access to the Sub-Contract Works and to the places on the Site where any work or materials therefor are being executed prepared or stored and the Sub-Contractor shall also permit or procure reasonable access for the Engineer, his servants and agents and for the Contractor, his servants and agents to such places off the Site where work is being executed or prepared by or on behalf of the Sub-Contractor in connection with the Sub-Contract Works.

Clause 5(1)—site hours

Security, safety and nuisance caused to neighbours are significant features of modern civil engineering construction. These are conveniently dealt with where the Sub-Contractor works the same hours as the Main Contractor and complies with his security and safety procedures. The expression "reasonable rules and regulations" seems directed at issues such as security and safety and noise etc.[15] The parties may expressly agree for the Sub-Contractor to work different hours.

Clause 5(2)—access and possession

Where the Contractor fails to make available so much of the Site as will enable him to execute the Works, this will be a breach of contract.

15 Note in this connection the Construction (Design and Management) Regulations 1994; see the commentary to Clause 71 of the ICE Conditions of Contract in Chapter 8.

Exclusive possession or control need not be provided. Where, however, the Contractor persistently fails to grant proper and sufficient access or space to undertake operations with reasonable efficiency using equipment etc. of a type which the Contractor should have foreseen being used, this is a failure to make the Site available.

Clause 5(3)—access for the Engineer and the Contractor

This is limited to working hours.

6 Commencement and completion

(1) Within 10 days, or such other period as may be agreed in writing, of receipt of the Contractor's written instructions so to do, the Sub-Contractor shall enter upon the Site and commence the execution of the Sub-Contract Works and shall thereafter proceed with the same with due diligence and without any delay, except such as may be expressly sanctioned or ordered by the Contractor or be wholly beyond the control of the Sub-Contractor. Subject to the provisions of this clause, the Sub-Contractor shall complete the Sub-Contract Works within the Period for Completion specified in the Third Schedule hereto.

(2) If the Sub-Contractor shall be delayed in the execution of the Sub-Contract Works:

(a) by any circumstances or occurrence (other than a breach of this Sub-Contract by the Sub-Contractor) entitling the Contractor to an extension of his time for completion of the Main Works under the Main Contract; or

(b) by the ordering of any variation of the Sub-Contract Works to which paragraph (a) of this sub-clause does not apply; or

(c) by any breach of this Sub-Contract by the Contractor;

then in any such event the Sub-Contractor shall be entitled to such extension of the Period for Completion as may in all the circumstances be fair and reasonable.

Provided always that in any case to which paragraph (a) of this sub-clause applies it shall be a condition precedent to the Sub-Contractor's right to an extension of the Period for Completion that he shall have given written notice to the Contractor of the circumstances or occurrence which is delaying him within 14 days of such delay first occurring together with full and detailed particulars in justification of the period of extension claimed in order that the claim may be investigated at the time and in any such case the extension shall not in any event exceed the extension of time to which the Contractor is properly entitled under the Main Contract.

(3) Where differing Periods of Completion are specified in the Third Schedule for different parts of the Sub-Contract Works, then for the purposes of the preceding provisions of this Clause each such part shall be treated separately in accordance with sub-clause (2) above.

(4) Nothing in this Clause shall be construed as preventing the Sub-Contractor from commencing off the Site any work necessary for the execution of the Sub-Contract Works at any time before receipt of the Contractor's written instructions under sub-clause (1) of this Clause.

(5) The Contractor shall notify the Sub-Contractor in writing of all extensions of time obtained under the provisions of the Main Contract which affect the Sub-Contract.

Sub-Contract timing

Where the Sub-Contract Works require significant preceding or enabling work by others, it is frequently not possible to predict the dates when a sub-contractor will be required on site. Nevertheless, sub-contracts are frequently made at the outset of the project so that the Contractor knows that he has agreed prices/rates on agreed terms. For this reason it is common to agree a Sub-Contract duration (which is inserted in the Third Schedule) and then to call upon the Sub-Contractor with 10 days notice (or other agreed notice period).

Due diligence and without any delay

Ordinarily a sub-contractor will be entitled to plan his work as he wishes.[16] This provision, however, requires him to make proper progress. This is often a matter of practical importance, for the following reasons.

(*a*) Delays by the Sub-Contractor to critical activities may delay the duration of the project as a whole. Such overall delays are frequently expensive, both where liquidated damages are levied and because of the costs of maintaining a site presence and

16 *Greater London Council* v. *Cleveland Bridge Engineering Co. Ltd* (1984) 34 BLR 50.

meeting claims for delay by other sub-contractors. Accordingly, the Contractor will wish to ensure that the rate of progress is adequate to ensure completion within the Sub-Contract period and to be able to take appropriate action in the event that it is not.

(*b*) Where there is non-exclusive possession, the Contractor will wish to manage access arrangements. This cannot be done where individual sub-contractors can leave work idle for periods and then attend in force during other periods, disrupting other operatives (e.g. by excessive use of hoists, access routes, storage facilities in the vicinity of the workfront, etc.).

Any failure to progress the Sub-Contract Works diligently and without delay will thus be a breach of contract and if sufficiently serious will entitle the Contractor to treat the Sub-Contractor's failure as a repudiation, so that the Contractor may terminate the Sub-Contract (see also Clause 17). "'Diligence' in this context means it seems to me, not only the personal industriousness of the [sub-contractor] himself but his efficiency and that of those who worked for him ...".[17] In the absence of a specific agreement, or where it is clear from the factual circumstances known at the time of the agreement, there is no obligation for the Sub-Contractor to progress work to keep pace with any other works being undertaken by the Contractor or his other sub-contractors.[18]

The Sub-Contractor is not obliged to progress the Sub-Contract where the Contractor sanctions a delay where there are reasons "wholly beyond his control".[19] It is thought that this latter phrase does not include such things as lack of any equipment or material which the Sub-Contractor is to provide, since its absence could have been covered by contingency supplies and so is not "wholly" beyond the Sub-Contractor's control.

---

17 *Hooker Constructions* v. *Chris's Engineering Contracting Company* [1970] ALR 821. See also *West Faulkner Associates* v. *London Borough of Newham* (1994) 71 BLR 1 (CA).

18 *Piggott Foundations Ltd* v. *Shepherd Construction Ltd* (1993) 67 BLR 48 (a case on the DOM/1 building sub-contract).

19 *Scott Lithgow Limited* v. *Secretary of State for Defence* (1989) 45 BLR 6.

Extensions of time

Clause 6(2) sets out the circumstances which entitle the Sub-Contractor to an extension of time. Three separate sources of entitlement are given. It is a condition precedent of the first (circumstances entitling the contractor to an extension under the Main Contract) that the Sub-Contractor shall give notice with full particulars within 14 days. This gives rise to the interesting prospect that the contractor may receive an extension for an event at the Employer's risk, but, where the Sub-Contractor fails to give the requisite notice, the Contractor is not obliged to pass on the extension to the Sub-Contractor. However, it is rare for liquidated damages to be agreed under the sub-contract and so damages for delay will be unliquidated and the Contractor's loss must be demonstrated; where the Contractor receives an extension from the Employer, he will be at least partly compensated for the delay and, to that extent, will not be able to claim against the Sub-Contractor for it. By Clause 6(5) the Contractor shall notify the Sub-Contractor of extensions of time obtained under the provisions of the Main Contract.

Off site work

Subject to any terms agreed, the Sub-Contractor may commence any work off site as soon as the Sub-Contract is agreed—Clause 6(4).

7 Instructions and decisions

(1) Subject to Clause 8 (Variations), the Sub-Contractor shall in relation to the Sub-Contract Works comply with all instructions and decisions of the Engineer and of the Engineer's Representative which are notified and confirmed in writing to him by the Contractor, irrespective of whether such instructions and decisions were validly given under the Main Contract. The Sub-Contractor shall have the like rights, (if any), to payment against the Contractor in respect of such compliance as the Contractor has against the Employer under the Main Contract. Further if any such instruction or decision notified and confirmed as aforesaid is invalidly or incorrectly given by the Engineer under the Main Contract, then the Sub-Contractor shall be entitled to recover such costs as may be reasonable (if any) from the Contractor of complying therewith provided that such costs were not caused or contributed to by any breach of this Sub-Contract by the Sub-Contractor.

(2) The Contractor shall have the like powers in relation to the Sub-Contract Works to give instructions and decisions as the Engineer has in relation to the Main Works under the Main Contract and the Sub-Contractor shall have the like obligations to abide by and comply therewith and the like rights in relation thereto as the Contractor has under the Main Contract. The said powers of the Contractor shall be exercisable in any case irrespective of whether the Engineer has exercised like powers in relation thereto under the Main Contract.

Instructions—general

In civil engineering contracts proper control must be exercisable by an identified individual. Under the Main Contract this control is generally exercised by the Engineer. He will ordinarily be entitled to issue instructions with which the contractor is obliged to comply. Where a Sub-Contractor is employed, means must be devised whereby the Engineer's decisions become applicable to and binding on the Sub-Contractor. In addition, the Contractor will need to exercise some control and will wish to be entitled to issue independent instructions. In either of these cases, the Sub-Contractor will need to be entitled to the costs of giving effect to instructions, where such instructions do not fall within the ambit of the Sub-Contract agreement. Clause 7 deals with such matters.

Instructions and decisions of the Engineer

By Clause 7(1) the Engineer's instructions and decisions are not automatically binding on the Sub-Contractor. The Sub-Contractor is only required to comply with all instructions and decisions of the Engineer which are notified and confirmed in writing by the Contractor. Where an Engineer's instruction or decision is notified and confirmed in writing, the Sub-Contractor is entitled as follows.

(*a*) Where the Engineer's instructions or decision is validly given, the Sub-Contractor shall have "the like rights, (if any), to payment ... as the Contractor has against the Employer ...". Thus where the Engineer's instruction does not entitle the Contractor to additional payment, the Sub-Contractor will not be entitled. Where the Contractor is entitled to claim, the Sub-Contractor will be entitled on the same basis (though not necessarily to the same amount).

(*b*) Where the Engineer's instruction is not validly given, the Sub-Contractor shall be entitled "to recover such costs as may be reasonable". The exact meaning of this is not clear. It is thought that this means that the Sub-Contractor is generally entitled to a *quantum meruit* since compliance with the instruction is work falling outside the agreed scope of the Sub-Contract.

The test as to which instructions/decisions are "validly given" and which are "invalidly or incorrectly given" is not stated. It is submitted that a valid instruction/decision is one which is given in circumstances where the Engineer had authority to give the instruction or make the decision. It is thought that a later finding by an arbitrator that, for example, the decision was wrong does not prevent it from having been given "validly".

Instructions and decisions issued by the Contractor

Clause 7(2) provides the Contractor with the same power to give instructions and decisions as the Engineer has in relation to the Main Works. The Sub-Contractor will have the same rights to payments as where the instructions or decisions emanate from the Engineer. Situations where the Contractor's instructions are invalid include where the Sub-Contractor is instructed to execute work which does not form part of the Main Works.

8 Variations

(1) The Sub-Contractor shall make such variations of the Sub-Contract Works, whether by way of addition, modification or omission, as may be:

(a) ordered by the Engineer under the Main Contract and confirmed in writing to the Sub-Contractor by the Contractor; or

(b) agreed to be made by the Employer and the Contractor and confirmed in writing to the Sub-Contractor by the Contractor; or

(c) ordered in writing by the Contractor.

Any order relating to the Sub-Contractor Works which is validly given by the Engineer under the Main Contract and constitutes a variation thereunder shall for the purposes of this Clause be deemed to constitute a variation of the Sub-Contract Works, if confirmed by the Contractor in accordance with paragraph (a) hereof.

(2) The Sub-Contractor shall not act upon an unconfirmed order for the variation of the Sub-Contract Works which is directly

received by him from the Employer or the Engineer. If the Sub-Contractor shall receive any such direct order, he shall forthwith inform the Contractor's agent or foreman in charge of the Main Works thereof and shall supply him with a copy of such direct order, if given in writing. The Sub-Contractor shall only act upon such order as directed in writing by the Contractor, but the Contractor shall give his directions thereon with all reasonable speed.

(3) Save as aforesaid the Sub-Contractor shall not make any alteration in or modification of the Sub-Contract Works.

(4) Variations carried out in accordance with this Clause shall be valued as provided in Clause 9 and payment made in accordance with Clause 15.

The scheme of Clause 8 is broadly similar to that of Clause 7. Variations ordered by the Engineer are not automatically binding on the Contractor unless confirmed in writing by the Contractor. The valuation of variations is in accordance with Clause 9.

9 Valuation of variations

(1) All authorised variations of the Sub-Contract Works shall be valued in the manner provided by this clause and the value thereof shall be added to or deducted from the price specified in the Third Schedule hereto or as the case may require.

(2) The value of all authorised variations shall be ascertained by reference to the rates and prices (if any), specified in this Sub-Contract for the like or analogous work, but if there are no such rates and prices, or if they are not applicable, then such value shall be such as is fair and reasonable in all the circumstances. In determining what is a fair and reasonable valuation, regard shall be had to any valuation made under the Main Contract in respect of the same variation.

(3) Where an authorised variation of the Sub-Contract Works, which also constitutes an authorised variation under the Main Contract is measured by the Engineer thereunder, then provided that the rates and prices in this Sub-Contract permit such variation to be valued by reference to measurement, the Contractor shall permit the Sub-Contractor to attend any measurement made on behalf of the Engineer and such measurement made under the Main Contract shall also constitute the measurement of the variation for the purposes of this Sub-Contract and it shall be valued accordingly.

(4) Save where the contrary is expressly stated in any bill of quantities forming part of this Sub-Contract, no quantity stated therein shall be taken to define or limit the extent of any work to be done by the Sub-Contractor in the execution and completion of the Sub-Contract Works, but any difference between the quantity so billed and the actual quantity executed shall be ascertained by measurement, valued under the clause as if it were an authorised variation and paid in accordance with the provisions of the Sub-Contract.

(5) Where the Sub-Contractor has been ordered in writing by the Contractor to carry out any additional or substituted work on a daywork basis the Sub-Contractor shall be paid for such work under the conditions set out in the Daywork Schedule included in the Bill of Quantities under the Main Contract (if any) and at the rates and prices affixed to his Tender or as otherwise agreed. Failing the provision of a Daywork Schedule the Sub-Contractor shall be paid in accordance with the "Schedules of Dayworks carried out incidental to Contract Work" issued by the Federation of Civil Engineering Contractors current at the date of the execution of the daywork.

(6) Without prejudice to the generality of Clause 3(1) where the Sub-Contractor is to be paid for dayworks at the rates provided in the Daywork Schedule included in the Bill of Quantities under the Main Contract the Main Contractor shall provide the Sub-Contractor with a copy of the said Daywork Schedule which should be referred to in Schedule 2.

The Scheme of Valuation

The valuation process is as follows.

(*a*) The primary basis of valuation is by reference to rates and/or prices for the like or analogous work.

(*b*) Where there are no equivalent or similar agreed rates and/or prices, the value shall be such as is fair and reasonable. In this case regard should be had to any valuations made under the Main Contract. Clearly, the Main Contract may have agreed rates and/or prices which may be "weighted" and hence may not bear any resemblance to what is fair and reasonable.

Dayworks

Clause 9(5) provides the Contractor's authority to order work on a daywork basis.

10 Notices and claims

(1) Without prejudice to the generality of Clause 3 hereof whenever the Contractor is required by the terms of the Main Contract to give any return, account, notice or other information to the Engineer or to the Employer, the Sub-Contractor shall in relation to the Sub-Contract Works give a similar return, account or notice or such other information in writing to the Contractor as will enable the Contractor to comply with such terms of the Main Contract and shall do so in sufficient time to enable the Contractor to comply with such terms punctually.

Provided always that the Sub-Contractor shall be excused any non-compliance with this sub-clause for so long as he neither knew nor ought to have known of the Contractor's need of any such return, account, notice or information from him.

(2) Subject to the Sub-Contractor's complying with this sub-clause, the Contractor shall take all reasonable steps to secure from the Employer such contractual benefits, if any, as may be claimable in accordance with the Main Contract on account of any adverse physical conditions or artificial obstructions or any other circumstances that may affect the execution of the Sub-Contract Works and the Sub-Contractor shall in sufficient time afford the Contractor all information and assistance that may be requisite to enable the Contractor to claim such benefits. On receiving any such contractual benefits from the Employer (including any extension of time) the Contractor shall in turn pass on to the Sub-Contractor such proportion thereof as may in all the circumstances be fair and reasonable. Save as aforesaid the Contractor shall have no liability to the Sub-Contractor in respect of any condition, obstruction or circumstance that may affect the execution of the Sub-Contract Works and the Sub-Contractor shall be deemed to have satisfied himself as to the correctness and sufficiency of the Price to cover the provision and doing by him of all things necessary for the performance of his obligations under the Sub-Contract. Provided always that nothing in this Clause shall prevent the Sub-Contractor claiming for delays in the execution of the Sub-Contract Works solely by the act or default of the Main Contractor on the ground only that the Main Contractor has no remedy against the Employer for such delay.

(3) If by reason of any breach by the Sub-Contractor of the provisions of sub-clause (1) of this Clause the Contractor is prevented from recovering any sum from the Employer under the Main Contract in respect of the Main Works, then without prejudice to any other remedy of the Contractor for such breach,

the Contractor may deduct such sums from monies otherwise due to the Sub-Contractor under this Sub-Contract.

Synopsis

Clause 10 deals with two distinct matters. First, it deals with information which the Contractor is obliged to supply under the terms of the Main Contract. Second, it provides that, in specified circumstances, the Contractor should take reasonable steps to obtain benefits for the Sub-Contractor.

Information

Civil engineering contracts frequently require that information such as labour, plant, etc., returns, notices, accounts, etc. be provided either as part of a general data collection programme or in order to substantiate particular claims. Clauses 10(1) and (3) deal with this. They provide that where the Main Contractor is obliged to provide information to the Employer, the Sub-Contractor shall provide such information as is required to enable the Main Contractor to comply. The Sub-Contractor is not obliged to do this where he does not know, nor ought to know, of the requirement. Clause 10(3) provides that where the Sub-Contractor fails to provide this information and the Contractor is thereby prevented from recovering any sum from the Employer, the Contractor may deduct such monies from the Sub-Contractor's account.

Adverse physical conditions, artificial obstructions or other circumstances that may affect the execution of the Sub-Contract works

Clause 10(2) primarily covers situations entitling the Contractor to recover under Clause 12 of the ICE Conditions of Contract. While the phrase "or any other circumstances..." appears to widen this, the Court of Appeal has held[20] that claims outside the scope of Clause 12 do not fall within the scope of Clause 10(2).

20 *Balfour Beatty Construction Ltd* v. *Kelston Sparkes Contractors Ltd*, CA 10 July 1996.

Obligations provided for in Clause 10(2)

Clause 10(2) provides that the Contractor shall take "all reasonable steps" to recover under the Main Contract amounts for adverse physical conditions etc. that affect the Sub-Contract works and pass on to the Sub-Contractor "such proportion thereof as may in all the circumstances be fair and reasonable". Clause 10(2) provides that, except in the case of delay, this is the only entitlement available to the Sub-Contractor. The Sub-Contractor may claim for delays where the delay has been caused by a default of the Contractor. It is thought, however, that where the Sub-Contractor's loss is due to a misrepresentation or other default by the Contractor this provision will not limit the Contractor's liability. Frequently Sub-Contractors frame their claims for adverse conditions etc. as a breach of Clause 10(2) by the Contractor; the allegation is made that the Contractor has failed to take all reasonable steps to recover for adverse physical conditions on their behalf.

11 Property in Materials and Plant

(1) Where it is provided by the Main Contract that the property in any Sub-Contractor's Equipment, Temporary Works, materials or things whatsoever shall in certain events vest in the Employer or revest in the Contractor, then in so far as such Sub-Contractor's Equipment, Temporary Works, materials or things are to be provided by the Sub-Contractor in connection with the Sub-Contract Works, the property therein shall pass from the Sub-Contractor to the Contractor immediately before it is due to vest in the Employer in pursuance of the Main Contract and shall re-pass from the Contractor to the Sub-Contractor immediately after it has revested in the Contractor in pursuance of the Main Contract.

(2) Without prejudice to the generality of Clause 3 (Main Contract), the Sub-Contractor shall comply with the requirements of the Main Contract as to the bringing on to and removal from the Site of Sub-Contractor's Equipment, Temporary Works, materials and other things and in so far as any items thereof are hired by the Sub-Contractor, he shall comply with all the requirements of the Main Contract as to the terms of such hirings and as to the giving of information and certificates in relation to thereto.

This clause operates to allow property in equipment, materials, etc. to pass through the Contractor to the Employer. Sub-Contractors should appreciate that they may lose their equipment where the Employer is entitled to take over the Contractor's equipment under the terms of the Main Contract.

12 Indemnities

(1) The Sub-Contractor shall at all times indemnify the Contractor against all liabilities to other persons (including the servants and agents of the Contractor or Sub-Contractor) for bodily injury, damage to property or other loss which may arise out of or in consequence of the execution, completion or maintenance of the Sub-Contract Works and against all costs, charges and expenses that may be occasioned to the Contractor by the claims of such persons.

Provided always that the Contractor shall not be entitled to the benefit of this indemnity in respect of any liability or claim if he is entitled by the terms of the Main Contract to be indemnified in respect thereof by the Employer.

Provided further that the Sub-Contractor shall not be bound to indemnify the Contractor against any such liability or claim if the injury, damage or loss in question was caused solely by the wrongful acts or omissions of the Contractor, his servants or agents.

(2) The Contractor shall indemnify the Sub-Contractor against all liabilities and claims against which the Employer by the terms of the Main Contract undertakes to indemnify the Contractor and to the like extent, but no further.

The clause provides for both the Contractor and the Sub-Contractor to be indemnified by the other in specified circumstances. The Contractor has an indemnity in respect of "all liabilities to other persons ..." unless (*a*) the Contractor is entitled to be indemnified by the Employer in which case the Contractor must look to the Employer, or (*b*) where the injury, damage in question was caused solely by the Contractor. It is thought that the expression "solely" will not entitle the Sub-Contractor to this defence where the default of the Contractor is merely the dominant cause.[20]

13 Outstanding work and defects

(1) If the Sub-Contractor shall complete the Sub-Contract Works as required by Clause 2(1) before the substantial completion of the Main Works, or where under the Main Contract the Main Works are to be completed by sections before the substantial completion of the section or sections in which the Sub-Contract Works are comprised, the Sub-Contractor shall maintain the Sub-Contract Works in the condition required by the Main

Contract (fair wear and tear excepted) to the satisfaction of the Engineer and shall make good every defect and imperfection therein from whatever cause arising until such substantial completion of the Main Works or section thereof is achieved and subject to Clause 14 (Insurance), shall not be entitled to any additional payment for so doing unless such defect or imperfection is caused by the act, neglect or default of the Employer, his servants or agents under the Main Contract or of the Contractor, his servants or agents under the Sub-Contract.

(2) After completion of the Main Works or of the section or sections thereof in which the Sub-Contract Works are comprised, as the case may be, the Sub-Contractor shall maintain the Sub-Contract Works and shall make good such defects and imperfections therein as the Contractor is liable to make good under the Main Contract for the like period and otherwise upon the like terms as the Contractor is liable to do under the Main Contract.

Provided always that if any defect or imperfection made good by the Sub-Contractor under this sub-clause is caused by the act, neglect or default under the Sub-Contract of the Contractor, his servant or agents, then notwithstanding that the Contractor may have no corresponding right under the Main Contract, the Sub-Contractor shall be entitled to be paid by the Contractor his reasonable costs of making good such defect or imperfection.

This Clause requires the Sub-Contractor to do the following.

(*a*) Keep the Sub-Contract works in good condition from the date of its completion until the date of Substantial Completion of the Works. This obligation exists even though defects are caused by the Contractor; in this latter case, however, the Sub-Contractor shall be entitled to "additional payment". The amount of additional payment will, it is submitted, be a reasonable amount in all the circumstances.

(*b*) Maintain the works[21] as required by the Main Contract.

21 Note that the expression "maintain" is not used in the ICE Conditions Main Contract, though it is used here and in Clause 2(1).

14 Insurances

(1) The Sub-Contractor shall effect insurance against such risks as are specified in Part I of the Fifth Schedule hereto and in such sums and for the benefit of such persons as are specified therein and unless the said Fifth Schedule otherwise provides, shall maintain such insurance from the time that the Sub-Contractor first enters upon the Site for the purpose of executing the Sub-Contract Works until he has finally performed his obligations under Clause 13 (Outstanding Work and Defects).

(2) The Contractor shall maintain in force until such time as the Main Works have been substantially completed or ceased to be at his risk under the Main Contract, the policy of insurance specified in Part II of the Fifth Schedule hereto. In the event of the Sub-Contract Works, or any Sub-Contractor's Equipment, Temporary Works, materials or other things belonging to the Sub-Contractor being destroyed or damaged during such period in such circumstances that a claim is established in respect thereof under the said policy, then the Sub-Contractor shall be paid the amount of such claim, or the amount of his loss, whichever is the less, and shall apply such sum in replacing or repairing that which was destroyed or damaged. Save as aforesaid the Sub-Contract Works shall be at the risk of the Sub-Contractor until the Main Works have been substantially completed under the Main Contract, or if the Main Works are to be completed by sections, until the last of the sections in which the Sub-Contract Works are comprised has been substantially completed, and the Sub-Contractor shall make good all loss of or damage occurring to the Sub-Contract Works prior thereto at his own expense.

(3) Where by virtue of this Clause either party is required to effect and maintain insurance, then at any time until such obligation has fully been performed, he shall if so required by the other party produce for inspection satisfactory evidence of insurance and in the event of his failing to do so, the other party may himself effect such insurance and recover the cost of so doing from the party in default.

This clause requires the Sub-Contractor to maintain the insurance specified in Part 1 of the Fifth Schedule and the Contractor to maintain insurances specified in Part 2 of the Fifth Schedule. By Clause 14(3), where either party fails to provide satisfactory evidence of insurance the other may effect such insurance at the cost of the other party.

15 Payment

(1) (a) The Sub-Contractor shall not less than 7 days before the date specified in the First Schedule (the "Specified Date") or otherwise as agreed submit to the Contractor a written statement of the value of all work properly done under the Sub-Contract and of all materials delivered to the Site for incorporation in the Sub-Contract Works and if allowable under the Main Contract the value of off-site materials for incorporation in the Sub-Contract Works at the date of such statement. The statement shall be in such form and contain such details as the Contractor may reasonably require and the value of work done shall be calculated in accordance with the rates and prices, if any, specified in the Sub-Contract, or if there are no such rates and prices, then by reference to the Price.

(b) The statement submitted by the Sub-Contractor as provided in the preceding sub-clause shall constitute a "valid statement" for the purposes of this Clause but not otherwise.

(2) (a) The Contractor shall make applications for payment in accordance with the Main Contract and subject to the Sub-Contractor having submitted a valid statement shall include in such applications claims for the value of work and materials set out in such statement.

(b) In any proceedings instituted by the Contractor against the Employer to enforce payment of monies due under any certificate issued by the Engineer in accordance with the provisions of the Main Contract there shall be included all sums certified and unpaid in respect of the Sub-Contract Works.

(3) (a) Within 35 days of the Specified Date or otherwise as agreed but subject as hereinafter provided, there shall be due to the Sub-Contractor in respect of the value of the work and materials if included in a valid statement payment of a sum calculated in accordance with the rates and prices specified in this Sub-Contract, or by reference to the Price, as the case may require, but subject to a deduction of previous payments and of retention monies at the rate(s) specified in the Third Schedule hereto until such time as the limit of retention (if any) therein specified has been reached.

(b) Subject to Clauses 3(4), and 10(3) and 17(3) and as hereinafter provided and without prejudice to any rights which exist at Common Law the Contractor shall be entitled to withhold or defer payment of all or part of any sums otherwise due pursuant to the provisions hereof where:—

(i) The amounts of quantities included in any valid statement together with any other sums to which the Sub-Contractor might otherwise be entitled in the opinion of the Contractor will not justify the issue of an interim certificate, or

(ii) the amounts or quantities included in any valid statement together with any sums which are the subject of an application by the Contractor in accordance with Clause 15(2) are insufficient to justify the issue of an interim certificate by the Engineer under the Main Contract, or

(iii) the amounts or quantities included in any valid statement are not certified in full by the Engineer, providing such failure to certify is not due to the act or default of the Contractor, or

(iv) the Contractor has included the amounts or quantities set out in the valid statement in his own statement in accordance with the Main Contract and the Engineer has certified but the Employer has failed to make payment in full to the Contractor in respect of such amounts or quantities, or

(v) a dispute arises or has arisen between the Sub-Contractor and the Contractor and/or the Contractor and the Employer involving any question of measurement or quantities or any matter included in any such valid statement.

(c) Any payment withheld under the provisions of sub-clauses (b)(iii), (iv) or (v) above shall be limited to the extent that the amounts in any valid statement are not certified, not paid by the Employer or are the subject of a dispute as the case may be.

(d) In the event of the Contractor withholding any payment he shall notify the Sub-Contractor of his reasons in writing as soon as is reasonably practicable but not later than the date when such payment would otherwise have been payable.

(e) The provisions of this Clause with regard to the time for payment shall not apply to the amounts or quantities in any valid statement by the Sub-Contractor which are included in the Contractor's statement of final account to the

Employer under the provisions of the Main Contract. In respect of any such amounts or quantities payment shall be due 7 days after receipt by the Contractor of any payment which includes a sum in respect of such amounts or quantities.

(f) In the event of the Contractor failing to make payment of any sum properly due and payable to the Sub-Contractor or in the event of payment being withheld pursuant to sub-clause 15(3)(b)(iv) the Contractor shall, upon receiving a written claim for interest from the Sub-Contractor within 7 days of the date when such sum became payable, pay to the Sub-Contractor interest on such overdue sum at the rate payable by the Employer to the Contractor under the provisions of the Main Contract. Provided always that in the event of the Contractor not receiving written notice of a claim for interest within 7 days of the date when such sum became payable aforesaid, interest shall be payable by the Contractor on such sum from the date of receipt of the said written claim.

(g) Notwithstanding sub-clause (f) the Sub-Contractor shall be paid any interest actually received by the Contractor from the Employer which is attributable to monies due to the Sub-Contractor.

(4) (a) Within 35 days of the issue by the Engineer of a Certificate including an amount in respect of payment to the Contractor of the first half of the retention monies or where the Main Works are to be completed by sections for any section in which the Sub-Contract Works are comprised the Contractor shall pay to the Sub-Contractor the first half of the retention monies under this Sub-Contract.

(b) Within 7 days of the Contractor's receipt of any payment under the Main Contract which is by way of release of the second half of the retention monies the Contractor shall pay the Sub-Contractor the second half of the retention monies under this Sub-Contract.

(5) Within three months after the Sub-Contractor has finally performed his obligations under Clause 13 (Outstanding Work and Defects), or within 14 days after the Contractor has recovered full payment under the Main Contract in respect of the Sub-Contract Works, whichever is the sooner and provided that one month has expired since the submission by the Sub-Contractor of his valid statement of final account to the Contractor, the Contractor shall pay to the Sub-Contractor the

Price and/or any other sums that may have become due under the Sub-Contract, less such sums as have already been received by the Sub-Contractor on account.

Provided always that if the Contractor shall have been required by the Main Contract to give to the Employer or to procure the Sub-Contractor to give to the Employer any undertaking as to the completion or maintenance of the Sub-Contract Works, the Sub-Contractor shall not be entitled to payment under this Sub-Contract until he has given a like undertaking to the Contractor or has given the required undertaking to the Employer, as the case may be.

(6) The Contractor shall not be liable to the Sub-Contractor for any matter or thing arising out of or in connection with this Sub-Contract or the execution of the Sub-Contract Works unless the Sub-Contractor has made written claim in respect thereof to the Contractor before the Engineer issues the Defects Correction Certificate in respect of the Main Works, or, where under the Main Contract the Main Works are to be completed by sections, the Defects Correction Certificate in respect of the last of such sections in which the Sub-Contract Works are comprised.

Synopsis

The broad scheme of Clause 15 is as follows.

(*a*) The Sub-Contractor submits a statement giving his opinion of the value of the work done to date.

(*b*) The Contractor submits the statement for payment in accordance with the Main Contract.

(*c*) The Contractor is generally obliged to pay the amount of the statement (less appropriate retentions), but this obligation is subject to several exceptions. In essence, these exceptions establish a partial "pay-when-paid" regime.[22] Where the Employer does not pay monies to the Contractor in respect of the Sub-Contract Works, the Contractor is not obliged to pay the Sub-

22 See Hevey G., Pay when paid clauses (1995) 11 Constr. L.J. 79 and McCosker B., Pay when paid clauses — the Australian experience, (1995) CILL 1029. Section 113 of the Housing Grants, Construction and Regeneration Act 1996, when in force, will render conditional payment provisions ineffective.

Contractor. There are other exceptions which are declaratory of the set-off rights of the Contractor.

(*d*) Where the Contractor fails to make proper payment the Sub-Contractor is entitled to recover interest at the Main Contract rate.

The Sub-Contractor's statement

A "Specified Date" is set out in the First Schedule. Seven or more days before this date, the Sub-Contractor submits a statement of:

(*a*) the value of all work properly done under the Sub-Contract;
(*b*) all materials delivered to the site for incorporation in the Sub-Contract Works; and
(*c*) if allowable under the Main Contract, the value of off-site materials for incorporation in the Sub-Contract Works.

The statement relates to work and materials done or delivered at the date of the statement; work, materials, etc. which may be done between the date of the statement and the date of payment should not be included.

This statement should comply with the following in regard to its format.

(*a*) It shall be in such format as the Contractor may reasonably require. Thus where the Main Contract requires statements to be submitted in a particular form, the Contractor may require the Sub-Contractor to supply statements in a corresponding form.
(*b*) It shall contain such details as the Contractor shall reasonably require.
(*c*) The value of work shall be calculated in accordance with the rates and prices specified; or if none by reference to the price. Where there is simply a price, "by reference to" it means, it is submitted, by estimating the proportion of the work completed and applying for that proportion of the price.

The Contractor's obligation to make applications for payment

By Clause 15(2)(a) the Contractor is obliged to apply for payment on behalf of the Sub-Contractor where the Sub-Contractor has submitted a statement which complies with Clause 15(1)(a). The Contractor is not entitled to moderate or recast or delete any of the Sub-Contractor's claims. Clause 15(2)(b) provides an analogous obligation to enforce certificates on the Sub-Contractor's behalf.

Payment

The provisions in Clause 15(3) set out a detailed scheme for payment. The Contractor's basic obligation is to make payment within 35 days of the Specified Date[23] (or other agreed date); but this is subject to a significant number of exceptions. In practice, Clause 15(3) establishes a pay-when-paid regime except where the Employer's failure to pay is attributable to the breach by the Contractor.

Notifying the Sub-Contractor of reasons

By Clause 15(3)(d), where the Contractor withholds payment he must notify the Sub-Contractor of his reasons. This notification must not occur later than the date when such payment would otherwise be payable. The reasons might properly include (*a*) a clear statement of the facts which it is alleged found the reason, (*b*) the provisions of Clause 15(3) and/or other terms of the Sub-Contract which it is alleged entitle the Contractor to withhold payment, and (*c*) a statement of the sum which is being withheld and its breakdown in sufficient detail to enable the Sub-Contractor to make representations on the sum. Where the Contractor fails to notify the Sub-Contractor of his reasons it is not thought that the Sub-Contractor can set up this failure as a condition precedent to an entitlement to continue to withhold the money; it is thought that a claim for interest will normally be the Sub-Contractor's only remedy.

Interest

Clause 15(3)(f), (g) deals with interest. It seems that in order to recover interest under the Sub-Contract, the Sub-Contractor must make a claim not only for the principal sum but also a specific claim for interest. Otherwise the Sub-Contractor is only entitled to interest which the Contractor actually receives. It is thought, however, that this provision does not preclude the Sub-Contractor claiming for statutory interest[24] where the Contractor fails to pay over money due.

---

23 Unless the provisions of Clause 15(3)(e) apply, where the money is paid following an application by way of statement of final account.

24 For example under section 19A of the Arbitration Act 1950.

Retentions

By Clause 15(3)(a) the Contractor is entitled to deduct retention monies as specified in the Third Schedule. Upon retention monies being released under the Main Contract, Clause 15(4) provides time limits for the release of retentions by the Contractor.

Longstop for Sub-Contractor's claims

A party to a contract is ordinarily entitled to recover for breaches by the other for six years after the breach.[25] Clause 15(6) provides, however, that a claim is barred unless the claim is made prior to the issue of the Defects Correction Certificate. It is thought that this limitation is effective, save where the Contractor is unjustly enriched as a result of the provision (e.g. where the Contractor actually recovers sums on the Sub-Contractor's behalf) or there is fraud (e.g. where the Contractor deceives the Sub-Contractor about monies which have been paid over to him). The position where the Sub-Contractor's cause of action has not arisen at the time of the Defects Correction Certificate is also an area of uncertainty.

16 Determination of the Main Contract

(1) If the Main Contract is determined for any reason whatsoever before the Sub-Contractor has fully performed his obligations under this Sub-Contract, then the Contractor may at any time thereafter by written notice to the Sub-Contractor forthwith determine the Sub-Contractor's employment under the Sub-Contract and thereupon the Sub-Contractor shall, subject to Clause 11 (Property in Materials), with all reasonable speed remove his men and Sub-Contractor's Equipment from the Site.

(2) Upon such a determination of the Sub-Contractor's employment, the other provisions of this Sub-Contract shall cease to have effect and subject to sub-clause (3) hereof, the Sub-Contractor shall be entitled to be paid the full value, calculated by reference to the Price and to the rates and prices contained in any bill of quantities or schedule forming part of this Sub-Contract, of all work properly done on the Site by the Sub-

25 Twelve years in the case of a contract under seal.

Contractor and of all materials properly bought and left on the Site by the Sub-Contractor, together with his reasonable costs of removing his Sub-Contractor's Equipment from the Site, but less such sums as the Sub-Contractor has already received on account. Furthermore if at the date of such determination the Sub-Contractor has properly prepared or fabricated off the Site any goods for subsequent incorporation in the Sub-Contract Works and he shall deliver such goods to the Site or to such other place as the Contractor may reasonably direct, then he shall be paid for such goods as for materials properly brought and left on the Site by him.

Provided always that nothing herein shall affect the rights of either party in respect of any breach of this Sub-Contract committed by the other prior to such determination, nor any right which accrued to the Sub-Contractor prior to such determination to receive any payment which is not in respect or on account of the Price.

(3) If the Main Contract is determined by the Employer in consequence of any breach of this Sub-Contract by the Sub-Contractor, then the provisions of the preceding sub-clause as to payment shall not apply, but the rights of the Contractor and the Sub-Contractor hereunder shall be the same as if the Sub-Contractor had by such breach repudiated this Sub-Contract and the Contractor had by his notice of determination under sub-clause (1) of this Clause elected to accept such repudiation.

The scope of Clause 16

This clause provides that the Sub-Contract is determined automatically upon the determination of the Main Contract. No distinction is drawn between a determination at the Employer's or the Main Contractor's instance. There seems to be no requirement that the determination be under the express determination provisions of the Main Contract; thus it is thought that Clause 16 operates where the Contractor elects to determine the Main Contract following the Employer's repudiation.

Financial consequences

Where the Main Contract is determined and the Contractor determines the Sub-Contractor's employment, the Sub-Contractor is entitled to be paid for work done to date, unless he is responsible for the default which caused the determination—see Clause 16(3). The Sub-Contractor is entitled to the costs of removing his equipment from Site.

He will not, however, be entitled to recover loss of profit for the uncompleted work; this loss is simply a risk which the Sub-Contractor takes.

17 Sub-Contractor's default

(1) If:

(a) the Sub-Contractor fails to proceed with the Sub-Contract Works with due diligence after being required in writing so to do by the Contractor; or

(b) the Sub-Contractor fails to execute the Sub-Contract Works or to perform his other obligations in accordance with the Sub-Contract after being required in writing so to do by the Contractor; or

(c) the Sub-Contractor refuses or neglects to remove defective materials or make good defective work after being directed in writing so to do by the Contractor; or

(d) the Sub-Contractor becomes bankrupt or has a receiving order or administration order made against him or presents his petition in bankruptcy or makes an arrangement with or assignment in favour of his creditors or agrees to carry out the Sub-Contract under a committee of inspection of his creditors or (being a corporation) goes into liquidation (other than a voluntary liquidation for the purposes of amalgamation or reconstruction); or

(e) the Contractor is required by the Engineer to remove the Sub-Contractor from the Works after due notice in writing from the Engineer to the Contractor in accordance with the Main Contract.

then in any such event and without prejudice to any other rights or remedies, the Contractor may by written notice to the Sub-Contractor forthwith determine the Sub-Contractor's employment under this Sub-Contract and thereupon the Contractor may take possession of all materials, Sub-Contractor's Equipment and other things whatsoever brought on to the Site by the Sub-Contractor and may use them for the purpose of executing, completing and maintaining the Sub-Contract Works and may, if he thinks fit, sell all or any of them and apply the proceeds in or towards the satisfaction of monies otherwise due to him from the Sub-Contractor.

(2) Upon such a determination, the rights and liabilities of the Contractor and the Sub-Contractor shall, subject to the

preceding sub-clause, be the same as if the Sub-Contractor had repudiated this Sub-Contract and the Contractor had by his notice of determination under the preceding sub-clause elected to accept such repudiation.

(3) The Contractor may in lieu of giving a notice of determination under this Clause take part only of the Sub-Contract Works out of the hands of the Sub-Contractor and may by himself, his servants or agents execute, complete and maintain such part and in such event the Contractor may recover his reasonable costs of so doing from the Sub-Contractor, or deduct such costs from monies otherwise becoming due to the Sub-Contractor.

Clause 17 provides the Contractor with wide-ranging power to determine the Sub-Contract.

Written notice

Clause 17(1) requires that the determination takes effect by written notice. Such notice should specify the grounds upon which the Sub-Contract is being determined. Where the notice does not specify the grounds it is not thought that this will render it ineffective. It is not thought that the Court will order the Contractor to specify the grounds. However, a notice of determination must, it is submitted, be made in good faith and a notice which does not disclose its reasoning may suggest a lack of good faith.

Partial take-over

Clause 17(3) provides the Contractor with an option other than suffering the Sub-Contractor to continue or determining his employment entirely. Here, the Contractor takes part only of the work "out of the hands" of the Sub-Contractor. This right is not unrestricted. It is in lieu of giving a notice of determination and hence must be supported by one or more of the reasons set out in Clause 17(1).

The Sub-Contractor's liability for work completed by others

Where any work is taken out of the Sub-Contractor's hands, it is thought that the Sub-Contractor bears no liability for defects or delays attributable to events occurring after the take-over. Where, however, the Sub-Contractor has partially completed the works and those works are then completed by another employee of the Contractor, difficult questions of causation may need to be answered should defects arise.

18 Disputes

(1) If any dispute or difference of any kind whatsoever shall arise between the Contractor and the Sub-Contractor in connection with or arising out of the Sub-Contract, or the carrying out of the Sub-Contract Works (excluding a dispute concerning VAT but including a dispute as to any act or omission of the Engineer) whether arising during the progress of the Works or after their completion it shall be settled in accordance with the following provisions:

(2) For the purposes of sub-clauses 18(3) to 18(5) inclusive, a dispute is deemed to arise when one party serves on the other a notice in writing (herein called the Notice of Dispute) stating the nature of the dispute. Provided that no Notice of Dispute may be served unless the party wishing to do so has first taken any step or invoked any procedure available elsewhere in the Sub-Contract in connection with the subject matter of such dispute, and the other party has:

(a) taken such step as may be required or

(b) been allowed a reasonable time to take any such action.

(3) In relation to any dispute notified under sub-clause 18(2) and in respect of which no notice to refer under sub-clause 18(5) has been served, either party may, within 28 days of the service of the Notice of Dispute, give a notice in writing under the Institution of Civil Engineers' Conciliation Procedure (1988) or any amendment or modification thereof being in force at the date of such notice and the dispute shall thereafter be referred and considered in accordance with the said Procedure.

(4) Where a dispute has been referred to a conciliator under the provisions of sub-clause 18(3), either party may, within 28 days of the receipt of the conciliator's recommendation, refer the dispute to arbitration if a person can be agreed upon by the parties by serving on the other party a written Notice to Refer. Where a written Notice to Refer is not served within the said period of 28 days, the recommendation of the conciliator shall be deemed to have been accepted in settlement of the dispute.

(5) Where a dispute has not been referred to a conciliator under the provisions of sub-clause 18(3), then either party may within 28 days of service of the Notice of Dispute under sub-clause 18(2) refer the dispute to the arbitration of a person to be agreed upon by the parties by serving on the other party a written Notice to Refer. Where a Notice to Refer is not served within the said period of 28 days, the Notice of Dispute shall be deemed to have been withdrawn.

(6) If the parties fail to appoint an arbitrator within 28 days of either party serving on the other party a written Notice to Concur in the appointment of an arbitrator, the dispute shall be referred to a person to be appointed on the application of either party by the President (or if he is unable to act, by any Vice-President) for the time being of the Institution of Civil Engineers.

(7) Any such reference to arbitration shall be conducted in accordance with the Institution of Civil Engineers' Arbitration Procedure (1983) or any amendment or modification thereof being in force at the time of the appointment of the arbitrator.

(8) If any dispute arises in connection with the Main Contract and the Contractor is of the opinion that such dispute touches or concerns the Sub-Contract Works and the dispute is referred to a conciliator or an arbitrator under the Main Contract the Contractor may by notice in writing require that the Sub-Contractor provide such information and attend such meetings in connection therewith as the Contractor may request. The Contractor may also by notice in writing require that any such dispute under this Sub-Contract be dealt with jointly with the dispute under the Main Contract and in like manner. In connection with any such joint dispute the Sub-Contractor shall be bound in like manner as the Contractor by any recommendation of the conciliator or any award by an arbitrator.

Clause 18 provides for arbitration and much of the wording is related to and intended to mesh with Clause 66 of the ICE Conditions of Contract.[26]

Synopsis

Where a dispute has arisen, either party may serve a Notice of Dispute upon the other. The Notice of Dispute has a lifetime of 28 days. During its life, either party may serve:

(*a*) a notice of conciliation; or
(*b*) a notice to refer to arbitration.

26 Reference should be made to the commentary on Clause 66 in Chapter 8.

Where more than one notice is served, the first notice is the effective notice. Thus a claimant who is determined not to have the matter conciliated can serve a Notice to Refer close on the heels of the Notice of Dispute, or even in the same document.

Connected disputes

It is convenient that connected disputes under the Sub-Contract and the Main Contract be resolved together. This ensures that the evidence and submissions are heard only once, thus saving costs. It also avoids the possibility of inconsistent findings. In order to ensure that connected disputes are heard together, Clause 18(8) entitles the Contractor to require the Sub-Contractor to join in the Main Contract arbitration. A prerequisite of the Contractor's notice is his opinion that the Main Contract dispute touches or concerns the Sub-Contract Works; this opinion is not wholly in the Contractor's discretion but must be based on some proper ground.[27] Whether or not there is a dispute under the Main Contract is a question of fact; such dispute may arise before the Contractor serves a Notice of Dispute.[28] Where such a dispute exists and the Contractor serves a notice under Clause 18(8), any subsequent appointment of an arbitrator will be invalid unless the Contractor concurs in the appointment.[29]

19 Value Added Tax

(1) The Sub-Contractor is deemed not to have allowed in the Price for any tax payable by him as a taxable person to the Commissioners of Customs and Excise.

(2) Tax Invoices shall not be submitted to the Contractor by the Sub-Contractor.

(3) Where the Sub-Contractor is registered as a taxable person under the Value Added Tax Act 1983 as amended from time to time there shall be added to the amount of every payment made by the Contractor to the Sub-Contractor pursuant to Clause 15

27 *Erith Contractor Ltd* v. *Costain Civil Engineering Ltd* (1993) (unreported), cited in Hevey G., Pay when paid clauses (1995) 11 Constr. L.J. 79 at 83.

28 *M. J. Gleeson Group plc* v. *Wyatt of Snetterton Ltd* (1994) 72 BLR 15 (CA).

29 *M. J. Gleeson Group plc* v. *Wyatt of Snetterton Ltd* (1994) 72 BLR 15 (CA).

and paid by the Contractor a separately identified amount equal to the amount of tax properly payable by the Sub-Contractor in respect of the taxable supply to which the payment made under Clause 15(3) relates.

(4) Where the Contractor operates the Authenticated Receipt System the Sub-Contractor will, upon receipt of the Contractor's Certificate of Payment and the payment detailed thereon, within 7 days return to the Contractor an Authenticated Receipt as required by the relevant VAT Regulations detailing the net payment and the Value Added Tax on the said payment.

(5) Where the Contractor operates Self-Billing procedures no Authenticated Receipt shall be required.

(6) Where the Contractor makes any provision pursuant to Clause 4 (Contractor's Facilities) and that provision is a taxable supply to the Sub-Contractor by the Contractor the Sub-Contractor shall pay tax to the Contractor in accordance with the requirements of the Contractor at the rate properly payable by the Contractor to the Commissioners of Customs and Excise.

Clause 19 provides that invoices will be net of VAT.

20 Law of Sub-Contract

The Law of the Country applying to the Main Contract shall apply to this Sub-Contract.

Schedules

The form of contract is published in a booklet which contains proformas for five schedules, as follows.

First Schedule	Particulars of Main Contract	
Second Schedule	(A)	Further documents forming part of the Sub-Contract
	(B)	Sub-Contract Works
	(C)	Fluctuation Provisions (if any)
Third Schedule	(A)	The Price/Measure and Value
	(B)	Retentions
	(C)	Period for Completion
Fourth Schedule	Contractor's Facilities	
	Part I	Common Facilities

	Part II	Exclusive Facilities
Fifth Schedule	Insurances	

Guidance notes

Following the form for signatures and/or sealing etc., there follows "Notes for the guidance of contractors on the completion of the schedules". These notes do not form part of the Sub-Contract—see Clause 1(3).

10

Design and construct contracts and the ICE Design and Construct Conditions of Contract

A design and construct contract is one in which the contractor agrees not only to be responsible for the construction of the works to be provided, but also to be responsible[1] for their design. Contracts of this sort are becoming increasingly common in civil engineering.

1. Introduction to design and construct contracts

A contract where the contractor designs the works as well as constructing them provides a number of benefits, particularly to the employer. These include the following.

Single point liability

Where a defect arises in works procured using traditional arrangements, it is frequently not clear whether the fault lies with the designer or with the contractor. Under a design and construct scheme the responsibility clearly lies with the contractor. Thus an employer is able to seek redress more readily.

1 Note that, strictly speaking, the key factor is not whether the contractor carries out the design, but whether he is responsible for it. Frequently, as described below, the design is physically carried out by consultants chosen by the employer; nevertheless, by the terms of the contract, the contractor takes over the design and becomes fully responsible for it.

Constructability

Using traditional arrangements, the designer's major preoccupation is to produce a design which is safe and able to meet any technical requirements in its completed condition. A design and construct contractor, however, works closely with the designers to produce a scheme which may also be constructed with efficiency. This produces cheaper construction which benefits the employer in the guise of lower tender prices.

Simpler organisation from the employer's perspective

Using traditional contracts, the employer engages a number of consultants (designer, quantity surveyors, etc.) as well as the contractor. Using design and construct arrangements, a firm of consultants is generally appointed to draw up the "employer's requirements" and to watch out for the employer's interests; however, this relationship tends to be less demanding on the employer's organisation.

Other factors which need to be addressed by parties considering entering into a design and construct contract include the following.

The standard of design care

Where a contractor agrees to design and construct works W for £*x*, without expressly agreeing terms as to the standard of design care, the agreement is ordinarily construed to mean that the works as completed will be fit for their known purpose.[2] Accordingly, where the works fail to meet reasonable performance, accommodation, etc. criteria the contractor will be in breach of contract. This will be so even though the defect arises from the design and the design meets the ordinary test of "reasonable skill and care". Many contractors are unhappy to take on a responsibility to provide works which are fit for their purpose,

2 *Greaves Contractors* v. *Baynham Meikle & Partners* [1975] 1 WLR 1095 (CA); *Hawkins* v. *Chrysler (UK) Ltd and Byrne Associates* (1986) 38 BLR 36.

which leaves them exposed to the employer and deprives them of a similar cause of action against their designers.[3] Accordingly, many design and construct standard form contracts[4] specifically limit the contractor's responsibility to the provision of a design which is done with reasonable skill and care. The effect of this, however, is, to create a risk for the employer in two senses. First, the employer carries the risk of work not being fit for its purpose where it has, nevertheless, been designed with reasonable skill and care. Second, where it is unclear whether defects result from the design or from the construction, there is a litigation risk; where it transpires that the defect is in the design and the design was done with reasonable skill and care, the employer's action fails and hence not only does he get no redress but he has to pay the contractor's costs.

Imposed design requirements

The employer will require that the works comply with specified performance criteria. He may, however, also stipulate that they meet other specified requirements. For instance the "employer's requirements" may specify a cable stay bridge with a specified cable arrangement, with cables of a specific type. Clearly such requirements impinge on the designer's function. This raises the question of the contractor's responsibility for a design where defects are attributable (partly or wholly) to the specified requirements. It is submitted that the matter is one of construing the contract. Where it seems in the light of all the circumstances that the employer has not placed the responsibility for these elements of design upon the contractor, the contractor is not responsible for them.[5] Where, however, the

---

3 It may be possible to require the designer to contract on a fitness for purpose basis; however, many are unwilling to do this either for professional or for insurance reasons. Note also that in some cases, a designer may owe a fitness for purpose design duty: see *Greaves Contractors* v. *Baynham Meikle & Partners* [1975] 1 WLR 1095 (CA).

4 Including the ICE Design and Construct Conditions of Contract.

5 There is a point of general principle, namely that where an employer requires his contractor to use specific techniques or material, the contractor will not warrant that the result will be fit for its purpose unless the terms of the contract clearly show that a different result was intended. See e.g. *Young & Marten* v. *McManus Childs* [1969] 1 AC 454 (HL); *Independent Broadcasting Authority* v. *EMI Electronics Ltd and BICC Construction Ltd* (1980) 14 BLR 1 (HL).

contractor has clearly agreed to take over the design, and constructs the works in accordance with that design, he will be responsible for it.[6] This will be so whether or not the specified requirements made it impossible for it to function safely or properly.[7] It is a matter of "presumed intention".[8] Note, however, that a contractor who recognises the difficulty prior to construction design cannot be required to construct it as currently specified;[9] it is thought that a term will ordinarily be implied that the employer will allow such reasonable modification of any of his requirements so as to enable a safe design to be effected.

Novation[10]

Employers frequently commence a design and construct project by taking on design consultants who develop a scheme. Where there are any novel features in the design, including structural innovations, unusual lighting schemes, mechanical plant, etc., the designer may work these elements up into a near-completed design. It is, in such a case, inconvenient for the designer's experience to be lost when the

6 In some cases, it may be possible to find an express or implied term that the specified requirements are not inimical to the proper design, construction and functioning of the works. In such a case, the employer is in breach where the specified requirements are incompatible with other contract requirements. It is submitted, however, that such terms will only be implied where necessary to give the contract business efficacy. Where the terms are clear in placing the responsibility upon the contractor, this presumed intention of both parties will prevail. The ICE Design and Construct Conditions of Contract are not altogether clear on this question: see the commentary below, and Clauses 8(1) and 5(1)(b) etc.

7 But see *H. R. & S. Sainsbury Ltd* v. *Street* [1972] 3 All ER 1127 where a possible doctrine of "partial excuse" was suggested.

8 *McRae* v. *Commonwealth Disposals Commission* (1950) 84 CLR 377 (High Court of Australia) per Dixon and Fullagar JJ. "The common law has generally been true to its theory of simple contract, and it has always regarded the fundamental question as being: 'what did the promisor really promise?' Did he promise to perform his part in all events ...? So questions of intention or 'presumed intention' arise, and these must be determined in the light of the words used by the parties and reasonable inferences from all the surrounding circumstances ... the problem is fundamentally one of construction."

9 In other words, there is no possibility of the court ordering specific performance.

10 See generally McNicholas, Novation of consultants to design build contractors, (1993) Constr. L.J. 263.

contractor takes over the design. The designer may, therefore, continue as the employer's agent for supervising the works or he may be "novated" to the contractor.[11] Novation[12] involves the dissolution of the contract between the employer and the designer and the creation of a new contract between the contractor and the designer.[13] This is achieved in practice by the inclusion of a term in the original contract between the designer and the employer that both will agree to the novation; and it will be a term of the contractor's contract that he will adopt the design and the designer in this way.[14] The novation agreement usually requires that, as between the employer and contractor, the contractor takes responsibility for the design from the moment of the conception of the design.

Collateral warranties

Where the contractor engages a designer, either by novation or as a domestic sub-contractor, the employer will have an interest in the performance of the designer. He may, therefore, wish to agree a collateral warranty so that whenever the designer is in breach of his design obligation to the contractor,[15] he is also in breach of the collateral warranty. The employer's right to take such a warranty is ordinarily included in the original contract between the designer and the employer (if there was one); alternatively, it will be a term in the contract between the employer and the contractor that no designer will be engaged who does not agree to the warranty.

11 There are strong reasons for the employer to prefer the latter. For instance, should a dispute arise he needs impartial advice, which the designer may not be able to give if there is criticism of the design.

12 See Chapter 3, section 6.

13 Generally, it is in the employer's interest to maintain a contract between himself and the consultant. This should be done expressly by the execution of a new agreement.

14 The similarity with the nomination of sub-contractors should be noted. In appropriate cases, the law relating to nominated sub-contractors will apply. Thus where, for example, the contractor is obliged to take on designer X and X is then dissolved (e.g. goes into liquidation if a company) the contractor will not be liable for any delay while the employer organises a renomination—*Bickerton* v. *North West Metropolitan Regional Hospital Board* [1970] 1 WLR 607 (HL).

15 The terms of the warranty are generally agreed in advance; otherwise the arrangement may be construed as an unenforceable agreement to agree: *May & Butcher* v. *R.* [1934] 2 KB 17 (HL). Frequently, one sees warranties which are more than collateral; they appear to give the employer a cause of action in a wider range of situations than provided for by the main design contract; likewise, the damages which flow from the breach are frequently not limited to the amount that the contractor could recover from the designer.

The Construction (Design and Management) Regulations 1994[16]

Construction projects are now subject to these regulations. Where a contractor takes on a design and construct role, he will need to satisfy himself as to the effect of these regulations. By Regulation 13 the designer shall ensure that any design he prepares pays proper regard to risks during and after construction and gives priority to measures which will protect people. He must ensure that information is available as to the design and he must cooperate with the planning supervisor. Where the designer fails in any of these regards, the planning supervisor may in appropriate circumstances stop the work until proper amendments are made. This suspension will be at the contractor's risk. Where the employer's requirements contain a provision that the design will accord with the 1994 Regulations[17] (as will ordinarily be the case) and the works are not safe, e.g. for cleaning, as required by the Regulations, this may amount to a "defect" entitling the employer to undertake "remedial works", the cost of which may be claimed from the contractor.

2. Introduction to the ICE Design and Construct Conditions of Contract

The ICE Design and Construct Conditions of Contract were drawn up by the Conditions of Contract Standing Joint Committee (CCSJC) and were published in 1992. The wording, numbering and layout of the clauses were designed to be as similar as possible to the "standard"[18] ICE Conditions of Contract. The provenance of the majority of clauses is readily apparent, and many are identical to those found in the standard Conditions of Contract. Consequently, this chapter focuses on aspects of difference and no clause-by-clause commentary is given.

16 See also the commentary on Clause 71 of the ICE Conditions of Contract in Chapter 8.

17 Note that the breaches of the Regulations, except in two specific cases, do not confer civil liability—Reg. 21. Hence the need for this term to be agreed.

18 The expression "standard" will be used here to describe the traditional form of the Conditions of Contract.

Principal points of difference

The ICE Design and Construct Conditions of Contract provide a design and construct obligation and a lump sum payment regime. The standard Conditions of Contract, on the other hand, is a construct only contract with a measure and value payment regime.[19]

Terminology compared with the standard ICE Conditions of Contract

The basic terms used are similar to those used in the standard ICE Conditions of Contract. The parties are the Employer and the Contractor. The obligation is to provide the Works on the Site. The basic performance certificates are the Certificate of Substantial Completion and the Defects Correction Certificate. However, there are also notable differences. For example, the Engineer is replaced by an Employer's Representative and the Drawings and Specification give way to the Employer's Requirements and the Contractor's Submission.

The Contract

The Contract and a number of related terms are defined in Clause 1(1):

1 Definitions

(1) (g) "Contract" means the Conditions of Contract the Employer's Requirements the Contractor's Submission and the written acceptance thereof together with such other documents as may be expressly agreed between the parties and the Contract Agreement (if completed).

(e) "Employer's Requirements" means the requirements which are identified as such at the date of the award of the Contract and any subsequent variations thereto and which may describe the standards performance and/or objectives that are to be achieved by the Works or parts thereof.

19 Although, as noted in Chapter 8, the CCSJC is considering whether to broaden the ambit of the standard Conditions of Contract to allow a variety of design and payment options.

(f) "Contractor's Submission" means the tender and all documents forming part of the Contractor's offer together with such modifications and additions thereto as may be agreed between the parties prior to the award of the Contract.

Thus the two principal documents which set out the scope of the Works are the Employer's Requirements and the Contractor's Submission. The Employer's Requirements set out whatever requirements the Employer wishes to specify. For general civil engineering works, which tend to be functional, these may include layouts, performance requirements, interfacing with other works, features, e.g. for maintenance and upgrading, materials specifications, etc. Where the Works are visible (e.g. bridges) the Employer's Requirements may prescribe the style, dimensions and other aesthetic characteristics. Where the Works include specialist features, such as a cable stayed structure, the Employer may even stipulate many of the design details.[20] The Contractor's Submission will be in a form required by the invitation to tender and will form, together with his prices, the basis of the Contractor's tender. It will set out the means by which the Contractor proposes to design and construct the Works, including such conceptual and detailed design ideas as are called for in the invitation to tender.

DOCUMENTATION AND INFORMATION

5 Contract documents

(1) (a) The several documents forming the Contract are to be taken as mutually explanatory of one another.

(b) If in the light of the several documents forming the Contract there remain ambiguities or discrepancies between the Employer's Requirements and the Contractor's Submission the Employer's Requirements shall prevail.

(c) (i) Any ambiguities or discrepancies within the Employer's Requirements shall be explained and adjusted by the Employer's Representative who shall thereupon issue to the Contractor appropriate instructions in writing.

20 See below for a discussion of Clause 8 on this point.

(ii) Should such instructions involve the Contractor in delay or disrupt his arrangements or methods of construction so as to cause him to incur cost beyond that reasonably to have been foreseen by an experienced contractor at the time of the award of the Contract then the Employer's Representative shall take such delay into account in determining any extension of time to which the Contractor is entitled under Clause 44 and the Contractor shall subject to Clause 53 be paid in accordance with Clause 60 the amount of such cost as may be reasonable. Profit shall be added thereto in respect of the design and construction of any additional permanent or temporary work.

(d) Any ambiguities or discrepancies within the Contractor's Submission shall be resolved at the Contractor's expense.

Supply of Contract documents

(2) Upon the award of the Contract the Employer shall assemble two complete copies of the Contract of which one copy shall be supplied to the Contractor free of charge.

This purports to be cast in similar terms to Clause 5 of the Standard Contract. However, an order of precedence is given; the Employer's Requirements prevail. It is thought, however, that this is subject to the general rules of construction of contracts. Thus, where the Contractor's Submission is specifically amended during the negotiations prior to the finalisation of the agreement, these amended items may take priority over the Employer's Requirements, notwithstanding Clause 5.

The Employer's Representative

1 Definitions

(c) "Employer's Representative means the person appointed by the Employer to act as such for the purposes of the Contract or any other person so appointed from time to time by the Employer and notified in writing as such to the Contractor.

Consent and approval

(7) (a) The giving of any consent or approval by or on behalf of the Employer's Representative shall not in any way relieve the Contractor of any of his obligations under the Contract or of his duty to ensure the correctness accuracy or suitability of the matter or thing which is the subject of the consent or approval.

(b) Failure by the Employer's Representative or any of his assistants to disapprove or object to any matter or thing shall not prejudice his power subsequently to take action under the Contract in connection therewith.

(c) Acceptance by the Employer's Representative of any programme or revised programme in accordance with Clause 14 shall not in any way imply that the programme so accepted is feasible suitable or appropriate and the Employer's rights under the Contract as to time shall in no way be impaired thereby.

2 Duties and authority of Employer's Representative

(1) (a) The Employer's Representative shall carry out the duties and may exercise the authority specified in or necessarily to be implied from the Contract.

(b) Except as expressly stated in the Contract the Employer's Representative shall have no authority to amend the Contract nor to relieve the Contractor of any of his obligations under the Contract.

Named individual

(2) Within 7 days of the award of the Contract and in any event before the Commencement Date the Employer shall notify to the Contractor in writing the name of the Employer's Representative. The Employer shall similarly notify the Contractor of any replacement of the Employer's Representative.

Delegation by the Employer's Representative

(3) The Employer's Representative may from time to time delegate to any person (including assistants appointed under sub-clause (4) of this Clause) any of the duties and authorities vested in him and he may at any time revoke such delegation.

Any such delegation

(a) shall be in writing and shall not take effect until such time as a copy thereof has been delivered to the Contractor or a representative appointed under Clause 15

(b) shall continue in force until such time as the Employer's Representative shall notify the Contractor or his representative in writing that the same has been revoked and

(c) shall not be given in respect of any decision to be taken or certificate to be issued under Clause 12(6) 15(2)(b) 44 46(3) 48 60(4) 61 or 65.

The Employer's Representative is able to delegate his authority to any person, save for his authority "in respect of any decision to be taken or certificate to be issued under Clauses 12(6), 15(2)(b), 44, 46(3), 48, 60(4), 61 or 65". These are broadly equivalent to the powers which the Engineer may not delegate under the standard Conditions of Contract. The role of the Employer's Representative is analogous in many respects to that of the Engineer under the standard Conditions of Contract. However, the Engineer's Representative is more able to devote himself to pursuing his client's interests: there is no equivalent of Clause 2(8) which requires the Engineer to act impartially. Nevertheless, it is submitted that he must act fairly in giving decisions and certificates. For example, when deciding whether or not ground conditions could have been reasonably foreseen by an experienced contractor, or whether a certificate of Substantial Completion is due etc., he must act fairly; and any interference from the Employer will be a breach of contract.[21]

An important difference between the Engineer and the Employer's Representative relates to the dispute resolution procedure. Under the standard Conditions of Contract, the parties are required to refer disputes to the Engineer prior to taking the matter to arbitration. Under the present contract, the Employer's Representative has no such role. The matter proceeds to Conciliation instead. See "Dispute resolution" below.

The Contractor's obligations

These are set out in Clause 8.

8 Contractor's general obligations

(1) The Contractor shall subject to the provisions of the Contract and save in so far as it is legally or physically impossible

21 See *London Borough of Merton* v. *Stanley Hugh Leach* (1985) 32 BLR 51. Note that this case involved a consideration of the role of the Architect under the JCT Contract, but it is submitted that the broad principle remains the same.

(a) design construct and complete the Works and

(b) provide all design services labour materials Contractor's Equipment Temporary Works transport to and from and in or about the Site and everything whether of a temporary or permanent nature required in and for such design construction and completion so far as the necessity for providing the same is specified in or reasonably to be inferred from the Contract.

Contractor's design responsibility

(2) (a) In carrying out all his design obligations under the Contract including those arising under sub-clause (2)(b) of this Clause (and including the selection of materials and plant to the extent that these are not specified in the Employer's Requirements) the Contractor shall exercise all reasonable skill care and diligence.

(b) Where any part of the Works has been designed by or on behalf of the Employer and that design has been included in the Employer's Requirements the Contractor shall check the design and accept responsibility therefor having first obtained the approval of the Employer's Representative for any modifications thereto which the Contractor considers to be necessary.

Quality assurance

(3) To the extent required by the Contract the Contractor shall institute a quality assurance system.

The Contractor's quality plan and procedures shall be submitted to the Employer's Representative for his prior approval before each design and each construction stage is commenced.

Should the Employer's Representative fail within a reasonable period following the Contractor's submission or re-submission of the quality plan and procedures to notify the Contractor either that he approves thereto or that he is withholding his approval he shall take such failure into account in determining any extension of time to which the Contractor is entitled under Clause 44 and the Contractor shall subject to Clause 53 be paid in accordance with Clause 60 any reasonable cost which may arise from such failure.

Compliance with such approved quality assurance system shall not relieve the Contractor from any of his other duties obligations or liabilities under the Contract.

Statutory checks

(4) Where any Act of Parliament Regulation or Bye-law requires that a separate check of the design or a test shall be carried out prior to the construction or loading of any permanent and temporary works the Contractor shall arrange and pay for such check or test.

Contractor responsible for safety

(5) The Contractor shall take responsibility for the safety of the design and for the adequacy stability and safety of all site operations and methods of construction.

Clause 8 is an important clause. It sets out the basic obligation, namely to "design construct and complete the Works". The express provisions in Clause 8(1)(b)—to do all things necessary;[22] and in Clause 2(a)—to select materials and plant with all reasonable skill care and diligence[23]—will ordinarily arise by operation of law in any event.[24]

The detail of the clause is important whenever the Employer's Requirements do more than set out quantifiable performance criteria. Where they stipulate materials, suppliers of mechanical plant, structural layouts, aesthetic conditions, etc. these may be incompatible with the proper design, construction or functioning of the Works. Here the position appears to be as follows.

(*a*) Where Employer's Requirements specify materials or plant, the Contractor is responsible neither for their operation nor even for using skill and care in their selection. This seems to follow from Clause 8(2)(a) "to the extent that these are not specified in the Employer's Requirements". Where materials or plant are not "specified"[25] the Contractor must exercise all reasonable skill care and diligence in selecting them.

22 See *Williams* v. *Fitzmaurice* (1853) 3 H & N 844.

23 See *Young & Marten* v. *McManus Childs* [1969] 1 AC 454 (HL).

24 "Plant" presumably means mechanical plant which forms part of the Permanent Works, rather than constructional plant.

25 That is to the extent that the Contractor has some discretion in their selection.

(*b*) Where the Works cannot be designed, constructed or commissioned in accordance with the Employer's Requirements, the Contractor faces an apparently intractable problem. To fail to design in accordance with the Employer's Requirements seems to be a breach; and yet to design defective Works is also a breach. This problem is addressed in two distinct provisions in the clause, neither of which is happily worded. First, the Contractor is only obliged to execute the Works "in so far as it is legally or physically possible". It might reasonably be said that there is a legal impossibility where compliance with a term is both to comply with and to breach a contract; likewise, where compliance with any term produces Works which cannot physically meet the specified performance, this is a physical impossibility. The second provision is given in Clause 8(2)(b). The wording here is not clear, but seems to entitle the Contractor to "check" the design. This suggests that there is no obligation to take on a design for which a check discloses clear problems. This is reinforced by the provision which allows the Contractor to modify the design where the Employer's Representative gives his approval. It is thought that the proper construction of this provision is that where a check discloses a latent risk which a reasonable designer might properly decline to take, the Contractor is entitled to refuse to perform the design as it stands.

(*c*) Where the Employer's Requirements stipulate for impossible, impracticable, ambiguous or other defective design requirements, the provisions of Clause 5 may be important. This requires that "any ambiguities or discrepancies within the Employer's Requirements shall be explained and adjusted by the Employer's Representative". This suggests that where the Employer's Requirements are inimical to the proper completion of the Works, the Contractor is not obliged to complete the relevant aspects of the Works without having the matter clarified by the Employer's Representative.

Variations

A variation to the Employer's Requirements arises under Clause 51.

51 Ordered variations to Employer's Requirements

(1) The Employer's Representative shall have power after consultation with the Contractor's Representative to vary the Employer's Requirements. Such variations may include

additions and/or omissions and may be ordered at any time up to the end of the Defects Correction Period for the whole of the Works.

Ordered variations to be in writing

(2) All variations shall be ordered in writing but the provisions of Clause 2(5) in respect of oral instructions shall apply.

Variation not to affect Contract

(3) No variation ordered under this Clause shall in any way vitiate or invalidate the Contract but the fair and reasonable value (if any) of all such variations shall be taken into account in ascertaining the amount of the Contract Price except to the extent that such variation is necessitated by the Contractor's default.

A variation may be ordered because of some additional requirement of the Employer which comes to light, or because of some incompatibility in the current Employer's Requirements; this may be due to the Employer's Representative resolving ambiguities or discrepancies, or because unforeseeable ground conditions arise which make it convenient to order a variation.[26] Where the Employer's Requirements are varied, the Contractor will be obliged to take over the design responsibility. It is submitted, however, that a term must be implied into the agreement[27] that the Employer's Requirements may not be varied so as materially to increase the Contractor's exposure to design risk.[28] Thus, for example, the Employer's Representative is not entitled to vary non-specific Employer's Requirements for a roof structure to a specific innovative[29] structural arrangement.

---

26 See Clause 12(4) of the ICE Design and Construct Conditions of Contract.

27 As a matter of business efficacy.

28 This would seem to follow both from principle and by analogy with the extent of permitted variations under a traditional civil engineering contract: see *Blue Circle Industries* v. *Holland Dredging Co. (UK) Ltd* (1987) 37 BLR 40.

29 Innovation requires the designer to take especial care: *Independent Broadcasting Authority* v. *EMI Electronics Ltd and BICC Construction Ltd* (1980) 14 BLR 1 (HL).

The significance of the phrase "after consultation with the Contractor's Representative" is unclear. It is not thought that the Contractor's Representative has any right to veto a variation which otherwise falls within the scope of the Contract. Its practical significance is that the variation is not ordered until the Contractor has had an opportunity to discuss it; and it may in law act as a condition precedent to the issue of a valid variation order.

Payment

The ICE Design and Construct Conditions of Contract are designed to operate as a lump sum contract. While the Contract provides for the case where there is a bill of quantities, schedule of rates, etc. it does not presuppose that this is the case. Clause 56(1) is almost dismissive of the possibility of remeasurement, and Clause 57 (which in the standard Conditions of Contract specifies the method of measurement) is pointedly not used.

Perhaps the most common method of setting up the payment scheme is the insertion of a Schedule of Prices in the Contractor's Submission. This contains a list of activities or elements of work with prices for each. The submission will state whether payment is due for activities or elements which are partly complete (in proportion to the degree of completion) or alternatively whether payment is due in respect of completed activities or elements only. Furthermore, the Contractor's Submission may contain a list of rates and/or prices which can be used to value additional or omitted work.

The terminology used for payment under this contract differs from that used in the standard ICE Conditions of Contract.

1 Definitions

(1) (h) "Contract Price" means the sum to be ascertained and paid for the design construction and completion of the Works in accordance with the Contract.

(i) "Prime Cost Item" means an item in the Contract which contains (either wholly or in part) a sum referred to as Prime Cost (PC) which will be used for the supply of goods materials or services for the Works.

(j) "Contingency" means any sum included and so designated in the Contract as a specific contingency for the carrying out of work or the supply of goods materials or services which may be used in whole in part or not at all in accordance with the specific requirements stated therefor.

The price is adjusted in various ways. The Contractor becomes entitled to additional cost and profit in various situations (e.g. where the Employer's Representative fails to issue further necessary information—Clause 6(1)(b); where physical conditions are not reasonably foreseeable—Clause 12(6), etc.).

Variations are valued in accordance with Clause 52.

52 Valuation of ordered variations

(1) When requested by the Employer's Representative the Contractor shall submit his quotation for the work as varied and his estimate of any delay. Wherever possible the value and delay consequences (if any) of each variation shall be agreed before the order is issued or before work starts.

(2) In all other cases the valuation of variations ordered by the Employer's Representative in accordance with Clause 51 shall be ascertained as follows.

(a) As soon as possible after receipt of the variation order the Contractor shall submit to the Employer's Representative

(i) his quotation for any extra or substituted works necessitated by the variation having due regard to any rates or prices included in the Contract and

(ii) his estimate of any delay occasioned thereby and

(iii) his estimate of the cost of any such delay.

(b) Within 14 days of receiving the said submissions the Employer's Representative shall

(i) accept the same or

(ii) negotiate with the Contractor thereon.

(c) Upon reaching agreement with the Contractor the Contract Price shall be amended accordingly.

(d) In the absence of agreement the Employer's Representative shall notify the Contractor of what in his opinion is a fair and reasonable valuation and thereafter shall make such interim valuations for the purposes of Clause 60 as may be appropriate.

Daywork

(3) The Employer's Representative may if in his opinion it is necessary or desirable order in writing that any additional or substituted work shall be executed on a daywork basis in accordance with the provisions of Clause 56(3).

The procedure here clearly involves negotiation, and envisages—as will frequently be the case—that the Employer's Representative will conclude the negotiation by notice of his opinion of a fair valuation.

Contingencies and Prime Cost Items are used as follows.

58 Use of Contingency and Prime Cost Items

(1) The Contractor shall not commence work on any Contingency or Prime Cost Item until he has secured the consent thereto of the Employer's Representative which consent shall not unreasonably be withheld.

Valuation and payment

(2) Contingencies and Prime Cost Items shall be valued and paid for in accordance with Clause 52 or as the Contract otherwise provides. The percentage to be used for overheads and profit in adjusting the Prime Cost element of any such item shall be the figure stated therefor in the Appendix to the Form of Tender.

The procedure for payment is given by Clause 60. Clause 60(1) is set out below. The scheme for the remainder of Clause 60 is broadly the same as under the standard ICE Conditions of Contract.

60 Interim statements

(1) The Contractor shall submit to the Employer's Representative at such times and in such form as the Contract prescribes statements showing

(a) the amounts which in the Contractor's opinion are due under the Contract

and where appropriate showing separately

(b) each amount expended against Contingencies and Prime Cost Items

(c) a list of any goods or materials delivered to the Site for but not yet incorporated in the Permanent Works and their value

(d) a list of any goods or materials which have not yet been delivered to the Site but of which the property has vested in the Employer pursuant to Clause 54 and their value and

(e) the estimated amounts to which the Contractor considers himself entitled in connection with all other matters for which provision is made under the Contract

unless in the opinion of the Contractor such values and amounts together will not justify the issue of an interim certificate.

Dispute resolution

One of the major criticisms of the traditional civil engineering contracts is the procedure whereby the Engineer acts both as the Employer's agent and as an impartial decision-maker. In the present contract, the procedure for referring disputes to the Engineer has been modified.

66 Settlement of disputes

(1) Except as otherwise provided in these Conditions if a dispute of any kind whatsoever arises between the Employer and the Contractor in connection with or arising out of the Contract or the carrying out of the Works including any dispute as to any decision opinion instruction direction certificate or valuation of the Employer's Representative (whether during the progress of the Works or after their completion and whether before or after the determination abandonment or breach of the Contract) it shall be settled in accordance with the following provisions.

Notice of Dispute

(2) For the purpose of sub-clauses (2) to (5) inclusive of this Clause a dispute shall be deemed to arise when one party serves on the other a notice in writing (hereinafter called the Notice of Dispute) stating the nature of the dispute. Provided that no Notice of Dispute may be served unless the party wishing to do so has first taken any steps or invoked any procedure available elsewhere in the Contract in connection with the subject matter of such dispute and the other party or the Employer's Representative as the case may be has

(a) taken such step as may be required or

(b) been allowed a reasonable time to take any such action.

Conciliation

(3) Every dispute notified under sub-clause (2) of this Clause if not already settled shall after the period of one calendar month from service of the Notice of Dispute be referred to conciliation in accordance with the Institution of Civil Engineers' Conciliation Procedure (1988) or any modification thereof being in force at the date of such referral. The conciliator shall make his recommendation in writing and give notice of the same to the Employer and the Contractor within three calendar months of the service of the Notice of Dispute.

Effect on Contractor and Employer

(4) Unless the Contract has already been determined or abandoned the Contractor shall in every case continue to proceed with the Works with all due diligence.

The Contractor and the Employer shall both give effect forthwith to every recommendation of the conciliator. Such recommendations shall be final and binding upon the Contractor and the Employer unless and until the recommendation of the conciliator is revised by an arbitrator as hereinafter provided and an award made and published.

The remainder of Clause 66 is broadly the same as for the standard ICE Conditions of Contract.

The procedure involves conciliation, which may be invoked at any time ("whether during the progress of the Works or after their completion ..."). The subject matter is unlimited and includes, for instance, an interim valuation by the Employer's Representative. The only requirement is that a Notice of Dispute be served after internal contract procedures have been exhausted or have been given a reasonable time.

The Conciliator's recommendation is to be given within three months of the Notice of Dispute. Any recommendation is "final and binding upon the Contractor and the Employer unless and until the recommendation of the conciliator is revised by an arbitrator ...".[30]

Clause 66 provides for arbitration to be the means of formal dispute resolution. Indeed the phrase "unless ... revised by an arbitrator ..." may mean that a court is unable to revise the recommendation.[31]

30 Apart from the fact that the conciliator has three months to make a recommendation, the role of the conciliator is fairly similar to that of an adjudicator whose decision is binding until reviewed.

31 This form of words may create a *Scott* v. *Avery* (1856) 5 HLC 811 (HL) precondition or a reinforced argument along the line of *Northern Regional Health Authority* v. *Crouch* [1984] QB 644 (CA).

11

The Engineering and Construction Contract (The NEC, Second Edition)

1. Introduction to the Engineering and Construction Contract

The New Engineering Contract (NEC) was created and drawn up by an Institution of Civil Engineers working group. A consultative version was published in 1991. The first edition was published in 1993. The second edition was published as the Engineering and Construction Contract (ECC) in November 1995. The new title reflects the aspirations of the NEC Panel that the contract should be used for construction work in all sectors, including traditional building. Many of the changes in this new edition are derived from recommendations in the Latham Report.[1]

The suite

The contract is not a single contract but a suite of contracts which share a core body of definitions and interrelationships; there are Core Clauses, Main Option Clauses and Secondary Option Clauses, together with Contract Data. The contract is published as a system in a series of documents. These documents include (*a*) the "Black Book" which houses the full inventory of clauses, (*b*) merged versions which contain clauses relevant to each of the Main Options, (*c*) guidance notes and (*d*) flow charts. Ancillary documents include a harmonised sub-contract and an Adjudicator's form of appointment.

1 Sir Michael Latham, *Constructing the team*, HMSO, 1994.

Style of drafting

The contract is drawn in a manner which seems unfamiliar and hence disconcerting to lawyers experienced in traditional contract design. It is a series of short, crisp statements. These are in the present tense[2] and use defined terms. For example, Clause 20.1 reads: "The *Contractor* Provides the Works in accordance with the Works Information". Words with initial capitals are defined in the contract. Italicised words[3] are defined in the contract and identified in the Contract Data. A numbering system is used which immediately identifies the location and objective of each clause.[4] There are some "flourishes" which will surprise lawyers: Clause 10.1, for example, requires the Employer, Contractor etc. to act "in a spirit of mutual trust and cooperation". This perhaps imports no substantive obligation, but merely assists in the construction of any terms which require cooperation.

Core clauses

These clauses remain constant across the range of contracts. Thus the advantages accruing to any standard form, such as familiarity, even-handedness, thorough checking and consultation are retained.

Main Option clauses

The six basic options differ principally in the agreed method of remuneration and its attendant mechanisms and risks. The versions are: (A) Priced contract with activity schedule; (B) Priced contract with bill of quantities; (C) Target contract with activity schedule; (D) Target contract with bill of quantities; (E) Cost reimbursable contract; (F) Management contract. Other options, such as contractor's design obligations, are readily achieved by inserting the extent of the design obligation into the Contract Data.

---

2 Except Clause 10.1; this is an exhortation and is in the future tense. In addition, the term "may" is used for discretionary action.

3 Save for direct quotations, this convention is not preserved in the text of this chapter.

4 Thus the "30" series clauses (from 30.1 to 36.5) all relate to aspects of time.

Secondary Option clauses

These provide a wide range of additional possibilities and are to be used in conjunction with the core clauses and a Main Option from A to F. The secondary option clauses are: (G) Performance bond; (H) Parent company guarantee; (J) Advanced payment to the Contractor; (K) Multiple currencies; (L) Sectional completion; (M) Limitation of the Contractor's liability for his design to reasonable skill and care; (N) Price adjustment for inflation; (P) Retention; (Q) Bonus for early completion; (R) Delay damages; (S) Low performance damages; (T) Changes to the law; (U) The Construction Design and Management Regulations 1994; (V) Trust Fund; (Z) Additional conditions.

Contract data

This contains the specific data which relates to the project and contract. Matters such as: defined personnel; the identity and location of "The Works Information" (i.e. drawings, specifications, etc.); the identity and location of the "Site Information" (i.e. information about the Site); contract dates; payment assessment intervals and payment currencies; insurances; dispute resolution tribunal; and other data required to make the various options work.

2. An outline of the contract

The parties

The parties to the contract are the Contractor and the Employer—Clause 11.2(1).

Administrators and decision-makers

Three appointments are defined in the contract: the Project Manager, the Supervisor and the Adjudicator—Clause 11.2(2). The Project Manager is the person who is the senior administrator on the project. He acts as the agent of the Employer and as an independent decision-maker. The Supervisor is responsible for day-to-day supervision of the Works on behalf of the Employer. The Adjudicator is an independent person appointed to resolve disputes.

The Project Manager

He is appointed by the Employer. His principal role is to manage the project, thereby achieving the Employer's objectives. The Project

Manager has considerable authority. His role is not defined in any one place, but his functions are stated throughout the contract terms. As with an Engineer under a traditional contract, he has functions which are purely as an agent of the Employer, but some which require a degree of fairness. Thus where he assesses payments due, he is not entitled to under-assess simply because this suits the business objectives of the Employer.[5] Where the Contractor (and in some cases, the Employer) is dissatisfied with the Project Manager's decision, they may refer the matter to the Adjudicator. The Supervisor has his own distinct functions, and there is no appeal to the Project Manager from his decision.[6]

The obligations of the Contractor[7]

The Contractor agrees to Provide the Works in accordance with the Works Information. The Works Information is contained in the documents identified in the Contract Data as amended by any instruction given in accordance with the contract.[8] The Project Manager may give an instruction to the Contractor which changes the Works Information.[9]

Time[10]

The starting date, the possession date and the completion date are given in the Contract Data. The Contractor may start work on the starting date, which may be prior to possession (e.g. where plant etc. is constructed off-site). The programme is a key document. It includes a plan of time and is regularly updated. Completion occurs when the Contractor has done the work in the Works Information and corrected any defects which would have prevented the Employer from using the

5 It is submitted that the term "assess" requires that the computation be fair as between the parties.

6 Compare this with the reference from the Engineer's Representative to the Engineer under the ICE Conditions of Contract—see Chapter 8.

7 See generally clauses in the 20 series.

8 Clause 11.2(5).

9 Clause 14.3.

10 See generally clauses in the 30 series.

Works.[11] Where a Compensation Event occurs, which cannot be accommodated into the programme float, the Project Manager extends the time for completion.

Payment[12]

Assessment dates for payment are set out in the Contract Data. The Project Manager assesses payment at the appropriate dates. The payment due at any assessment date has several components. The principal component is The Price for Work Done to Date (PWDD). The computation of the PWDD depends on which of the six Main Options is being operated. Option A contains an activity schedule, that is a list of activities and correlated prices for each activity; the PWDD is the total of the prices for activities in the schedule which have been completed. In addition, there is an adjustment for Compensation Events (which are priced on the basis of Actual Cost, which in turn depends on a Cost Schedule which forms part of the contract). Furthermore, a retention is deducted. The Project Manager certifies payment within one week of the assessment date and payment is made within three weeks of the assessment date; interest is paid on any late payments.

Compensation events[13]

Compensation events are those events which entitle the Contractor to claim compensation. These include[14] where the Project Manager makes a change to the work, where the Employer fails to give possession on time, where the Employer fails to provide something which he is to provide or do some work which he is to do, where unforeseen physical conditions occur, where very exceptional weather conditions occur, etc. Where the event arises as a result of an instruction etc. of the Project Manager, the Project Manager notifies[15] the Contractor. Where the Contractor believes that a compensation

---

11 Clause 11.2(13).

12 See generally clauses in the 50 series.

13 See generally clauses in the 60 series.

14 They are listed out in Clause 60.1.

15 See generally Clause 61 as to notifications for compensation events.

event has occurred or anticipates that one will occur, he notifies the Project Manager. The Project Manager decides whether or not a compensation event has occurred or will occur; if so, he asks for a quotation or alternative quotations based on alternative ways of dealing with the compensation event.[16] The Project Manager assesses the change in Price and any delay to the Completion Date caused by the compensation event.[17]

3. Commentary on selected core clauses

The principal innovations of the Engineering and Construction Contract includes its use and promotion of management principles and its handling of disputes. In this selective commentary, matters such as payment details are, therefore, given less prominence than the management innovations.

Early warning

16 Early warning

16.1 The *Contractor* and the *Project Manager* give an early warning by notifying the other as soon as either becomes aware of any matter which could

- increase the total of the Prices,
- delay Competition or
- impair the performance of the *works* in use.

16.2 Either the *Project Manager* or the *Contractor* may instruct the other to attend an early warning meeting. Each may instruct other people to attend if the other agrees.

16.3 At an early warning meeting those who attend co-operate in

- making and considering proposals for how the effect of each matter which has been notified as an early warning can be avoided or reduced,

16 See generally Clause 62 as to quotations for compensation events.

17 See generally Clauses 63, 64, 65 as to assessment and incidental matters in respect of compensation events.

- seeking solutions that will bring advantage to all those who will be affected and
- deciding upon actions which they will take and who, in accordance with this contract, will take them.

16.4 The *Project Manager* records the proposals considered and the decisions taken at an early warning meeting and gives a copy of his record to the *Contractor*.

Early warning notice—the nature of the duty to notify Clause 16.1 is expressed in mandatory terms. Thus, where either the Contractor or the Project Manager becomes aware of any condition which requires the giving of an early warning notice, and he does not do so, this may rank as a breach of contract. Although a sanction is provided in the contract for the late notification by the Contractor,[18] the Project Manager's motivation is his interest in efficient completion. However, the question of the mandatory nature of Clause 16.1 may raise interesting questions. For instance where the Project Manager is aware of a notifiable defect which he fails to notify and this defect causes the Contractor serious loss at a later date, does this loss flow from the Project Manager's breach on behalf of the Employer? If so, this clause may reverse the position found on traditional contracts, namely that the employer's agent is not responsible for the contractor's defaults.[19]

The relevant conditions These are "any matter which could increase the total of the Prices, delay Completion or impair the performance of the *works* in use". A matter which could increase the total of the Prices includes compensation events (e.g. unforeseen ground conditions). Matters which could delay Completion range from those which are entirely due to the Contractor's default or inefficiency to those for which an extension of time is indubitably due. And matters which could impair the performance of the works in progress include design defects noticed by the Contractor, any adverse ground conditions, any defective work, etc.

18 Clause 63.4, see below.

19 See Chapter 3, section 3 as to performance. *East Ham Borough Council* v. *Bernard Sunley & Sons Ltd* [1966] AC 406 (HL); *AMF International Ltd* v. *Magnet Bowling Ltd* [1968] 1 WLR 1028.

Convening an early warning meeting Clause 16.2 is drawn in discretionary terms. Accordingly there is no obligation to call an early warning meeting. But if either the Contractor or the Project Manager wishes to do so, he may instruct the other to attend. It seems that the instruction to attend may be included with the early warning notice. The timing of the meeting is not specified, but a reasonable period should be allowed between notice/instruction and the meeting to enable the parties to collect such information as is reasonably required to comply with Clause 16.3, given the urgency, value and importance of the matter to be discussed. The final sentence of Clause 16.2 provides that other people may be instructed to attend. The term "other people" is not defined, but no doubt includes lawyers if necessary. However, any such attendance requires the consent of the other. On one reading of this provision it seems as if the only persons entitled to attend as of right are the Project Manager or his delegate and the Contractor's representative. It is thought, however, that the provision impliedly includes (at the instance of either the Contractor or Project Manager) the Supervisor and other persons directly involved in the management of the project as well as any directly concerned sub-contractor or designer.[20]

The meeting Clause 16(3) sets out the agenda for the meeting. In accordance with Clause 16.4, the proposals and decision are recorded by the Project Manager and a copy is provided to the Contractor.

61 Notifying compensation events

61.5 If the *Project Manager* decides that the *Contractor* did not give an early warning of the event which an experienced contractor could have given, he notifies this decision to the *Contractor* when he instructs him to submit quotations.

63 Assessing compensation events

63.4 If the *Project Manager* has notified the *Contractor* of his decision that the *Contractor* did not give an early warning of a compensation event which an experienced contractor could have given, the event is assessed as if the *Contractor* had given early warning.

20 Note that the defined term "Others" is not used here—see Clause 11.2(2).

These clauses set out the contractual sanction against the Contractor. It seems that a notice under Clause 61.5 is a condition precedent to the action under Clause 63.4. Note also that the Clause 16.1 requires actual awareness, while Clause 61.5 requires that "an experienced contractor" would have given the notice. The meaning of Clause 63.4 is not altogether clear; it is thought that it means that the time or cost which is assessed as being due to the Contractor is not the time and cost actually incurred, but the time or costs which would have been incurred had the early warning been given at the proper time; in short, the Contractor pays for his own failure to mitigate his loss.

Programme

31 The programme

31.1 If a programme is not identified in the Contract Data, the *Contractor* submits a first programme to the *Project Manager* for acceptance within the period stated in the Contract Data.

31.2 The *Contractor* shows on each programme which he submits for acceptance

- the *starting date*, *possession dates* and Completion Date,
- for each operation, a method statement which identifies the Equipment and other resources which the *Contractor* plans to use
- planned Completion,
- the order and timing of
  - the operations which the *Contractor* plans to do in order to Provide the Works and
  - the work of the *Employer* and Others either as stated in the Works Information or as later agreed with them by the *Contractor*,
- the dates when the *Contractor* plans to complete work needed to allow the *Employer* and Others to do their work,
- provisions for
  - float,
  - time risk allowances,
  - health and safety requirements and

    - the procedures set out in this contract

- the dates when, in order to Provide the Works in accordance with his programme, the *Contractor* will need
    - possession of a part of the Site if later than its *possession date*,
    - acceptances and
    - Plant and Materials and other things to be provided by the *Employer* and
- other information which the Works Information requires the *Contractor* to show on a programme submitted for acceptance

31.3 Within two weeks of the *Contractor* submitting a programme to him for acceptance, the *Project Manager* either accepts the programme or notifies the *Contractor* of his reasons for not accepting it. A reason for not accepting a programme is that

- the *Contractor*'s plans which it shows are not practicable,
- it does not show the information which this contract requires,
- it does not represent the *Contractor*'s plans realistically or
- it does not comply with the Works Information.

32 Revising the programme

32.1 The *Contractor* shows on each revised programme

- the actual progress achieved on each operation and its effect upon the timing of the remaining work,
- the effects of implemented compensation events and of notified early warning matters,
- how the *Contractor* plans to deal with any delays and to correct notified Defects and
- any other changes which the *Contractor* proposes to make to the Accepted Programme.

32.2 The *Contractor* submits a revised programme to the *Project Manager* for acceptance

- within the *period for reply* after the *Project Manager* has instructed him to,
- when the *Contractor* chooses to and, in any case,
- at no longer interval than the interval stated in the Contract Data from the *starting date* until Completion of the whole of the *works*.

The original and contractual status of the programme The programme is either (*a*) identified in the Contract Data or (*b*) submitted by the Contractor within a period stated in the Contract Data. Even where it is included in the Contract Data, it does not represent an invariate term of the contract.[21]

The content of the programme—general[22] Unlike programmes which are submitted under traditional standard form civil engineering contracts, the content of the programme is specified in some detail in Clause 31.2. Thus wherever a programme is called for under the contract, e.g. as part of a quotation for a compensation event, this includes a call for not only timing information, but also any necessary method statements and/or health and safety requirements, etc. Furthermore, revised programmes must show the matters set out in Clause 32.1, including progress to date and proactive control measures.

Float and time risk allowances Separate provisions are to be identified on the programme relating to (*a*) float and (*b*) time risk allowances. These are not defined anywhere in the contract. The normal meaning of "float" is the time slack, so that an activity may finish later than programmed by the float period and still not delay the completion. The apparent meaning of "time risk allowances" is the

21 Compare with the position where a programme is incorporated into a traditional contract, such as the ICE Conditions of Contract: see *Yorkshire Water Authority* v. *McAlpine* (1985) 32 BLR 114.

22 Note also that the Contractor may be the principal contractor under the Construction (Design and Management) Regulations 1994. If so, he will be required to maintain the health and safety plan which will contain much of the same information as required in this programme. See, generally, the commentary on Clause 71 of the ICE Conditions of Contract.

allowance to be made in addition to the nominal activity duration for uncertainty in the activity duration. The guidance notes (which form no part of the contract) state:[23] "The Contractor's [sic] time risk allowances ... are owned by the Contractor as part of his realistic planning to cover his risks". "Float is any spare time within the programme after the time risk allowances have been included. It is normally available to accommodate the time effects of a compensation event in order to mitigate or avoid any delay to planned Completion." There is no provision for what is frequently termed "overall float". Where the planned completion precedes the contract completion date, the difference in time is not available to the Project Manager; accordingly where a compensation event causes a delay to Completion, the Project Manager must also delay the Completion Date: see Clause 63.3.

Acceptance Clause 31.3 sets out the reasons for non-acceptance. These are not stated to be comprehensive, but they set a presumption of construction that any reason for non-acceptance must be of these general types. Where the Contractor is dissatisfied with the Project Manager's decision he may refer the matter to the Adjudicator. The Adjudicator will not, of course, require the Project Manager to accept the programme; but a compensation event will arise if the Adjudicator decides that the original programme complied with the contract work as then defined. When accepted, a programme becomes the Accepted Programme.[24]

Revising the programme The contract provides a liberal environment for submission of revised programmes. The interval contained in the Contract Data is the main constraint; but where either the Project Manager or Contractor wish it, a revised programme can be submitted at any time.

62 Quotations for compensation events

62.2 Quotations for compensation events comprise proposed changes to the Prices and any delay to the Completion Date

23 Page 39.

24 Defined in Clause 11.2(14).

assessed by the *Contractor*. The *Contractor* submits details of his assessment with each quotation. If the programme for remaining work is affected by the compensation event, the *Contractor* includes a revised programme in his quotation showing the effect.

Revised programme for compensation events Whenever a compensation event occurs, the Project Manager calls for a quotation and specifies what it is to include. The action proposed by the Contractor in his quotation may require a revision to the Accepted Programme. If so, he submits a revised programme, which must comply with the requirements of Clauses 31 and 32. This programme, if accepted, becomes the Accepted Programme and becomes an important element in the Project Manager's assessment of the compensation due.

Physical conditions

60 Compensation events

60.1 The following are compensation events ...

(12) The *Contractor* encounters physical conditions which

- are within the Site,
- are not weather conditions and
- which an experienced contractor would have judged at the Contract Date to have such a small chance of occurring that it would have been unreasonable for him to have allowed for them.

60.2 In judging the physical conditions, the Contractor is assumed to have taken into account

- the Site Information,
- publicly available information referred to in the Site Information,
- information available from a visual inspection of the Site and
- other information which a reasonably experienced contractor could reasonably be expected to have or to obtain.

60.3 If there is an inconsistency within the Site Information (including the information referred to in it), the *Contractor* is assumed to have taken into account the physical conditions more favourable to doing the work.

Clauses 11 and 12 of the ICE Conditions of Contract and construing the present clauses The wording of the present clauses differs from that found in Clauses 11 and 12 of the ICE Conditions of Contract. Nevertheless, the parallels are clear and it is tempting to construe the present terms in light of the traditional Clause 12. It is submitted, however, that the present contract stands alone and is to be construed within its own terms.[25] Consider, for instance, the wording of Clause 60.1(12), 2nd bullet: "are not weather conditions". The question may arise of whether flooding due to exceptionally heavy rainfall comes within this provision. It is tempting to refer to the corresponding wording in Clause 12 of the ICE Conditions of Contract (which includes the phrase "or conditions due to weather conditions") and to conclude that the ambit of Clause 60.2 has been deliberately expanded. It is submitted, however, that the wording of 60.1(12) permits the flooding claim within its own terms or it does not. This temptation to cross-refer between contracts is, of course, particularly difficult to resist in the case of the test in Clause 60.1(12), 3rd bullet.

The test It is a precondition of any claim under these provisions that the conditions referred to are within the Site, and that they are not weather conditions (which are dealt with specifically in Clause 60.1(13)).

Weather conditions Conditions such as low temperature causing delays to concreting, high winds causing inability to work on tower cranes, etc. are clearly weather conditions. On-site flooding, snow loading, wind loading, etc. on the Site are also thought to be "weather conditions", otherwise, the phrase "weather conditions" is largely emasculated. Where, however, there are heavy rainfalls off the site

25 See e.g. *Luxor* v. *Cooper* [1941] AC 108 (HL); *Mitsui* v. *AG of Hong Kong* (1986) 33 BLR 1 (PC) where Lord Bridge said at 18 "comparison of one contract with another can seldom be a useful aid to construction and may be ... positively misleading".

which cause flooding on the Site (as where a river breaks its banks) it is submitted that this is a physical condition which is not a weather condition and is, therefore, included. This is because the physical condition does not enter the Site as "weather" but as some other condition.

The test Clause 16.1(12), 3rd bullet provides the essential test. An experienced contractor is thought to be one who has average experience of the type of work envisaged at the Contract Date. The time at which the judgment is to be made is the Contract Date. The key question relates to the meaning of: "such a small chance of occurring that it would have been unreasonable for him to have allowed for them". This clearly cannot be reduced to precise probabilities. It seems that the Employer is not required to pay compensation unless the Contractor can show that the conditions which occurred are surprising and that any contractor who allowed for them in his prices would stand rightly accused of undue pessimism.

Information Clause 60.2 does not require the Employer to disclose information within his possession even if it has a clear bearing upon the works.[26] The information which the contractor is assumed to have or to be able to obtain is to be judged at the Contract Date; consequently, it is no defence to say that the Contractor could have obtained a piece of information if there was nothing to alert him to obtain it at that date.

Inconsistencies Clause 60.3 provides for the situation where there are inconsistencies in the Site Information. Many disputes under the traditional standard forms are generated by inconsistent site investigation data; the employer and contractor each hope to persuade the arbitrator that their interpretation is the more reasonable. The intention of this clause is to provide a test where the data suggest two or more possible interpretations; the Contractor is assumed to have taken the more favourable of them into account. However, the situation which generally gives difficulty is where data are not strictly inconsistent but where the totality of data gives more than one

---

26 Compare this with Clause 11 of the ICE Conditions of Contract; see the commentary in Chapter 8.

reasonable interpretation. This problem is not explicitly resolved. Thus, for example, where the majority of data are consistent with poor ground conditions and the remaining data suggest benign conditions, it is not clear whether or not the Contractor is entitled to take into account the benign conditions.

Disputes

90 Settlement of disputes

90.1 Any dispute arising under or in connection with this contract is submitted to and settled by the *Adjudicator* as follows.

ADJUDICATION TABLE

Dispute about:	**Which Party may submit it to the *Adjudicator*?**	**When may it be submitted to the *Adjudicator*?**
An action of the *Project Manager* or the *Supervisor*	The *Contractor*	Between two and four weeks after the *Contractor*'s notification of the dispute to the *Project Manager*, the notification itself being made not more than four weeks after the *Contractor* becomes aware of the action
The *Project Manager* or *Supervisor* not having taken an action	The *Contractor*	Between two and four weeks after the *Contractor*'s notification of the dispute to the *Project Manager*, the notification itself being made not more than four weeks after the *Contractor* becomes aware that the action was not taken
Any other matter	Either Party	Between two and four weeks after notification of the dispute to the other Party and the *Project Manager*

90.2 The *Adjudicator* settles the dispute by notifying the Parties and the *Project Manager* of his decision together with his reasons within the time allowed by this contract. Unless and until there is such a settlement, the Parties and the *Project Manager* proceed as if the action, inaction or other matter disputed were not disputed. The decision is final and binding unless and until revised by the *tribunal*.

91 The adjudication

91.1 The Party submitting the dispute to the *Adjudicator* includes with his submission information to be considered by the *Adjudicator*. Any further information from a Party to be considered by the *Adjudicator* is provided within four weeks from the submission. The *Adjudicator* notifies his decision within four weeks of the end of the period for providing information. The four-week periods in this clause may be extended if requested by the *Adjudicator* in view of the nature of the dispute and agreed by the Parties.

91.2 If a matter disputed under or in connection with a subcontract is also a matter disputed under or in connection with this contract, the *Contractor* may submit the subcontract dispute to the *Adjudicator* at the same time as the main contract submission. The *Adjudicator* then settles the two disputes together and references to the Parties for the purposes of the dispute are interpreted as including the Subcontractor.

92 The Adjudicator

92.1 The *Adjudicator* settles the dispute as independent adjudicator and not as arbitrator. His decision is enforceable as a matter of contractual obligation between the Parties and not as an arbitral award. The *Adjudicator*'s powers include the power to review and revise any action or inaction of the *Project Manager* or *Supervisor* related to the dispute. Any communication between a Party and the *Adjudicator* is communicated also to the other Party. If the *Adjudicator*'s decision includes assessment of additional cost or delay caused to the *Contractor*, he makes his assessment in the same way as a compensation event is assessed.

92.2 If the *Adjudicator* resigns or is unable to act, the Parties choose a new adjudicator jointly. If the Parties have not chosen a new adjudicator jointly within four weeks of the *Adjudicator* resigning or becoming unable to act, a Party may ask the person stated in the Contract Data to choose a new adjudicator and the Parties accept his choice. The new adjudicator is appointed as *Adjudicator* under the NEC Adjudicator's Contract. He has power to settle disputes that were currently submitted to his predecessor but had not been settled at the time when his predecessor resigned or became unable to act. The date of his appointment is the date of submission of these disputes to him as *Adjudicator.*

93 Review by the *tribunal*

93.1 If after the *Adjudicator*

- notifies his decision or
- fails to do so

within the time provided by this contract a Party is dissatisfied, that Party notifies the other Party of his intention to refer the matter which he disputes to the *tribunal.* It is not referable to the *tribunal* unless the dissatisfied Party notifies his intention within four weeks of

- notification of the *Adjudicator*'s decision or
- the time provided by this contract for this notification if the *Adjudicator* fails to notify his decision within that time

whichever is the earlier. The *tribunal* proceedings are not started before Completion of the whole of the *works* or earlier termination.

93.2 The *tribunal* settles the dispute referred to it. Its powers include the power to review and revise any decision of the *Adjudicator* and any action or inaction of the *Project Manager* or the *Supervisor* related to the dispute. A Party is not limited in the *tribunal* proceedings to the information, evidence or arguments put to the *Adjudicator.*

Matters which may be referred Almost any conceivable matter arising under the contract may be referred to the Adjudicator by the Contractor. All instructions, decisions, certificates, notices, etc. are included as are all alleged failures to issue any of the above. The Employer may refer any matter to the Adjudicator, except the alleged actions or inactions of the Project Manager or Supervisor.

Role of the Adjudicator The Adjudicator is not an arbitrator. Accordingly he does not need to provide the parties with the same opportunity to be heard as would an arbitrator. However, he is independent and any communication is communicated to the other party, so that there is transparency about what materials are before him. While no procedural rules are laid down, it is submitted that the Adjudicator will have the authority to request clarifications, further documents, meetings, etc. as he deems fit, notwithstanding the apparent limitation on the period for providing information.[27]

Time limits Time limits are prescribed for both the submission of a dispute to the Adjudicator and the decision by the Adjudicator.[28] All time limits may be waived by the other party in the normal way.[29] There seems to be no provision for extending the time limits for submission of disputes relating to actions or inactions by the Project Manager or Supervisor. It may be, therefore, that four weeks after an action or inaction comes to the notice of the Contractor, his right to claim will be barred unless he has given notice of the dispute, though it may be that any repeat of the action or inaction may revive the right to make a submission to the Adjudicator. Furthermore, Clause 93.2 seems to open up the possibility that the tribunal may be re-invested with authority to deal with "any action or inaction of the *Project Manager* or the *Supervisor* related to [another validly referred] dispute". As to the time limits for making the decision, the only scope for a decision outside the period is where the Adjudicator requests it and both Parties agree.[30]

Procedure When the time periods for submission of disputes become active,[31] the claimant submits his claim with information to be considered. A period of four weeks begins to run, during which any

27 Clause 91.1.

28 When the Housing Grants, Construction and Regeneration Act 1996 comes into force, these time limits will need to be shortened to comply with Section 108.

29 *W. J. Alan & Co.* v. *El Nasr Export and Import Co.* [1972] 2 All ER (CA). But any person waiving such rights must be authorised to do so by the relevant party. It is thought that the Project Manager, who is the Employer's primary agent, will not be authorised to act for the Employer in this matter.

30 Clause 91.1.

31 See the Adjudication Table in Clause 90.1.

further information is provided.[32] Within the next four weeks, the Adjudicator publishes his decision.

Review Where the Adjudicator fails to issue a decision or the decision is disputed and a notice to refer is served on the other party within four weeks of the time set out in Clause 93.1, either party may refer the matter to the Tribunal. The Tribunal is identified in the Contract Data. The parties may choose arbitration or a binding expert decision or Disputes Review Panel. Where they do not specify the Tribunal, the matter will automatically be dealt with by the court. By Clause 93.2 the Tribunal will be entitled to review and revise decisions of the Adjudicator and of the Project or Supervisor.[33]

32 It seems that where the respondent files his submission and information at the end of the four week period, the claimant will have no right to reply. However, the Adjudicator will as suggested above have authority to receive additional information up until the expiry of the time for making his decision.

33 Hence avoiding the risk that court will not be empowered to operate the contractual machinery: see *Northern Regional Health Authority* v. *Crouch* [1984] QB 644 (CA) and section 100 of the Courts and Legal Services Act 1990.

12

The Institution of Civil Engineers' Arbitration and Conciliation Procedures

1. Introduction to the ICE Arbitration Procedure (England and Wales) (1983)

The Arbitration Procedure was approved by the Institution of Civil Engineers in 1983. It is designed for use with the ICE Conditions of Contract. It is also used for other engineering arbitrations, though several of its provisions refer directly to Clause 66 of the ICE Conditions of Contract.

New legislation and revised rules of procedure

The Arbitration Act 1996 was enacted in June 1996. Following consultation, it may come into force in the first half of 1997. It will apply to all arbitrations commenced after that date.[1] This new legislation is designed primarily to consolidate existing statutes, although it contains new powers,[2] which are, in the main, already contained in the ICE Arbitration Procedure. The new Act adopts the guiding principle that the parties are empowered to make agreements as to procedural rules. Arbitrations conducted under the ICE Arbitration Procedure will not, therefore, be significantly affected. The main bodies who publish rules of procedure are currently considering

1 Section 84(2) of the Arbitration Act 1996.

2 See Sections 34 to 41 of the Arbitration Act 1996.

whether or not a set of arbitration rules to cover the entire construction industry is warranted.

Purpose of an arbitration procedure

In the absence of an agreed procedure, the Arbitrator controls the proceedings.[3] However, he has no inherent authority to order security for costs,[4] to make reviewable interim awards[5] and the like; the Arbitration Procedure gives him power to exercise these useful powers. Furthermore, it is convenient that the Arbitrator's directions are related to well-established written procedures so that all parties can fully understand the process.

The nature of the rules set out in the Arbitration Procedure

Many of the rules are merely declaratory of powers which an arbitrator already possesses or are formulations based on common arbitral practice. Other rules apply only where the parties specifically agree that they shall. There are, however, a core of rules to which frequent reference is made. These include Rules 1, 4, 6, 7 and 14.

The binding effect of the Arbitration Procedure[6]

The Arbitration Procedure is binding on the parties wherever they have agreed this in their contract (as parties to the ICE Conditions and FCEC Sub-Contract will have done through Clauses 66(8) and 18(7) respectively[7]).

---

3 *Bremer Vulkan Schiffbau und Maschinenfabrik* v. *South India Shipping Corporation* [1981] AC 909 at 985 (HL).

4 This is the position under the 1950 Act: *Unione Stearinerie Lanza* v. *Wiener* [1917] 2 KB 558. Under Section 38(3) of the 1996 Act the arbitrator may order security for costs unless the parties agree otherwise.

5 *Fidelitas Shipping Co.* v. *Exportchleb* [1966] 1 QB 630 (CA). See also Section 39(4) of the 1996 Act.

6 See also the commentary on Rule 26.1.

7 Note that in each case the reference is to the ICE Arbitration Procedure or any amendment or modification thereof being in force at the time of the appointment of the arbitrator.

2. Commentary on the ICE Arbitration Procedure (England and Wales) (1983)

PART A. REFERENCE AND APPOINTMENT

Rule 1. Notice to Refer

1.1 A dispute or difference shall be deemed to arise when a claim or assertion made by one party is rejected by the other party and that rejection is not accepted. Subject only to Clause 66(1) of *the ICE Conditions of Contract* (if applicable), either party may then invoke arbitration by serving a *Notice to Refer* on the other party.

1.2 The Notice to Refer shall list the matters which the issuing party wishes to be referred to arbitration. Where Clause 66 of the ICE Conditions of Contract applies the Notice to Refer shall also state the date when the matters listed therein were referred to the Engineer for his decision under Clause 66(1) and the date on which the Engineer gave his decision thereon or that he has failed to do so.

Disputes and the service of a Notice to Refer

Generally speaking, the only precondition for the commencement of arbitration under the rules is the existence of a dispute. Where, however, the ICE Conditions of Contract apply, the provisions of Clause 66 prevail. Those provisions require that a Notice of Dispute be served setting out the matters referred to the Engineer; the Engineer is then allowed a period of time in which to make a formal decision. Where the Engineer fails to make a decision or where he does so and either party is dissatisfied with it, a Notice to Refer may be served.

Commencement and conduct of the arbitration

A conflict may arise between the primary arbitration clause in a civil engineering contract and Rule 1.2 which purports to require that the Notice to Refer shall give specified details. It is submitted that the primary clause prevails. Clause 66 of the ICE Conditions of Contract requires that the arbitration be commenced by a Notice to Refer, but does not specify any detailed format or content for this notice. Clause 66(8) provides that the reference is to be "conducted in accordance with the ICE Arbitration Procedure". Thus the Procedure applies only to the "conduct" and not to the "commencement" of the reference. Where the Notice to Refer omits some detail required by the Arbitration Procedure, e.g. the dates upon which disputes were referred to the Engineer, this will not invalidate the Notice to Refer.[8] Where, however, the primary clause requires that the arbitration is to be "commenced and conducted" with the ICE Arbitration Procedure the result may be different. Where the notice is defective only as to details, the Court may exercising its discretion to extend time to allow a revised Notice to Refer to be served.[9]

The Notice to Refer—suggested format

Rule 1.2 provides that the Notice to Refer should contain the following.

(*a*) A list of matters to be referred. Where any item on the list is narrowly defined, other related matters which do not fall within its terms are not properly submitted.[10] It is advisable, therefore, to set out the matters to be referred in broad terms; and hence it is important to have referred the matter to the Engineer in the same broad terms. It is important, however, that any dispute is not so broadly drawn that the other party is prejudiced; thus where the Notice of Dispute purports to refer "other matters" this is not effective to bring all other matters into the reference.[11]

(*b*) A list of dates corresponding to the matters referred in (*a*) above.

Disputes over jurisdiction

Wherever the Arbitrator's jurisdiction in respect of any matter is challenged (e.g. where one party alleges that a matter has not been properly referred to the Engineer), difficult questions arise. The fundamental problem is that an arbitrator may not rule upon his own

---

8 *Christiani & Nielsen Ltd* v. *Birmingham City Council* (1994) CILL 1014.

9 Section 27 of the Arbitration Act 1950 and Section 12 of the Arbitration Act 1996. This power is used sparingly: see *Comdel Commodities Ltd* v. *Siporex Trade SA* [1990] 2 All ER 552; *The Joceleyne* [1977] 2 Lloyd's Rep. 121; *Crown Estate Commissioners* v. *John Mowlem & Co.* (1994) 70 BLR 1 (CA).

10 Rule 4 provides that additional disputes may be referred to the same Arbitrator before the Arbitrator's appointment is completed; or that any other dispute which has not been referred to the Engineer may yet be referred to the Arbitrator where it is necessary for the determination of matters already referred.

11 *Wigan Metropolitan Borough Council* v. *Sharkey Bros Ltd* (1987) 4 Constr. L.J. 162.

jurisdiction, for he must logically have jurisdiction in the first place in order to give such a ruling any effect. Accordingly, an arbitrator whose jurisdiction is seriously challenged faces a difficulty. This is a matter which arises also in the context of Rule 4 and is dealt with in detail in the commentary to that Rule.

Rule 2. Appointment of sole Arbitrator by agreement

2.1 After serving the Notice to Refer either party may serve upon the other a *Notice to Concur* in the appointment of an Arbitrator listing therein the names and addresses of any persons he proposes as Arbitrator.

2.2 Within 14 days thereafter the other party shall

(a) agree in writing to the appointment of one of the persons listed in the Notice to Concur

or

(b) propose a list of alternative persons.

2.3 Once agreement has been reached the issuing party shall write to the person to selected inviting him to accept the appointment enclosing a copy of the Notice to Refer and documentary evidence of the other party's agreement.

2.4 If the person so selected accepts the appointment he shall notify the issuing party in writing and send a copy to the other party. The date of posting or service as the case may be of this notification shall be deemed to be the date on which the Arbitrator's appointment is completed.

The parties frequently agree the appointment of an Arbitrator. The timetable given here is rarely adhered to, but this does not affect the validity of the appointment.

Rule 3. Appointment of sole Arbitrator by the President

3.1 If within one calendar month from service of the Notice to Concur the parties fail to appoint an Arbitrator in accordance with Rule 2 either party may then apply to the President to appoint an Arbitrator. The parties may also agree to apply to the President without a Notice to Concur.

3.2 Such application shall be in writing and shall include copies of the Notice to Refer, the Notice to Concur (if any) and any other relevant documents. The application shall be accompanied by the appropriate fee.

3.3 The Institution will send a copy of the application to the other party stating that the President intends to make the appointment on a specified date. Having first contacted an appropriate person and obtained his agreement the President will make the appointment on the specified date or such later date as may be appropriate which shall then be deemed to be the date on which the Arbitrator's appointment is completed. The Institution will notify both parties and the Arbitrator in writing as soon as possible thereafter.

This reflects the procedure laid down in Clause 66 of the ICE Conditions of Contract, but in rather more procedural detail.[12] It specifies the documents to be supplied in support of the application. It also sets out the procedure to be adopted by the President, namely that he contacts and obtains the consent of an appropriate person before making an appointment. This procedure means that the appointment is made before the Arbitrator has contact with the parties. He is therefore unable to impose upon them his standard scale of fees or cancellation charges.[13]

Rule 4. Notice of further disputes or differences

4.1 At any time before the Arbitrator's appointment is completed either party may put forward further disputes or differences to be referred to him. This shall be done by serving upon the other party an additional Notice to Refer in accordance with Rule 1.

4.2 Once his appointment is completed the Arbitrator shall have jurisdiction over any issue connected with and necessary to the determination of any dispute or difference already referred to him whether or not the connected issue has been first referred to the Engineer for his decision under Clause 66(1) of the ICE Conditions of Contract.

12 Where a conflict arises between the primary clause and the Arbitration Procedure, the primary clause will prevail. In the case of the ICE Conditions of Contract, this means that the terms of Clause 66 will prevail: see *Christiani & Nielsen Ltd* v. *Birmingham City Council* (1994) CILL 1014.

13 Where an arbitrator attempts to impose cancellation charges or his own standard scale of charges against the wishes of any party, he may misconduct himself: see *K/S Norjarl* v. *Hyundai Heavy Industries Co. Ltd* [1992] 1 QB 863.

Additional Notice to Refer

A further Notice to Refer may be served upon the other party setting out additional matters to be referred. Where it is valid, the Arbitrator has jurisdiction over those additional matters. There are two conditions for its validity.

(*a*) It must be served upon the other party before the Arbitrator's appointment is completed. Where the Arbitrator is appointed by the parties, the appointment is completed on the date on which both parties and arbitrator have each indicated agreement—see Rule 2.4. Where an appointment is made by the President, the appointment is made after obtaining the consent of a prospective arbitrator and when the President signifies that he makes the formal appointment.

(*b*) Any dispute or difference referred under Rule 4.1 must comply with any condition laid down in Rule 1. Accordingly, where the ICE Conditions of Contract apply, the issue must have been referred to the Engineer and the Engineer must either have given a decision or have failed to give his decision in the stipulated period.

Jurisdiction over any matter connected with and necessary to the determination of any dispute already referred—Rule 4.2

An issue which has not passed through the regular process of reference to the Engineer can only be referred to the Arbitrator upon two conditions being satisfied. First, the issue must be connected to a dispute already referred; this can frequently be demonstrated. Second, the issue must be necessary to the determination of a matter already referred. It is thought that such issues may be classified as:

(*a*) issues whose resolution forms a necessary step in the determination of matters already referred

(*b*) claims which overlap with any matter referred to such a degree that it could reasonably be said that a decision on one raises a serious possibility of *res judicata* in relation to the other[14]

14 Similar reasoning is frequently used for allowing a claimant to make a very late amendment to plead his case on a different basis: otherwise a potentially valid claim may not be lost on account of the *res judicata* rule. See e.g. *Beoco Ltd* v. *Alfa Laval Co. Ltd* [1994] 3 WLR 1179.

(*c*) defences (including set-offs) to any claim.

In practice, additional matters are frequently brought into the arbitration by consent of the parties to avoid the same parties having to run two parallel arbitrations relating to the same general facts. Where a separate arbitration is started and the same Arbitrator is appointed, there is frequently no objection to his dealing with both arbitrations concurrently.[15]

Disputes over jurisdiction

In arbitrations governed by the Arbitration Act 1950, the Arbitrator may not rule definitively on the initial existence of the agreement from which his jurisdiction derives.[16] Logically, a ruling by an "arbitrator" without jurisdiction can be of no effect; and an arbitrator who has jurisdiction cannot simply divest himself of it. The 1996 Act, however, allows the Arbitrator to break this circle. It entitles him to make a valid award as to his own jurisdiction, which award becomes binding unless challenged by appeal.[17] Even under the 1950 Act, however, there is scope for the Arbitrator to make a ruling on some aspects of his jurisdiction. The doctrine of the separability of the arbitration agreement[18] means that providing the arbitration agreement itself has come into existence, he may rule as to:[19]

(*a*) whether or not the dispute falls within the scope of the arbitration agreement
(*b*) whether or not a condition precedent to making a claim has been fulfilled
(*c*) whether or not a procedure for referring disputes has not been complied with.

15 Note that Rule 7.1 does not cover this since there are not "two or more contracts".

16 *Ashville Investments* v. *Elmer Contractors* [1989] QB 488 (CA).

17 Section 30 of the Arbitration Act 1996.

18 *Heyman* v. *Darwins Ltd* [1942] AC 365 (HL).

19 See e.g. *Harbour Assurance* v. *Kansa* [1993] QB 701 (CA) in relation to the illegality of the contract.

Nevertheless, it is thought that it is in the best interests of the parties for the Arbitrator to treat all matters which are essential to his jurisdiction as matters which he cannot conclusively decide. It is appropriate, therefore, for him to form a provisional view as to the merits of any challenge to his jurisdiction or any application to add new matters (e.g. under Rule 4.2). He should act with caution. He should be slow to accept that he has no jurisdiction in respect of a matter which has apparently been referred to him. It is always open to either party to seek a declaration of the court to the effect that the Arbitrator does, or does not, have jurisdiction in respect of any matter. It may be appropriate for the Arbitrator to direct a short adjournment so that such an application may be made to the court.

PART B. POWERS OF THE ARBITRATOR

Rule 5. Power to control the proceedings

5.1 The Arbitrator may exercise any or all of the powers set out or necessarily to be implied in this Procedure on such terms or conditions as he thinks fit. These terms or conditions may include orders as to payment of expenses, time for compliance and the consequences of non-compliance.

5.2 Powers under this Procedure shall be in addition to any other powers available to the Arbitrator.

An Arbitrator has substantial power to control the proceedings and to require the parties to do all things which he shall require.[20] Rule 5(1) reinforces this and Rule 5(2) counters the possibility that any rule may inadvertently derogate from any general power which the Arbitrator may have.

Rule 6. Power to order protective measures

6.1 The Arbitrator shall have power

(a) to give directions for the detention storage sale or disposal of the whole or any part of the subject matter of the dispute at the expense of one or both of the parties

20 Section 12(1) of the Arbitration Act 1950; *Bremer Vulkan Schiffbau und Maschinenfabrik* v. *South India Shipping Corporation* [1981] AC 909 at 985 (HL). See also Sections 34 to 41 of the 1996 Act.

(b) to give directions for the preservation of any document or thing which is or may become evidence in the arbitration

(c) to order the deposit of money or other security to secure the whole or any part of the amount(s) in dispute

(d) to make an order for security for costs in favour of one or more of the parties

and

(e) to order his own costs to be secured.

6.2 Money ordered to be paid under this Rule shall be paid without delay into a separate bank account in the name of a stakeholder to be appointed by and subject to the directions of the Arbitrator.

The powers given in Rule 6 are analogous to a number of powers which the Court may exercise under the Arbitration Act.[21]

Rules 6.1(a), (b), (c)

The Arbitrator must exercise these powers judicially. He should have proper regard to the principles laid down by the Court in equivalent cases and ask to be addressed as to the applicable principles. Thus, for example, where a party applies to the Arbitrator for an order for the detention or storage of any subject matter which is allegedly the subject matter of the dispute, he must satisfy himself that it is bona fide the subject matter of the dispute.[22] Where appropriate there must be some element of risk in not making the order and the Arbitrator shall consider alternatives, such as creating a video recording rather than preserving "things" which may become evidence.[23] An order under Rule 6.1(c) may overlap with the Court's jurisdiction to grant a Mareva[24] injunction, and where appropriate, regard should be had to the practice and principles of the Court. Once the Arbitrator has given

---

21 Section 12(6) of the Arbitration Act 1950. For the exercise by the Court of its power see *Richo International Ltd* v. *Industrial Food Co. SAL, The Fayrouz III* [1989] 2 Lloyd's Rep. 10. Section 38 of the 1996 Act provides that an arbitrator will have similar powers to those provided in Rule 6 unless the parties otherwise agree.

22 *Scott* v. *Mercantile Accident Insurance Co.* (1892) 8 TLR.

23 *Ash* v. *Buxted Poultry Ltd, The Times* 29 November 1989.

24 See e.g. *Z Ltd* v. *A–Z* [1982] QB 558.

the direction he is in no position to enforce it; however, either party may seek a court order to enforce the Arbitrator's direction.

Security for costs—Rule 6.1(d)[25]

A successful respondent will ordinarily be awarded the costs which he has reasonably incurred in running his defence.[26] Such an award of costs is, however, of little value to him if the claimant has no funds to pay those costs. In order to provide protection against such an eventuality, a respondent may, in some situations, obtain an order during the course of the proceedings whereby the claimant is required to provide security for some or all of the respondent's projected costs. If such an order is made, and the claimant fails to comply, his claims will normally be stayed.

General principles for an Arbitrator dealing with an application for security for costs

The question of whether an order for security for costs/fees is appropriate may conveniently be dealt with in two stages.

(*a*) *Jurisdiction*. The Arbitrator's jurisdiction is determined by construing Rule 6.1(d) and is not limited to the special categories of claimant defined in Order 23.[27]

(*b*) *Exercise of the power*. The exercise of the power is discretionary and must therefore be exercised judicially. The principal matters to which an Arbitrator may have regard are the general matters set out by Lord Denning in *Sir Lindsay Parkinson* v. *Triplan*.[28]

---

25 In arbitrations under the 1950 Act, the Arbitrator has no power to order security for costs unless clothed with such power by agreement of the parties. Section 38(3) of the 1996 Act, however, provides the Arbitrator with general power to order security unless the parties otherwise agree.

26 *Smeaton Hanscombe & Co.* v. *Sassoon I. Setty, Son & Co. (No. 2)* [1953] 1 WLR 1481; *Tramountana Armadora* v. *Atlantic Shipping Co.* [1978] 1 Lloyd's Rep 391.

27 Order 23 of the Rules of the Supreme Court establishes the regime for security for costs in the High Court. The Order 23 categories may be used for guidance. Under the 1996 Act, unless the parties otherwise agree, the arbitrator must not make such an order on the ground that the claimant is resident outside the UK — Section 38(3).

28 [1973] 1 QB 609 (CA). While this case addresses the specific situation of an application under Order 23 of the Rules of the Supreme Court against a limited company, the principles outlined are, it is submitted, of general application.

> The principal factors to be weighed include (1) the bona fides of the claim, (2) the prospects of success, if these are readily apparent,[29] (3) any admissions or offers which have been made which are not wholly privileged,[30] (4) any oppressive features of the application for security, (5) the effect of the respondent's behaviour on the claimant's want of means and (6) the timing of the application.[31] While these are the principal factors which are normally considered, the court is not constrained in the factors which it may take into account when exercising its discretion.[32]

Dealing with applications for security for costs

The Arbitrator may entertain an application in any format which he deems acceptable, providing he abides by the rules of natural justice and acts judicially. In practice, the application should be made after the claimant has been given an opportunity to provide voluntary security. The applicant should show that he has asked for reasonable security with supporting reasons and that he has given the claimant a reasonable time to provide it. The application must be served on the claimant, giving him a proper and adequate opportunity to respond. The claimant must be given a proper opportunity to address the Arbitrator as to (*a*) whether the Arbitrator has jurisdiction to order security for costs, (*b*) whether the application raises an appropriate case for security,[33] (*c*) whether the quantum requested is reasonable and (*d*) whether the form in which the security is requested is appropriate. While applications for security for costs and the

29 The court will not investigate the merits unless they are clear: *Porzelack K. G.* v. *Porzelack UK Ltd* [1987] 1 WLR 420, per Sir Nicolas Browne-Wilkinson V.-C. at 423. To do so would not only require the court to hear evidence but would also pre-empt the eventual outcome which is a matter solely within the Arbitrator's province: see *Flender Werft* v. *Aegean Maritime* [1990] 2 Lloyd's Rep. 27, per Saville J at 30.

30 *Simaan General Contracting Co.* v. *Pilkington Glass Ltd* [1987] 1 WLR 516.

31 An application may in principle be made at any time: see *Re Smith* (1896) LT 46. The application in *Sir Lindsay Parkinson Ltd* v. *Triplan Ltd* [1973] 1 QB 609 was brought very late; this was considered a strong reason for rejecting it.

32 *Sir Lindsay Parkinson Ltd* v. *Triplan Ltd* [1973] 1 QB 609, per Lawton LJ at 629.

33 Unless the powers vested in the Arbitrator are unusual and are very different from those which the court might exercise, this will involve submissions on whether the criteria established by the courts (in *Sir Lindsay Parkinson Ltd* v. *Triplan Ltd* and subsequent cases) have been met.

responses thereto may be dealt with solely by written submissions, it is usually appropriate to hold a short oral hearing. Wherever the respondent makes out a prima facie case for the ordering of security, the likelihood that the claimant will be unable to pay the security arises, almost by definition. Since an order for security may result in a terminal stay of the claimant's case, the parties should normally be allowed an opportunity to make oral submissions, unless the Arbitrator considers the application to be hopeless. Where the Arbitrator considers that security should be given, he should draw up a formal order, giving directions as to the provision of the security and clearly setting out the effects of non-compliance.

Security for costs where there is a counterclaim

Where an application for security is made by a respondent whose counterclaim overlaps in its subject matter with the claim, or where a claimant seeks an order for security for costs against a respondent in respect of a counterclaim, difficult questions may arise as to whether or not security should be ordered and as to the form of the order which should be made.[34]

Security for the arbitrator's costs—Rule 6.1(e)

It is submitted that the Arbitrator should only exercise the power to order security if there is some reason to believe that his fees are at risk.[35] In all cases the Arbitrator must take the interests of the parties fully into account as well as his own. It is thought that the tests normally applicable where the respondent seeks an order for security do not govern this situation: for example, the Arbitrator cannot undertake company searches in the same way as might a respondent or make any other enquiries without the full cooperation of the parties. Where the Arbitrator does order security for his own fees, he must be careful to make the quantum reasonable and capable of being

34 See *B. J. Crabtree (Insulation) Ltd* v. *GPT Communication Systems Ltd* (1989) 59 BLR 43 (CA); *Flender Werft* v. *Aegean Maritime* [1990] 2 Lloyd's Rep. 27; *New Fenix Compagnie Anonyme* v. *General Accident* [1911] 2 KB 619 (CA); *Neck* v. *Taylor* [1893] 1 QB 560 at 562 (CA); *Cathery* v. *Lithodmos Ltd* (1987) 41 BLR 76 (CA).

35 If the reference is likely to span a significant period of time, this may of itself be sufficient.

substantiated. The question of who should provide the security for the Arbitrator's fees is not always easy to decide. Where the respondent's case is purely defensive, it is suggested that the claimant alone should be required to provide the security; but where the respondent advances an independent counterclaim both parties should be required to provide security in proportion to the likely costs which are likely to be consumed on the respective claims. It is submitted that the better solution in any event is for the Arbitrator to submit regular accounts for work done, payable on the above basis and to be adjusted at the conclusion of the reference; this provides the Arbitrator with reasonable "security".

Rule 7. Power to order concurrent Hearings

7.1 Where disputes or differences have arisen under two or more contracts each concerned wholly or mainly with the same subject matter and the resulting arbitrations have been referred to the same Arbitrator he may with the agreement of all the parties concerned or upon the application of one of the parties being a party to all the contracts involved order that the whole or any part of the matters at issue shall be heard together upon such terms or conditions as the Arbitrator thinks fit.

7.2 Where an order for concurrent Hearings has been made under Rule 7.1 the Arbitrator shall nevertheless make and publish separate Awards unless the parties otherwise agree but the Arbitrator may if he thinks fit prepare one combined set of Reasons to cover all the Awards.

This rule provides for two or more arbitrations to be heard together. Where appropriate, the rule must be read together with Clause 18 of the FCEC Form of Sub-Contract.

The operation of the rule is conditional upon the following.

(*a*) All the arbitrations being "concerned wholly or mainly with the same subject matter". It is thought that this means that the same facts or alleged facts must be material in each arbitration.

(*b*) Both arbitrations being referred to the same Arbitrator. This may come about where the appointing authority (ordinarily the President of the ICE) appoints the same arbitrator. It may arise also where a Contractor and the Sub-Contractor combine to recover monies under the Main Contract. For example, where the Contractor seeks benefits for his Sub-Contractor under Clause 10(2) of the FCEC Form of Sub-Contract, the Contractor and Sub-

Contractor may request the Arbitrator appointed under the Main Contract to determine their dispute also.

(c) The application of a party who is a party to all the arbitrations. An alternative is the agreement of all the parties, but such an agreement necessarily entails the consent of a party who is a party to all the arbitrations.

Such terms and conditions as the Arbitrator thinks fit

Where the above conditions are met, the Arbitrator may issue such directions as he considers proper. The following considerations may arise.

(a) *Evidence:* The Arbitrator may direct that both arbitrations proceed fully in parallel, with all parties being able to examine all witnesses. Or he may direct that each arbitration be concurrent but evidentially separate. Where a Contractor and Sub-Contractor combine against the Employer, there may be very simple pleadings in the Sub-Contract arbitration and it may be "adjourned *sine die*": in other words, it does not feature until the Main Contract arbitration is completed at which point the Arbitrator decides how any recovered sum is to be divided between the Main Contractor and Sub-Contractor.

(b) *Discovery:* Where there are more than two parties, the question arises of whether a party is entitled to seek discovery against a third party who is nevertheless a party to the arbitration. It is thought that the Arbitrator may issue directions that any party must disclose any relevant documents to any other party.

Separate awards

There should be separate awards, including awards as to costs.[36]

Rule 8. Powers at the Hearing

8.1 The Arbitrator may hear the parties their representatives and/or witnesses at any time or place and may adjourn the arbitration for any period on the application of any party as he thinks fit.

36 *The Catherine L.* [1982] 1 Lloyd's Rep. 484.

8.2 Any party may be represented by any person including in the case of a company or other legal entity a director officer employee or beneficiary of such company or entity. In particular, a person shall not be presented from representing a party because he is or may be also a witness in the proceedings. Nothing shall prevent a party from being represented by different persons at different times.

8.3 Nothing in these Rules or in any other rule custom or practice shall prevent the Arbitrator from starting to hear the arbitration once his appointment is completed or at any time thereafter.

8.4 Any meeting with or summons before the Arbitrator at which both parties are represented shall if the Arbitrator so directs be treated as part of the hearing of the arbitration.

Representatives

In arbitration proceedings there are no restrictions as to who may represent the parties and "lay advocates" frequently appear.[37] Rule 8.2 is merely declaratory.

Rule 9. Power to appoint assessors or to seek outside advice

9.1 The Arbitrator may appoint a legal or technical or other assessor to assist him in the conduct of the arbitration. The Arbitrator shall direct when such assessor is to attend hearings of the arbitration.

9.2 The Arbitrator may seek legal technical or other advice on any matter arising out of or in connection with the proceedings.

9.3 Further and/or alternatively the Arbitrator may rely upon his own knowledge and expertise to such extent as he thinks fit.

Assessors

The Court may appoint an assessor who assists and advises the judge in technical matters. Rule 9 provides that an Arbitrator may do likewise.[38] An Arbitrator who intends to appoint an assessor should

---

37 As to the costs of lay advocates and claims consultants generally see: *Piper Double Glazing Ltd* v. *DC Contracts* [1994] 1 WLR 777.

38 Under the 1996 Act an arbitrator is entitled to appoint an assessor even in the absence of such an agreed rule: Section 37(1)(a)(ii).

first give the parties an opportunity to make representations. He must not delegate his decision to the assessor. Furthermore, where the assessor advances a line of reasoning which is not contended for by either party, that reasoning should be put to the parties for them to comment and, if necessary, to bring further evidence.

Advice

It is generally accepted that an Arbitrator may take legal advice[39] providing he does not delegate his decision to the lawyer.[40] The costs of this advice is part of the costs of the award. It is thought that an Arbitrator may take technical advice also, though there is no firm guidance from the Court on this point and Rule 9.2 is a welcome clarification.

Arbitrator's own knowledge

Subject to a contrary agreement, a technically qualified Arbitrator is always able to use his own general knowledge and expertise.[41] Where he has, by virtue of his particular experience, special knowledge he should advise the parties so that they may comment and make submissions.[42]

---

39 This has now been enacted in Section 37(1)(a)(i) of the 1996 Act.

40 *Ellison* v. *Bray* (1864) 9 LT 730 where Shee J said: "It is clear that an arbitrator or umpire is allowed to consult others if he wishes to inform his own mind, but he must not substitute the opinion of another for his own".

41 *Mediterranean & Eastern Export Co. Ltd* v. *Fortress Fabrics (Manchester) Ltd* [1948] 2 All ER 186; *Fox* v. *Wellfair* [1981] 2 Lloyd's Rep. 514 (CA); *F. R. Waring (UK) Ltd* v. *Administracao Geral do Acucar* [1983] 1 Lloyd's Rep. 45.

42 *Fox* v. *Wellfair* [1981] 2 Lloyd's Rep. 514 (CA).

PART C. PROCEDURE BEFORE THE HEARING

Rule 10. The Preliminary Meeting

10.1 As soon as possible after accepting the appointment the Arbitrator shall summon the parties to a Preliminary Meeting for the purpose of giving such directions about the procedure to be adopted in the arbitrations as he considers necessary.

10.2 At the Preliminary Meeting the parties and the Arbitrator shall consider whether and to what extent

(a) Part F (Short Procedure) or Part G (Special Procedure for Experts) of these rules shall apply

(b) the arbitration may proceed on documents only

(c) progress may be facilitated and costs saved by determining some of the issues in advance of the main Hearing

(d) the parties should enter into an exclusion agreement (if they have not already done so) in accordance with S.3 of the Arbitration Act 1979 (where the Act applies to the arbitration)

and in general shall consider such other steps as may minimise delay and expedite the determination of the real issue between the parties.

10.3 If the parties so wish they may themselves agree directions and submit them to the Arbitrator for his approval. In so doing the parties shall state whether or not they wish Part F or Part G of these Rules to apply. The Arbitrator may then approve the directions as submitted or (having first consulted the parties) may vary them or substitute his own as he thinks fit.

Preliminary Meeting

A meeting shortly after the Arbitrator's appointment is useful in a number of respects. The Arbitrator meets the parties and any challenge to his jurisdiction can be raised. At this meeting the Arbitrator will give "directions", which will include:

(*a*) directions as to the form which the pleadings will take
(*b*) directions as to the timetable for pleadings

(*c*) directions as to expert witnesses (including the maximum number permitted,[43] their disciplines and any arrangements for them to prepare reports and/or to meet to attempt to settle technical issues[44])

(*d*) directions as to other witnesses (including whether their primary evidence is to be reduced to writing and if so whether the document containing it is to be released to the other party in advance of the hearing[45])

(*e*) directions as to the hearing (whether there will be a formal hearing, whether advocates will make their primary submissions in writing, whether each party will have limited time to present its case, etc.)

(*f*) directions as to the award (whether it is to be supported by reasons etc.[46]).

Directions agreed by the parties

There is some academic debate as to whether an Arbitrator is bound by directions agreed by the parties. There is a spectrum of views, the extreme positions being: (1) the Arbitrator is master of the proceedings and need not heed the agreements of the parties made after his appointment; (2) the Arbitrator is the servant of the parties and bound by their agreement. In practice problems rarely arise because agreed directions are generally reasonable and acceptable to the Arbitrator.

Rule 11. Pleadings and discovery

11.1 The Arbitrator may order the parties to deliver pleadings or statements of their cases in any form he thinks appropriate. The Arbitrator may order any party to answer the other party's case and to give reasons for any disagreement.

11.2 The Arbitrator may order any party to deliver in advance of formal discovery copies of any documents in his possession custody or

43 Note that a direction entitling a party to call *X* experts does not mean that the costs of *X* witnesses are necessarily recoverable: *Atwell* v. *Ministry of Public Buildings and Works* [1969] 1 WLR 1074.

44 See also Rules 13.1(c) and 16.2.

45 See also Rules 16.3 and 16.5.

46 See Rule 18.

power which relate either generally or specifically to matters raised in any pleading statement or answer.

11.3 Any pleading statement or answer shall contain sufficient detail for the other party to know the case he has to answer. If sufficient detail is not provided the Arbitrator may of his own motion or at the request of the other party order further and better particulars to be delivered.

11.4 If a party fails to comply with any order made under this Rule the Arbitrator shall have power to debar that party from relying on the matters in respect of which he is in default and the Arbitrator may proceed with the arbitration and make his Award accordingly. Provided that the Arbitrator shall first give notice to the party in default that he intends to proceed under this Rule.

Pleadings

In litigation, formal pleadings are served to define the issues. The rules of formal pleading provide that facts shall be pleaded in terse numbered paragraphs; evidence is not pleaded. Such restrictions are generally inappropriate in technical arbitrations. Rule 11.1 allows the Arbitrator to order that pleadings be delivered in any form which he considers appropriate. He may direct that pleadings similar to those in a High Court action be served; at the other extreme he may direct that a statement of case be served, with all technical arguments, documents and drawings, etc. to be appended.

The sequence of pleadings

Pleadings are ordinarily served as follows.

(*a*) Points of Claim—served by the Claimant, setting out his case.
(*b*) Points of Defence (and Counterclaim)—served by the Respondent, setting out his answer to the Points of Claim. Where a Counterclaim is served, this contains a claim by the Respondent against the Claimant.
(*c*) Points of Reply (and Defence to Counterclaim)—served by the Claimant answering the Respondent's Points of Defence and where appropriate setting out the Claimant's defence to the Points of Counterclaim.
(*d*) Points of Reply to Counterclaim—served by the Respondent where there is a Counterclaim.

In complex arbitrations there may be additional pleadings in order to narrow the issues yet further. The terms used above are the traditional names of pleadings. Arbitrators frequently prefer to use alternative terms to indicate that they require "non-traditional" pleadings. Thus where the Arbitrator requires that supporting documents be served he may describe the Points of Claim as "Claimant's Statement of Case" or even "Claimant's Statement of Case with Supporting Materials" in order the better to describe the format and content of the document which he expects. None of these descriptions are terms of art.

Discovery

Rule 11 is entitled "Pleadings and discovery". Curiously, the only reference to "discovery" is in respect to the advance copies of documents referred to in Rule 11.2. The Arbitration Act[47] provides that the Arbitrator may require the parties to produce any document in their possession or power.[48] Hence no formal rule is required. The terminology used is as follows: (*a*) discovery is the process whereby a party advises the other of the documents which he has or has had in his possession; (*b*) inspection is where the other party is given sight of the documents. "Discovery by list" means that each and every relevant document in a party's possession is to be listed in a schedule. A term which is sometimes used, "Discovery by category", means that broad categories of document are to be set out in a schedule. It is for the Arbitrator to direct the extent to which discovery is required. Sometimes the Arbitrator will direct that parties need only disclose those documents upon which they specifically rely; in traditional arbitrations, every document which is or may be relevant is disclosed.

---

47 Section 12(1) of the Arbitration Act 1950; Section 34(2)(d) of the 1996 Act.

48 Unless it is privileged. Categories of privileged documents include (1) correspondence between parties and their lawyers, (2) reports compiled with a view to giving advice concerning contemplated or pending litigation or arbitration and (3) documents relating to negotiation with a view to settlement of the dispute: *South Shropshire District Council* v. *Amos* [1986] 1 WLR 1271 (CA).

Documents relating generally or specifically to any pleading—Rule 11.2

It is a matter of fairness that where a party refers to a document in his possession, the other party (who is to answer the pleading) shall be able to see the document. Rule 11.2 provides for this.

Sufficient detail—Rule 11.3

Frequently a party will advance a contention without giving sufficient detail for the other party to know what case he has to meet. In such a case a "Request for Further and Better Particulars" may be made by the disadvantaged party. Where there is a dispute as to whether the particulars furnished are adequate the Arbitrator will give a ruling.

Rule 12. Procedural meetings

12.1 The Arbitrator may at any time call such procedural meetings as he deems necessary to identify or clarify the issues to be decided and the procedures to be adopted. For this purpose the Arbitrator may request particular persons to attend on behalf of the parties.

12.2 Either party may at any time apply to the Arbitrator for leave to appear before him on any interlocutory matter. The Arbitrator may call a procedural meeting for the purpose or deal with the application in correspondence or otherwise as he thinks fit.

12.3 At any procedural meeting or otherwise the Arbitrator may give such directions as he thinks fit for the proper conduct of the arbitration. Whether or not formal pleadings have been ordered under Rule 11 such directions may include an order that either or both parties shall prepare in writing and shall serve upon the other party and the Arbitrator any or all of the following

(a) a summary of that party's case

(b) a summary of that party's evidence

(c) a statement or summary of the issues between the parties

(d) a list and/or summary of the documents relied upon

(e) a statement or summary of any other matters likely to assist the resolution of the disputes or differences between the parties.

It is frequently useful for procedural meetings to be convened. On occasions, one party wishes to make a formal application. Typical matters dealt with are:

(*a*) applications for security for costs or for the exercise of one of the other powers set out in Rule 6.1

(*b*) applications for directions or revised directions in respect of the format of pleadings, discovery, etc.

(*c*) applications for Summary Awards under Rule 14.

Rule 13. Preparation for the Hearing

13.1 In addition to his powers under Rules 11 and 12 the Arbitrator shall also have power

(a) to order that the parties shall agree facts as facts and figures as figures where possible

(b) to order the parties to prepare an agreed bundle of all documents relevant to the arbitration. The agreed bundle shall thereby be deemed to have been entered in evidence without further proof and without being read out at the Hearing. Provided always that either party may at the Hearing challenge the admissibility of any document in the agreed bundle

(c) to order that any experts whose reports have been exchanged before the Hearing shall be examined by the Arbitrator in the presence of the parties or their legal representatives and not by the parties or their legal representatives themselves. Where such an order is made either party may put questions whether by way of cross-examination or re-examination to any party's expert after all experts have been examined by the Arbitrator provided that the party so doing shall first give notice of the nature of the questions he wishes to put.

13.2 Before the Hearing the Arbitrator may and shall if so requested by the parties read the documents to be used at the Hearing. For this or any other purpose the Arbitrator may require all such documents to be delivered to him at such time and place as he may specify.

Agreeing facts as facts and figures as figures

This simply means achieving such agreement as to facts and valuations as may be achieved. These "agreements" may be contingent upon the Arbitrator's making specific findings in relation to liability. Thus "quantum" experts may agree upon the valuation of a Clause 12 claim, subject to the arbitrator finding that physical conditions could not reasonably have been foreseen. Clearly the Arbitrator cannot order that an agreement be reached.

Agreed bundle

The term "bundle" means a file of documents. In traditional litigation a document must be read into court before it carries any evidential weight. Rule 13.1(b) provides that the bundle is to be taken as evidence. This does not mean that a statement in a document is to be taken as proved. It simply means that unless specifically challenged, documents are to be taken as valid. Where a party wishes to rely upon a statement in a document as evidence of some fact he must still put the statement to the Arbitrator. In litigation, such statements are "hearsay" and their admissibility is governed by complex technical rules. It is submitted that the law of evidence applicable to litigation does not apply to arbitrations[49] and hence such "hearsay" is admissible. It is for the Arbitrator to decide what weight is to be given to any such statement.

Experts to be examined by the Arbitrator

Where the Arbitrator is an engineer and the parties put forward engineering experts it is sensible that the Arbitrator should be able to question them without the "shield" provided by the advocates.[50] Where the power in Rule 13.1(c) is exercised, the term "examination" rarely gives a proper impression of the proceedings. The term "learned discussion" is often more accurate; the expert witnesses frequently arrive at a common view or at least at a common perspective on the factors which will settle the dispute.

Rule 14. Summary Awards

14.1 The Arbitrator may at any time make a *Summary Award* and for this purpose shall have power to award payment by one party to another of a sum representing a reasonable proportion of the final nett amount which in his opinion that party is likely to be ordered to pay after determination of all the issues in the arbitration and after

49 Except the rule in relation to privileged documents. The rationale for this rule is equally strong whatever the tribunal. Section 34(2)(f) of the 1996 Act empowers the arbitrator to decide whether or not to apply strict rules of evidence.

50 Under Section 34(2)(g) of the 1996 Act the arbitrator will be entitled, subject to an agreement of the parties, to "take the initiative in ascertaining the facts...".

taking into account any defence or counterclaim upon which the other party may be entitled to rely.

14.2 The Arbitrator shall have power to order the party against whom a Summary Award is made to pay part or all of the sum awarded to a stakeholder. In default of compliance with such an order the Arbitrator may order payment of the whole sum in the Summary Award to the other party.

14.3 The Arbitrator shall have power to order payment of costs in relation to a Summary Award including power to order that such costs shall be paid forthwith.

14.4 A Summary Award shall be final and binding upon the parties unless and until it is varied by any subsequent Award made and published by the same Arbitrator or by any other arbitrator having jurisdiction over the matters in dispute. Any such subsequent Award may order repayment of monies paid in accordance with the Summary Award.

The Arbitration Act provides that an Arbitrator may issue an interim award.[51] Where the Arbitrator takes the view that an irreducible minimum is due to a party he may award that sum immediately.[52] However, the Courts have decided that any interim award is final and binding in relation to the findings it makes.[53] The Arbitrator is not entitled to revisit the findings and to revise them. Rule 14 allows the Arbitrator to make an "award" and later to revise it.[54] This valuable power mirrors the Court's power to make an interim payment;[55] while cases on the application of the Court's power may shed light on the

51 Section 14 of the 1950 Act. Section 47 of the 1996 Act does not use the expression "interim award", but entitles the arbitrator to make "more than one award at different times on different aspects of the matters to be determined".

52 *The Kostas Melas* [1981] 1 Lloyd's Rep. 18; *The Trade Fortitude* [1992] 1 Lloyd's Rep. 169.

53 *Fidelitas Shipping Co. Ltd* v. *V/O Exportchleb* [1966] 1 QB 630 (CA). See also Section 39 of the 1996 Act which supports this.

54 It may be a matter of semantics whether or not a Rule 14 award is an "award" at all or merely a procedural step. However, the distinction is important where either party wishes to challenge the "award". The Court is extremely reluctant to interfere with an Arbitrator's procedural decisions: see *The Trade Fortitude* [1992] 1 Lloyd's Rep. 169.

55 Under Order 29 of the Rules of the Supreme Court.

principles which are to be applied, it is submitted that Rule 14 is less demanding on an applicant.[56]

PART D. PROCEDURE AT THE HEARING

Rule 15. The Hearing

15.1 At or before the Hearing and after hearing representations on behalf of each party the Arbitrator shall determine the order in which the parties shall present their cases and/or the order in which the issues shall be heard and determined.

15.2 The Arbitrator may order any submission or speech by or on behalf of any party to be put into writing and delivered to him and to the other party. A party so ordered shall be entitled if he so wishes to enlarge upon or vary any such submission orally.

15.3 The Arbitrator may on the application of either party or of his own motion hear and determine any issue or issues separately.

15.4 If a party fails to appear at the Hearing and provided that the absent party has had notice of the Hearing or the Arbitrator is satisfied that all reasonable steps have been taken to notify him of the Hearing the Arbitrator may proceed with the Hearing in his absence. The Arbitrator shall nevertheless take all reasonable steps to ensure that the real issues between the parties are determined justly and fairly.

General

An Arbitrator is entitled to give directions as to the hearing and its arrangement; however, in the absence of agreement between the parties, the limits of this power are unclear. It has been decided that an Arbitrator may direct that representative issues may be advanced instead of all issues.[57] But it seems also to have been decided that the

---

56 Order 29 rule 11(1) of the Rules of the Supreme Court requires that the Court be "satisfied ... that if the action proceeded to trial, the plaintiff would obtain judgment for substantial damages ...". In *British & Commonwealth Holdings* v. *Quadrex Holdings* [1989] QB 842 (CA) it was decided that this was in essence the same standard required to obtain summary judgment under Order 14. Rule 14 (which it is suggested is infelicitously numbered) requires the claimant to overcome a lower hurdle: "a reasonable proportion of the final nett amount which in his opinion that party is likely to be ordered to pay ...".

57 See *Carlisle Place Investments Ltd* v. *Wimpey Construction (UK) Ltd* (1980) 15 BLR 109.

Arbitrator may not conduct the proceedings inquisitorially.[58] Rule 15 provides additional and welcome clarity.

The order in which the parties shall present their cases

In litigation the ordinary sequence of presentation is: (*a*) the Plaintiff (Claimant) opens his case and calls his witnesses; (*b*) the Defendant (Respondent) opens his case, calls his witnesses and then makes final submissions; (*c*) the Plaintiff (Claimant) makes final submissions; (*d*) the judge gives his judgment; (*e*) submissions are made as to costs. The right to open is considered valuable since ordinarily it carries with it the right to make the last submissions.[59] Rule 15, however, allows the Arbitrator to determine the order of presentations. This enables him to take into account:

(*a*) the availability of any witnesses
(*b*) the convenience of dealing with discrete issues before moving on to others[60]
(*c*) questions as to which party must prove what (the burden of proof), which may differ from issue to issue
(*d*) how to arrange proceedings where there are more than two parties.[61]

Speeches in writing

In traditional litigation, all submissions, speeches, etc. are made orally. It is becoming increasingly common for submissions to be reduced to writing, with brief elaborations and restricted oral argument. In arbitrations, speeches reduced to writing have long been common. Rule 15.2 reinforces this practice.

58 *Town & City Properties (Development) Ltd* v. *Wiltshier Southern Ltd and Gilbert Powell* (1989) 44 BLR 109.

59 This may have been important when juries decided cases in civil litigation, but it is probably of no importance in modern litigation. There is one exception: where the Defendant chooses not to call any evidence, he is entitled to the "last word".

60 See Rule 15.3.

61 See Rule 7.

Proceeding in the absence of a party

Where the claimant wilfully fails to appear to prosecute the claim, the respondent is entitled to an award dismissing the claim. Where the respondent fails to appear, the claimant must still prove his case on a balance of probabilities. Where the Arbitrator considers the claimant's case to be misconceived he must give the claimant proper notice.[62]

Rule 16. Evidence

16.1 The Arbitrator may order a party to submit in advance of the Hearing a list of the witnesses he intends to call. That party shall not thereby be bound to call any witness so listed and may add to the list so submitted at any time.

16.2 No expert evidence shall be admissible except by leave of the Arbitrator. Leave may be given on such terms and conditions as the Arbitrator thinks fit. Unless the Arbitrator otherwise orders such terms shall be deemed to include a requirement that a report from each expert containing the substance of the evidence to be given shall be served upon the other party within a reasonable time before the Hearing.

16.3 The Arbitrator may order disclosure or exchange of proofs of evidence relating to factual issues. The Arbitrator may also order any party to prepare and disclose in advance a list of points or questions to be put in cross-examination of any witness.

16.4 Where a list of questions is disclosed whether pursuant to an order of the Arbitrator or otherwise the party making disclosure shall not be bound to put any question therein to the witness unless the Arbitrator so orders. Where the party making disclosure puts a question not so listed in cross-examination the Arbitrator may disallow the costs thereby occasioned.

16.5 The Arbitrator may order that any proof of evidence which has been disclosed shall stand as the evidence in chief of the deponent provided that the other party has been or will be given an opportunity to cross-examine the deponent thereon. The Arbitrator may also at any time before such cross-examination order the deponent or some other identified person to deliver written answers to questions arising out of the proof of evidence.

62 *Fox* v. *Wellfair* [1981] 2 Lloyd's Rep. 514 (CA).

16.6 The Arbitrator may himself put questions to any witness and/or require the parties to conduct enquiries tests or investigations. Subject to his agreement the parties may ask the Arbitrator to conduct or arrange for any enquiry test or investigation.

Rule 16 is concerned both with evidence of fact and expert evidence. The Arbitrator will have wide powers as to evidence in any event, and the rule is probably entirely declaratory, save for some aspects of Rule 16.6.

Evidence and examination

Witnesses are examined in three stages. First, the witness is examined by the advocate calling him. This is called examination-in-chief. Then the witness's evidence is tested by cross-examination. Finally the witness is re-examined by the advocate calling him.

Evidence of fact

The general rule is that a witness is entitled to give evidence only of what he perceived. Rule 16.3 provides that the Arbitrator may "order disclosure or exchange of proofs of evidence". Where the evidence is reduced to writing this is called a "proof". Disclosure is the process whereby the proof is published to the other party. Exchange is a procedure whereby proofs are exchanged simultaneously, presumably to prevent a party altering his story upon sight of the opponent's evidence.[63] The fact that evidence is contained in a proof does not put it into evidence in the arbitration until the person whose evidence it is (sometimes called the deponent[64]) declares it to be his evidence. By Rule 16.3, a witness must present himself for cross-examination.

Expert evidence

The general rule that a witness is entitled to give evidence only of what he perceived has a number of exceptions. For present purposes

63 It is often advantageous for the proofs to be exchanged a little while before the Arbitrator sees them in case any privileged matter has been included by mistake.

64 See Rule 16.5.

the most important exception concerns expert evidence. Where a witness has a special skill or experience, he may give his opinion in relation to that skill or experience as expert evidence. This is common in civil engineering arbitrations. The expert witness need not have any formal qualifications[65] and it is for the Arbitrator to decide whether he is sufficiently skilled or experienced to be treated as an expert. Expert witnesses may advise their clients as to the clients' best case during preliminary stages; but when they give evidence (either in their report or at the hearing) they must be strictly impartial.[66]

Arbitrator may put questions or order tests etc.

Rule 16.6 provides the Arbitrator with the right to conduct the proceedings in a partly inquisitorial style. It has been held that an Arbitrator is constrained to conduct the proceedings in an adversarial manner[67] unless the parties agree otherwise. It has also been held that an Arbitrator is not to gather information.[68] The express right to do these things is thus important.

PART E. AFTER THE HEARING

Rule 17. The Award

17.1 Upon the closing of the Hearing (if any) and after having considered all the evidence and submissions the Arbitrator will prepare and publish his Award.

17.2 When the Arbitrator has made and published his Award (including a Summary Award under Rule 14) he will so inform the parties in

65 *R.* v. *Silverlock* [1894] 2 QB 766 (Crown Cases Reserved).

66 A number of commentators have argued that an expert witness is entitled not to volunteer information even though he knows the questioner will be misled by his failure to do so. This is not thought to be correct. An expert witness is given a special right to testify as to things he has not personally witnessed. He owes a corresponding special duty to ensure that he does not mislead. See generally *Whitehouse* v. *Jordan* [1981] 1 WLR 246 (HL); *University of Warwick* v. *Sir Robert McAlpine* (1988) 42 BLR 1.

67 *Town & City Properties (Development) Ltd* v. *Wiltshier Southern Ltd and Gilbert Powell* (1989) 44 BLR 109. But see Section 34(2)(g) of the 1996 Act.

68 *Owen* v. *Nicholl* [1948] 1 All ER 707 (CA); *Top Shop Estates* v. *Danino* (1984) 273 EG 197.

writing and shall specify how and where it may be taken up upon due payment of his fee.

Publication of the Award

An Award is published when the parties are advised that it is available to be collected. Where the Arbitrator makes collection conditional upon his fee being paid, this does not affect the date of publication. The date may be important since the Award cannot be challenged later than 21 days after publication.[69]

Payment of the Arbitrator

It is common practice for the Arbitrator to insist on being paid prior to releasing the Award. This is known as the arbitrator's lien. Where the party wishing to collect the Award disputes the amount of fees claimed by the Arbitrator, he may pay the money into court and the court will tax the Arbitrator's fees and order the Arbitrator to deliver up the Award.[70]

Rule 18. Reasons

18.1 Whether requested by any party to do so or not the Arbitrator may at his discretion state his Reasons for all or any part of his Award. Such Reasons may form part of the Award itself or may be contained in a separate document.

18.2 A party asking for Reasons shall state the purpose for his request. If the purpose is to use them for an appeal (whether under S.1 of the Arbitration Act 1979 or otherwise) the requesting party shall also specify the points of law with which he wishes the Reasons to deal.

69 See Order 73 of the Rules of the Supreme Court. The Arbitration Act 1996 allows a period of 28 days — Section 70(3).

70 Section 19 of the 1950 Act and Section 56 of the 1996 Act. But as to the difficulties in this procedure see *Government of Ceylon* v. *Chandris* [1963] 1 Lloyd's Rep. 214 and *Rolimpex Centrala Handlu Zagranicznego* v. *Haji E. Dossa & Sons Ltd* [1971] 1 Lloyd's Rep. 380.

In that event the Arbitrator shall give the other party an opportunity to specify additional points of law to be dealt with.

18.3 Reasons prepared as a separate document may be delivered with the Award or later as the Arbitrator thinks fit.

18.4 Where the Arbitrator decides not to state his Reasons he shall nevertheless keep such notes as will enable him to prepare Reasons later if so ordered by the High Court.

The benefits of giving reasons

Arbitrators agree that giving reasons serves three important functions.

(*a*) Principles of fairness suggest that the parties are "entitled" to know the reasoning that led to the result.
(*b*) The discipline of giving reasons helps the Arbitrator to clarify and to check his own reasoning.
(*c*) Arithmetic errors and other slips may be made and the parties are given an opportunity to check these. In "single issue" arbitrations this may not be a major problem. Civil engineering arbitrations, however, frequently involve many issues and calculation errors are not uncommon.

The effect of Rule 18

An Arbitrator is generally not required by law to give reasons for his award; an award which contains no reasons is valid.[71] However, the court may order that reasons be supplied where either party has requested them. In civil engineering arbitrations, the invariable practice is that Arbitrators give reasons wherever requested by one or more parties. Unless the parties have agreed to exclude any rights of appeal,[72] the reasons tend to form part of the Award, rather than being supplied in a separate document. It is thought that Rule 18 has no

71 Under Section 52 of the 1996 Act, the award shall, unless the parties otherwise agree, contain reasons.

72 See section 3 of the Arbitration Act 1979 and Rule 10.2(d). See Sections 69 and 87 of the 1996 Act.

practical importance save that an Arbitrator puts the parties on notice that he may supply reasons whether or not they have been requested—Rule 18.1. The Arbitrator's discretion to refuse to deliver reasons unless a purpose is stated (Rule 18.2) or to deliver them later (Rule 18.3) will not, it is thought, prevent the Court from ordering the provision of reasons.

Sufficient reasons

The reasons supplied do not have to be extremely detailed. They must allow the parties broadly to see and understand the Arbitrator's reasoning; they must be sufficient to allow the Court to see the legal reasoning involved.[73]

Content of a reasoned award

The content will include:[74]

(*a*) recitals as to the arbitration agreement, the mode and date of appointment of the Arbitrator, the procedure adopted and so forth
(*b*) a statement of the issues in the case
(*c*) the Arbitrator's summing up of submissions of law and his decisions on the law[75]

---

73 *The Nimenia* [1986] QB 802 per Sir John Donaldson MR at 807: a reasoned award is one which "states the reasons for the award in sufficient detail for the court to consider any question of law arising therefrom".

74 In *Bremer Handelgesellschaft mbH* v. *Westzucker GmbH* [1981] 2 Lloyd's Rep. 130 (CA): at 133 Donaldson LJ made a broad statement as to the matters which a reasoned award should contain. He said that the key was to tell a story "logically, coherently and accurately" and that an arbitrator's award differed from a judgment in that "arbitrators will not be expected to analyse the law. It will be quite sufficient that they should explain how they reached their conclusion". Nevertheless, he went on to say that any reasoning offered would not be unwelcome.

75 *Universal Petroleum Co. Ltd* v. *Handels und Transportgesellschaft mbH* [1987] 2 All ER 737 (CA) per Kerr LJ at 748: "A reasoned award is usually requested in order to lay the foundation for a possible application for leave to appeal. An arbitrator should therefore remember to deal in his reasoned award with all issues which may be described as having a 'conclusive' nature, in the sense that he should give reasons for his decisions on all issues which lead to conclusions on liability or other major matters in dispute on which leave to appeal may subsequently be sought."

(*d*) the Arbitrator's summing up of the evidence and his findings of fact[76]

(*e*) a statement of the Arbitrator's overall decision on the issues in the light of his findings of law and fact.

There is no standard format and the key is clear reasoning supported by clear language.

Rule 19. Appeals

19.1 If any party applies to the High Court for leave to appeal against any Award or decision or for an order staying the arbitration proceedings or for any other purpose that party shall forthwith notify the Arbitrator of the application.

19.2 Once any Award or decision has been made and published the Arbitrator shall be under no obligation to make any statement in connection therewith other than in compliance with an order of the High Court under S.1(5) of the Arbitration Act 1979.

Appeals

Unless both parties agree, an appeal against an Arbitrator's award is possible only in exceptional circumstances, namely where the Court can see from a brief perusal that the award is likely to be wrong and that the potential error may substantially affect the rights of the parties.[77] The right to appeal can be excluded by agreement after the dispute has arisen.[78]

76 *J. H. Rayner (Mincing Lane) Ltd* v. *Shaher Trading Co.* [1982] 1 Lloyd's Rep. 632 per Bingham J at 636: "... arbitrators should set out what, on their view of the evidence, did or did not happen, and should explain succinctly why, in the light of what happened, they reached their decision ...".

77 *The Nema* [1982] AC 724 (HL); *Antaios Compania Naviera SA* v. *Salen Rederierna* [1985] AC 191 (HL). The scheme in Section 69 of the 1996 Act is similar to that under the 1979 Act.

78 See section 3 of the Arbitration Act 1979 and Rule 10.2(d). See Sections 69 and 87 of the 1996 Act.

PART F. SHORT PROCEDURE

Rule 20. Short Procedure

20.1 Where the parties so agree (either of their own motion or at the invitation of the Arbitrator) the arbitration shall be conducted in accordance with the following Short Procedure.

20.2 Each party shall set out his case in the form of a file containing

(a) a statement as to the orders or awards he seeks

(b) a statement of his reasons for being entitled to such orders or awards

and

(c) copies of any documents on which he relies (including statements) identifying the origin and date of each document

and shall deliver copies of the said file to the other party and to the Arbitrator in such manner and within such time as the Arbitrator may direct.

20.3 After reading the parties' cases the Arbitrator may view the site or the Works and may require either or both parties to submit further documents or information in writing.

20.4 Within one calendar month of completing the foregoing steps the Arbitrator shall fix a day when he shall meet the parties for the purpose of

(a) receiving any oral submissions which either party may wish to make

and/or

(b) the Arbitrator's putting questions to the parties their representatives or witnesses.

For this purpose the Arbitrator shall give notice of any particular person he wishes to question but no person shall be bound to appear before him.

20.5 Within one calendar month following the conclusion of the meeting under Rule 20.4 or such further period as the Arbitrator may reasonably require the Arbitrator shall make and publish his Award.

Rule 21. Other matters

21.1 Unless the parties otherwise agree the Arbitrator shall have no power to award costs to either party and the Arbitrator's own fees and charges shall be paid in equal shares by the parties. Where one party has agreed to the Arbitrator's fees the other party by agreeing to this Short Procedure shall be deemed to have agreed likewise to the Arbitrator's fees.

21.2 Either party may at any time before the Arbitrator has made and published his Award under this Short Procedure require by written notice served on the Arbitrator and the other party that the arbitration shall cease to be conducted in accordance with this Short Procedure. Save only for Rule 21.3 the Short Procedure shall thereupon no longer apply or bind the parties but any evidence already laid before the Arbitrator shall be admissible in further proceedings as if it had been submitted as part of those proceedings and without further proof.

21.3 The party giving written notice under Rule 21.2 shall thereupon in any event become liable to pay

(a) the whole of the Arbitrator's fees and charges incurred up to the date of such notice

and

(b) a sum to be assessed by the Arbitrator as reasonable compensation for the costs (including any legal costs) incurred by the other party up to the date of such notice.

Payment in full of such charges shall be a condition precedent to that party's proceeding further in the arbitration unless the Arbitrator otherwise directs. Provided that non-payment of the said charges shall not prevent the other party from proceeding in the arbitration.

It is thought that this procedure is rarely used. Rule 21.1 provides that where the Short Procedure is used neither party may receive an award of costs unless otherwise agreed. Also note Rules 21.2, 21.3 which provide that either party may, upon payment of costs, revert to the standard procedure at any time.

PART G. SPECIAL PROCEDURE FOR EXPERTS

Rule 22. Special Procedure for Experts

22.1 Where the parties so agree (either of their own motion or at the invitation of the Arbitrator) the hearing and determination of any issues of fact which depend upon the evidence of experts shall be conducted in accordance with the following Special Procedure.

22.2 Each party shall set out his case on such issues in the form of a file containing

(a) a statement of the factual findings he seeks

(b) a report or statement from and signed by each expert upon whom that party relies

and

(c) copies of any other documents referred to in each expert's report or statement or on which the party relies identifying the origin and date of each document

and shall deliver copies of the said file to the other party and to the Arbitrator in such manner and within such time as the Arbitrator may direct.

22.3 After reading the parties' cases the Arbitrator may view the site or the Works and may require either or both parties to submit further documents or information in writing.

22.4 Thereafter the Arbitrator shall fix a day when he shall meet the experts whose reports or statements have been submitted. At the meeting each expert may address the Arbitrator and put questions to any other expert representing the party. The Arbitrator shall so direct the meeting as to ensure that each expert has an adequate opportunity to explain his opinion and to comment upon any opposing opinion. No other person shall be entitled to address the Arbitrator or question any expert unless the parties and the Arbitrator so agree.

22.5 Thereafter the Arbitrator may make and publish an Award setting out with such details or particulars as may be necessary his decision upon the issues dealt with.

Rule 23. Costs

23.1 The Arbitrator may in his Award make orders as to the payment of any costs relating to the foregoing matters including his own fees and charges in connection therewith.

23.2 Unless the parties otherwise agree and so notify the Arbitrator neither party shall be entitled to any costs in respect of legal representation assistance or other legal work relating to the hearing and determination of factual issues by this Special Procedure.

It is thought that this procedure is rarely used. By Rule 23.2 there is no recovery of costs for legal representation or assistance in relation to factual issues.

PART H. INTERIM ARBITRATION

Rule 24. Interim Arbitration

24.1 Where the Arbitrator is appointed and the arbitration is to proceed before completion or alleged completion of the Works then save in the case of a dispute arising under Clause 63 of the ICE Conditions of Contract the following provisions shall apply in addition to the foregoing Rules and the arbitration shall be called an Interim Arbitration.

24.2 In conducting an Interim Arbitration the Arbitrator shall apply the powers at his disposal with a view to making his Award or Awards as quickly as possible and thereby allowing or facilitating the timely completion of the Works.

24.3 Should an Interim Arbitration not be completed before the Works or the relevant parts thereof are complete the Arbitrator shall within 14 days of the date of such completion make and publish his Award findings of fact or Interim Decision pursuant to Rule 24.5 hereunder on the basis of evidence given and submissions made up to that date together with such further evidence and submissions made up to that date together with such further evidence and submissions as he may in his discretion agree to receive during the said 14 days. Provided that before the expiry of the said 14 days the parties may otherwise agree and so notify the Arbitrator.

24.4 For the purpose only of Rule 24.3 the Arbitrator shall decide finally whether and if so when the Works or the relevant parts thereof are complete.

24.5 In an Interim Arbitration the Arbitrator may make and publish any or all of the following

(a) a Final Award or an Interim Award on the matters at issue therein

(b) findings of fact

(c) a Summary Award in accordance with Rule 14

(d) an Interim Decision as defined in Rule 24.6.

An Award under (a) above or a Finding under (b) above shall be final and binding upon the parties in any subsequent proceedings. Anything not expressly identified as falling under either of headings (a) (b) or (c) above shall be deemed to be an Interim Decision under heading (d). Save as aforesaid the Arbitrator shall not make an Interim Decision without first notifying the parties that he intends to do so.

24.6 An Interim Decision shall be final and binding upon the parties and upon the Engineer (if any) until such time as the Works have been completed or any Award or decision under Rule 24.3 has been given. Thereafter the Interim Decision may be re-opened by another Arbitrator appointed under these Rules and where such other Arbitrator was also the Arbitrator appointed to conduct the Interim Arbitration he shall not be bound by his earlier Interim Decision.

24.7 The Arbitrator in an Interim Arbitration shall have power to direct that Part F (Short Procedure) and/or Part G (Special Procedure for Experts) shall apply to the Interim Arbitration.

The ICE Conditions of Contract at one time imposed an embargo on arbitration proceedings prior to substantial completion. This has now been replaced by provisions which allow either party to commence arbitration at any stage of the project. Rule 24 deals with the situation where the arbitration proceeds in advance of completion. The key to this Rule is speed. Rule 24.2 requires the Arbitrator to act quickly. Rules 24.3 and 24.5 require the Arbitrator to publish an award within 14 days of substantial completion on the basis of evidence and submissions received up until that date. This "award" will normally be an Interim Decision in accordance with Rule 24.6 which may be reopened by the same or another Arbitrator.

PART J. MISCELLANEOUS

Rule 25. Definitions

25.1 In these Rules the following definitions shall apply.

(a) 'Arbitrator' includes a tribunal of two or more Arbitrators or an Umpire.

(b) 'Institution' means The Institution of Civil Engineers.

(c) 'ICE Conditions of Contract' means the Conditions of Contract for use in connection with Works of Civil Engineering Construction published jointly by the Institution, the Association of Consulting Engineers and the Federation of Civil Engineering Contractors.

(d) 'Other party' includes the plural unless the context otherwise requires.

(e) 'President' means the President for the time being of the Institution or any Vice-President acting on his behalf.

(f) 'Procedure' means The Institution of Civil Engineers'

Arbitration Procedure (1983) unless the context otherwise requires.

(g) 'Award', 'Final Award' and 'Interim Award' have the meanings given to those terms in or in connection with the Arbitration Acts 1950 to 1979. 'Summary Award' means an Award made under Rule 14 hereof.

(h) 'Interim Arbitration' means a decision as defined in Rule 24.6 hereof.

Rule 26. Application of the ICE Procedure

26.1 This Procedure shall apply to the conduct of the arbitration if

(a) the parties at any time so agree

(b) the President when making an appointment so directs

or

(c) the Arbitrator so stipulates at the time of his appointment

Provided that where this Procedure applies by virtue of the Arbitrator's stipulation under (c) above the parties may within 14 days of that appointment agree otherwise in which event the Arbitrator's appointment shall terminate and the parties shall pay his reasonable charges in equal shares.

26.2 This Procedure shall not apply to arbitrations under the law of Scotland for which a separate ICE Arbitration Procedure (Scotland) is available.

26.3 Where an arbitration is governed by the law of a country other than England and Wales this procedure shall apply to the extent the applicable law permits.

Rule 27. Exclusion of liability

27.1 Neither the Institution nor its servants or agents nor the President shall be liable to any party for any act omission or misconduct in connection with any appointment made or any arbitration conducted under this Procedure.

Rule 26 purports to provide that the Arbitration Procedure applies where the President directs or the Arbitrator stipulates. It is thought that both of these provisions entail circular arguments. It is submitted that the Arbitration Procedure applies only where the parties have agreed this either in their arbitration agreement or by express subsequent agreement.[79]

3. Introduction to the ICE Conciliation Procedure (1994)

Conciliation is provided for in many standard form civil engineering contracts. It is used optionally in both the ICE Conditions of Contract and the FCEC Form of Sub-Contract. It is used as a mandatory step in the dispute resolution process provided for in the ICE Design and Construct Contract.

The legal status of conciliation is discussed in Chapter 7. The rules in the ICE Conciliation Procedure correspond with the typical model of conciliation, namely:

(*a*) proceedings are conducted on a "without prejudice" basis
(*b*) the conciliator may talk privately with either party
(*c*) there is no cross-examination or need to disclose all the information available
(*d*) the Conciliator's Recommendation need not accord with the exact legal position but may suggest a practical solution which takes account of not only the strict rights and obligations of the parties, but also their other interests.

In general, the recommendation of a conciliator is not enforceable. However, the ICE Conditions of Contract and the FCEC Form of Sub-Contract each provide that when the Recommendation is published it becomes binding unless a Notice to Refer is served within a short period of one month and 28 days respectively.[80] The ICE Design and Construct Conditions of Contract provide a period of three months, but in this latter contract the conciliator very much fulfils the role of the Engineer under the standard ICE and FCEC Contracts. Where the

79 See *Pratt* v. *Swanmore Builders Ltd and Baker* [1980] 2 Lloyd's Rep. 504.

80 ICE Conditions of Contract, Clause 66(5) provides: "The recommendation of the conciliator shall be deemed to have been accepted in settlement of the dispute unless a written Notice to Refer ... is served within one calendar month of its receipt". The FCEC Sub-Contract provides at Clause 18(4): "Where a Notice to Refer is not served within the said period of 28 days, the recommendation of the conciliator shall be deemed to have been accepted in settlement of the dispute".

conciliator's recommendation becomes binding through passage of time, it is thought that the only arguments available to overturn the recommendation are that:

(*a*) the conciliator lacked jurisdiction either because he was not properly appointed or because he made a recommendation on a matter which was not properly referred to him
(*b*) the other party fraudulently procured the recommendation.

It is thought that no defence may be raised that the parties or conciliator have failed to operate the provisions of the Conciliation Procedure, unless a party has raised and maintained his objection throughout. It is thought that an argument may not be maintained that the Recommendation lacks internal consistency providing the final result proposed is clear. Where, however, the Recommendation is hopelessly unclear in its terms it is submitted that it is unenforceable for want of certainty.

4. Commentary on the provisions of the Conciliation Procedure (1994)

CONCILIATION PROCEDURE (1994)

1 This Procedure shall apply whenever

(a) the Parties have entered into a contract which provides for Conciliation for any dispute which may arise between the Parties in accordance with the Institution of Civil Engineers' Conciliation Procedure, or

(b) where the Parties have agreed that the Institution of Civil Engineers' Conciliation Procedure shall apply.

2 This Procedure shall be interpreted and applied in the manner most conducive to the efficient conduct of the proceedings with the primary objective of achieving a settlement to the dispute by agreement between the Parties as quickly as possible.

3 Subject to the provisions of the Contract relating to Conciliation, any Party to the Contract may by giving to the other Party a written notice, hereafter called a Notice of Conciliation, request that any dispute in connection with or arising out of the Contract or the carrying out of the Works shall be referred to a Conciliator. Such Notice shall be accompanied by a brief statement of the matter or matters which it is desired to refer to the Conciliation, and the relief or remedy sought.

4 Save where a Conciliator has already been appointed, the Parties shall agree upon a Conciliator within 14 days of the Notice being given under Paragraph 3. In default of agreement any Party may request the President (or, if he is unable to act, any Vice President) for the time being of the Institution of Civil Engineers to appoint a Conciliator within 14 days of receipt of the request by him, which request shall be accompanied by a copy of the Notice of Conciliation.

5 If, for any reason whatsoever, the Conciliator is unable, or fails to complete the Conciliation in accordance with this Procedure, then any Party may require the appointment of a replacement Conciliator in accordance with the procedures of Paragraph 4.

6 The Party requesting Conciliation shall deliver to the Conciliator, immediately on his appointment, and at the same time to the other Party if this has not already been done, a copy of the Notice of Conciliation, or as otherwise required by the Contract, together with copies of all relevant Notices of Dispute and of any other notice or decision which is a condition precedent to Conciliation.

7 The Conciliator shall start the Conciliation as soon as possible after his appointment and shall use his best endeavours to conclude the Conciliation as soon as possible and in any event within any time limit as may be stated in the Contract, or two months from the date of his appointment, or within such other time as may be agreed between the Parties.

8 Any Party may, upon receipt of notice of the appointment of the Conciliator and within such period as the Conciliator may allow, send to the Conciliator and to the other Party a statement of its views on the dispute and any issues that it considers to be of relevance to the dispute, and any financial consequences.

9 As soon as possible after his appointment, the Conciliator shall issue instructions establishing, *inter alia*, the date and place for the conciliation meeting with the Parties. Each Party shall inform the Conciliator in writing of the name of its representative for the Conciliation, who shall have full authority to act on behalf of that Party, and the names of any other persons who will attend the Conciliation meeting. This information shall be given at least seven days before the Conciliation meeting with copies to the other Party.

10 The Conciliator may, entirely at his own discretion, issue such further instructions as he considers to be appropriate, meet and question the Parties and their representatives, together or separately, investigate the facts and circumstances of the dispute, visit the site and request the production of documents or the attendance of people whom he considers could assist in any way. The Conciliator may conduct the proceedings in any way that he wishes, and with the prior agreement of the Parties obtain legal or technical advice, the cost of which shall be met by the Parties, in accordance with Paragraph 17, or as may be agreed by the Parties and the Conciliator.

11 The Conciliator may consider and discuss such solutions to the dispute as he thinks appropriate or as may be suggested by any Party. He shall observe and maintain the confidentiality of particular information which he is given by any Party privately, and may disclose it only with the explicit permission of that Party. He will try to assist the Parties to resolve the dispute in any way which is acceptable to them.

12 Any Party may, at any time, ask that additional claims or disputes, or additional parties, shall be joined in the Conciliation. Such applications shall be accompanied by details of the relevant contractual facts, notices and decisions. Such joinder shall be subject to the agreement of the Conciliator and all other Parties. Any additional party shall, unless otherwise agreed by the Parties, have the same rights and obligations as the other Parties to the Conciliation.

13 If, in the opinion of the Conciliator, the resolution of the dispute would be assisted by further investigation by any Party or by the Conciliator, or by an interim agreement, including some action by any Party, then the Conciliator will, with the agreement of the Parties, give instructions and adjourn the proceedings as may be appropriate.

14 Once a settlement has been achieved of the whole or any part of the matters in dispute, the Conciliator may assist the Parties to prepare an Agreement incorporating the terms of the settlement.

15 If, in the opinion of the Conciliator, it is unlikely that the Parties will achieve an agreed settlement to their disputes, or if any Party fails to respond to an instruction by the Conciliator, or upon the request of any Party, the Conciliator may advise all Parties accordingly and will forthwith prepare his Recommendation.

16 The Conciliator's Recommendation shall state his solution to the dispute which has been referred for Conciliation. The

Recommendation shall not disclose any information which any Party has provided in confidence. It shall be based on his opinion as to how the Parties can best dispose of the dispute between them and need not necessarily be based on any principles of the Contract, law, or equity. The Conciliator shall not be required to give reasons for his Recommendation. Nevertheless should he choose to do so, his reasons shall be issued as a separate document, within 7 days of the giving of his Recommendation.

17 When a settlement has been reached or when the Conciliator has prepared his Recommendation, or at an earlier date solely at the discretion of the Conciliator, he shall notify the Parties in writing and send them an account of his fees and disbursements. Unless otherwise agreed between themselves each Party shall be responsible for paying and shall within 7 days of receipt of the account from the Conciliator pay an equal share save that the Parties shall be jointly and severally liable to the Conciliator for the whole of his account. Upon receipt of payment in full the Conciliator shall send his Recommendation to all the Parties. If any Party fails to make the payment due from him the other Party may pay the sum to the Conciliator and recover the amount from the defaulting Party as a debt due. Each Party shall meet his own costs and expenses.

18 The Conciliator may be recalled, by written agreement of the Parties and upon payment of an additional fee, to clarify, amplify or give further consideration to any provision of the Recommendation.

19 The Conciliator shall not be appointed arbitrator in any subsequent arbitration between the Parties whether arising out of the dispute, difference or other matter or otherwise arising out of the same Contract unless the Parties otherwise agree in writing. No Party shall be entitled to call the Conciliator as a witness in any subsequent arbitration or litigation concerning the subject matter of the Conciliation.

20 The confidential nature of the Conciliation shall be respected by every person who is involved in whatever capacity.

21 The Conciliator shall not be liable to the Parties or any person claiming through them for any matter arising out of or in connection with the Conciliation or the way in which it is or has been conducted, and the Parties will not themselves bring any such claims against him.

22 Any notice required under this Procedure shall be sent to the Parties by recorded delivery to the principal place of business or if a company to its registered office, or to the address which the Party has notified to the Conciliator. Any notice required by this Procedure to be sent to the Conciliator shall be sent by recorded delivery to him at the address which he shall notify to the Parties on his appointment.

23 In this Procedure where the context so required 'Party' shall mean 'Parties' and 'he' shall mean 'she'.

Terminology and interpretation

The individual sections of the Procedure are referred to as "paragraphs" rather than rules or articles. This indicates that the provisions are to be interpreted in the manner most conducive to the efficient conduct of the proceedings—Paragraph 2.

Where a party refuses to take part

The Procedure does not oblige either party to participate in the conciliation. Where either party refuses to take part, the Conciliator will normally consider, in accordance with Paragraph 15, that it is unlikely that the parties will reach an agreed settlement and he will then prepare his Recommendation based on the material supplied by the participating party.

The date of publication of the Recommendation

By Paragraph 17, the Conciliator notifies the parties in writing when his Recommendation is ready. He is not obliged to send it until his fees and disbursements etc. are paid. Where the contract provides that the Recommendation becomes binding a specified number of days after the date of the Conciliator's Recommendation, the date from which time begins to run will be the date set out in the contract.[81]

81 The ICE Design and Construct Conditions of Contract provide that the three months run upon "receiving notice of the conciliator's recommendation". The ICE Conditions of Contract and the FCEC Sub-Contract both refer to "receipt" of the recommendation.

Appendix

The text of Part II of the Housing Grants, Construction and Regeneration Act 1996 is reproduced below.

PART II CONSTRUCTION CONTRACTS

Introductory provisions

104 Construction contracts

(1) In this Part a "construction contract" means an agreement with a person for any of the following—

(a) the carrying out of construction operations;

(b) arranging for the carrying out of construction operations by others, whether under sub-contract to him or otherwise;

(c) providing his own labour, or the labour of others, for the carrying out of construction operations.

(2) References in this Part to a construction contract include an agreement—

(a) to do architectural, design, or surveying work, or

(b) to provide advice on building, engineering, interior or exterior decoration or on the laying-out of landscape,

in relation to construction operations.

(3) References in this Part to a construction contract do not include a contract of employment (within the meaning of the Employment Rights Act 1996).

(4) The Secretary of State may by order add to, amend or repeal any of the provisions of subsection (1), (2) or (3) as to the agreements which are construction contracts for the purposes of this Part or are to be taken or not to be taken as included in references to such contracts.

No such order shall be made unless a draft of it has been laid before and approved by a resolution of each of [sic] House of Parliament.

(5) Where an agreement relates to construction operations and other matters, this Part applies to it only so far as it relates to construction operations.

An agreement relates to construction operations so far as it makes provision of any kind within subsection (1) or (2).

(6) This Part applies only to construction contracts which—

(a) are entered into after the commencement of this Part, and

(b) relate to the carrying out of construction operations in England, Wales or Scotland.

(7) This Part applies whether or not the law of England and Wales or Scotland is otherwise the applicable law in relation to the contract.

105 Meaning of "construction operations"

(1) In this Part "construction operations" means, subject as follows, operations of any of the following descriptions—

(a) construction, alteration, repair, maintenance, extension, demolition or dismantling of buildings, or structures forming, or to form, part of the land (whether permanent or not);

(b) construction, alteration, repair, maintenance, extension, demolition or dismantling of any works forming, or to form, part of the land, including (without prejudice to the foregoing) walls, roadworks, power-lines, telecommunication apparatus, aircraft runways, docks and harbours, railways, inland waterways, pipe-lines, reservoirs, water-mains, wells, sewers, industrial plant and installations for purposes of land drainage, coast protection or defence;

(c) installation in any building or structure of fittings forming part of the land, including (without prejudice to the foregoing) systems of heating, lighting, air-conditioning, ventilation, power supply, drainage, sanitation, water supply or fire protection, or security or communications systems;

(d) external or internal cleaning of buildings and structures, so far as carried out in the course of their construction, alteration, repair, extension or restoration;

(e) operations which form an integral part of, or are preparatory to, or are for rendering complete, such operations as are previously described in this subsection, including site clearance, earth-moving, excavation, tunnelling and boring, laying of foundations, erection, maintenance or dismantling of scaffolding, site restoration, landscaping and the provision of roadways and other access works;

(f) painting or decorating the internal or external surfaces of any building or structure.

(2) The following operations are not construction operations within the meaning of this Part—

(a) drilling for, or extraction of, oil or natural gas;

(b) extraction (whether by underground or surface working) of minerals; tunnelling or boring, or construction of underground works, for this purpose;

(c) assembly, installation or demolition of plant or machinery, or erection or demolition of steelwork for the purposes of supporting or providing access to plant or machinery, on a site where the primary activity is—

(i) nuclear processing, power generation, or water or effluent treatment, or

(ii) the production, transmission, processing or bulk storage (other than warehousing) of chemicals, pharmaceuticals, oil, gas, steel or food and drink;

(d) manufacture or delivery to site of—

(i) building or engineering components or equipment,

(ii) materials, plant or machinery, or

(iii) components for systems of heating, lighting, air-conditioning, ventilation, power supply, drainage, sanitation, water supply or fire protection, or for security or communications systems,

except under a contract which also provides for their installation;

(e) the making, installation and repair of artistic works, being sculptures, murals and other works which are wholly artistic in nature.

(3) The Secretary of State may by order add to, amend or repeal any of the provisions of subsection (1) or (2) as to the operations and

work to be treated as construction operations for the purposes of this Part.

(4) No such order shall be made unless a draft of it has been laid before and approved by a resolution of each House of Parliament.

106 Provisions not applicable to contract with residential occupier

(1) This Part does not apply—

(a) to a construction contract with a residential occupier (see below), or

(b) to any other description of construction contract excluded from the operation of this Part by order of the Secretary of State.

(2) A construction contract with a residential occupier means a construction contract which principally relates to operations on a dwelling which one of the parties to the contract occupies, or intends to occupy, as his residence.

In this subsection "dwelling" means a dwelling-house or a flat; and for this purpose—

"dwelling-house" does not include a building containing a flat; and

"flat" means separate and self-contained premises constructed or adapted for use for residential purposes and forming part of a building from some other part of which the premises are divided horizontally.

(3) The Secretary of State may by order amend subsection (2).

(4) No order under this section shall be made unless a draft of it has been laid before and approved by a resolution of each House of Parliament.

107 Provisions applicable only to agreements in writing

(1) The provisions of this Part apply only where the construction contract is in writing, and any other agreement between the parties as to any matter is effective for the purposes of this Part only if in writing.

The expressions "agreement", "agree" and "agreed" shall be construed accordingly.

(2) There is an agreement in writing—

(a) if the agreement is made in writing (whether or not it is signed by the parties),

(b) if the agreement is made by exchange of communications in writing, or

(c) if the agreement is evidenced in writing.

(3) Where parties agree otherwise than in writing by reference to terms which are in writing, they make an agreement in writing.

(4) An agreement is evidenced in writing if an agreement made otherwise than in writing is recorded by one of the parties, or by a third party, with the authority of the parties to the agreement.

(5) An exchange of written submissions in adjudication proceedings, or in arbitral or legal proceedings in which the existence of an agreement otherwise than in writing is alleged by one party against another party and not denied by the other party in his response constitutes as between those parties an agreement in writing to the effect alleged.

(6) References in this Part to anything being written or in writing include its being recorded by any means.

Adjudication

108 Right to refer disputes to adjudication

(1) A party to a construction contract has the right to refer a dispute arising under the contract for adjudication under a procedure complying with this section.

For this purpose "dispute" includes any difference.

(2) The contract shall—

(a) enable a party to give notice at any time of his intention to refer a dispute to adjudication;

(b) provide a timetable with the object of securing the appointment of the adjudicator and referral of the dispute to him within 7 days of such notice;

(c) require the adjudicator to reach a decision within 28 days of referral or such longer period as is agreed by the parties after the dispute was referred;

(d) allow the adjudicator to extend the period of 28 days by up to 14 days, with the consent of the party by whom the dispute was referred;

(e) impose a duty on the adjudicator to act impartially; and

(f) enable the adjudicator to take the initiative in ascertaining the facts and the law.

(3) The contract shall provide that the decision of the adjudicator is binding until the dispute is finally determined by legal proceedings, by arbitration (if the contract provides for arbitration or the parties otherwise agree to arbitration) or by agreement.

The parties may agree to accept the decision of the adjudicator as finally determining the dispute.

(4) The contract shall also provide that the adjudicator is not liable for anything done or omitted in the discharge or purported discharge of this functions as adjudicator unless the act or omission is in bad faith, and that any employee or agent of the adjudicator is similarly protected from liability.

(5) If the contract does not comply with the requirements of subsections (1) to (4), the adjudication provisions of the Scheme for Construction Contracts apply.

(6) For England and Wales, the Scheme may apply the provisions of the Arbitration Act 1996 with such adaptations and modifications as appear to the Minister making the scheme to be appropriate.

For Scotland, the Scheme may include provision conferring powers on courts in relation to adjudication and provision relating to the enforcement of the adjudicator's decisions.

Payment

109 Entitlement to stage payments

(1) A party to a construction contract is entitled to payment by installments, stage payments or other periodic payments for any work under the contract unless—

(a) it is specified in the contract that the duration of the work is to be less than 45 days, or

(b) it is agreed between the parties that the duration of the work is estimated to be less than 45 days.

(2) The parties are free to agree the amounts of the payments and the intervals at which, or circumstances in which, they become due.

(3) In the absence of such agreement, the relevant provisions of the Scheme for Construction Contracts apply.

(4) References in the following sections to a payment under the contract include a payment by virtue of this section.

110 Dates for payment

(1) Every construction contract shall—

(a) provide an adequate mechanism for determining what payments become due under the contract, and when, and

(b) provide for a final date for payment in relation to any sum which becomes due.

The parties are free to agree how long the period is to be between the date on which a sum becomes due and the final date for payment.

(2) Every construction contract shall provide for the giving of notice by a party not later than five days after the date on which a payment becomes due from him under the contract, or would have become due if—

(a) the other party had carried out his obligations under the contract, and

(b) no set-off or abatement was permitted by reference to any sum claimed to be due under one or more other contracts,

specifying the amount (if any) of the payment made or proposed to be made, and the basis on which that amount was calculated.

(3) If or to the extent that a contract does not contain such provision as is mentioned in subsection (1) or (2), the relevant provisions of the Scheme for Construction Contracts apply.

111 Notice of intention to withhold payment

(1) A party to a construction contract may not withhold payment after the final date for payment of a sum due under the contract unless he has given an effective notice of intention to withhold payment.

The notice mentioned in section 110(2) may suffice as a notice of intention to withhold payment if it complies with the requirements of this section.

(2) To be effective such a notice must specify—

(a) the amount proposed to be withheld and the ground for withholding payment, or

(b) if there is more than one ground, each ground and the amount attributable to it,

and must be given not later than the prescribed period before the final date for payment.

(3) The parties are free to agree what that prescribed period is to be.

In the absence of such agreement, the period shall be that provided by the Scheme for Construction Contracts.

(4) Where an effective notice of intention to withhold payment is given, but on the matter being referred to adjudication it is decided that the whole or part of the amount should be paid, the decision shall be construed as requiring payment not later than—

(a) seven days from the date of the decision, or

(b) the date which apart from the notice would have been the final date for payment,

whichever is the later.

112 Right to suspend performance for non-payment

(1) Where a sum due under a construction contract is not paid in full by the final date for payment and no effective notice to withhold payment has been given, the person to whom the sum is due has the right (without prejudice to any other right or remedy) to suspend performance of his obligations under the contract to the party by whom payment ought to have been made ("the party in default").

(2) The right may not be exercised without first giving to the party in default at least seven days' notice of intention to suspend performance, stating the ground or grounds on which it is intended to suspend performance.

(3) The right to suspend performance ceases when the party in default makes payment in full of the amount due.

(4) Any period during which performance is suspended in pursuance of the right conferred by this section shall be disregarded in computing for the purposes of any contractual time limit the time taken, by the party exercising the right or by a third party, to complete any work directly or indirectly affected by the exercise of the right.

Where the contractual time limit is set by reference to a date rather than a period, the date shall be adjusted accordingly.

113 Prohibition of conditional payment provisions

(1) A provision making payment under a construction contract conditional on the payer receiving payment from a third person is ineffective, unless that third person, or any other person payment by whom is under the contract (directly or indirectly) a condition of payment by that third person, is insolvent.

(2) For the purposes of this section a company becomes insolvent—

(a) on the making of an administration order against it under Part II of the Insolvency Act 1986,

(b) on the appointment of an administrative receiver or a receiver or manager of its property under Chapter I of Part III of that Act, or the appointment of a receiver under Chapter II of that Part,

(c) on the passing of a resolution for voluntary winding-up without a declaration of solvency under section 89 of that Act, or

(d) on the making of a winding-up order under Part IV or V of that Act.

(3) For the purposes of this section a partnership becomes insolvent—

(a) on the making of a winding-up order against it under any provision of the Insolvency Act 1986 as applied by an order under section 420 of that Act, or

(b) when sequestration is awarded on the estate of the partnership under section 12 of the Bankruptcy (Scotland) Act 1985 or the partnership grants a trust deed for its creditors.

(4) For the purposes of this section an individual becomes insolvent—

(a) on the making of a bankruptcy order against him under Part IX of the Insolvency Act 1986, or

(b) on the sequestration of his estate under the Bankruptcy (Scotland) Act 1985 or when he grants a trust deed for his creditors.

(5) A company, partnership or individual shall also be treated as insolvent on the occurrence of any event corresponding to those specified in subsection (2), (3) or (4) under the law of Northern Ireland or of a country outside the United Kingdom.

(6) Where a provision is rendered ineffective by subsection (1), the parties are free to agree other terms for payment.

In the absence of such agreement, the relevant provisions of the Scheme for Construction Contracts apply.

Supplementary provisions

114 The Scheme for Construction Contracts

(1) The Minister shall by regulations make a scheme ("the Scheme for Construction Contracts") containing provision about the matters referred to in the preceding provisions of the Part.

(2) Before making any regulations under this section the Minister shall consult such persons as he thinks fit.

(3) In this section "the Minister" means—

(a) for England and Wales, the Secretary of State, and

(b) for Scotland, the Lord Advocate.

(4) Where any provisions of the Scheme for Construction Contracts apply by virtue of this Part in default of contractual provision agreed by the parties, they have effect as implied terms of the contract concerned.

(5) Regulations under this section shall not be made unless a draft of them has been approved by resolution of each House of Parliament.

115 Service of notices, &c

(1) The parties are free to agree on the manner of service of any notice or other document required or authorised to be served in pursuance of the construction contract or for any of the purposes of this Part.

(2) If or to the extent that there is no such agreement the following provisions apply.

(3) A notice or other document may be served on a person by any effective means.

(4) If a notice or other document is addressed, pre-paid and delivered by post—

(a) to the addressee's last known principal residence or, if he is or has been carrying on a trade, profession or business, his last known principal business address, or

(b) where the addressee is a body corporate, to the body's registered or principal office,

it shall be treated as effectively served.

(5) This section does not apply to the service of documents for the purposes of legal proceedings, for which provision is made by rules of court.

(6) References in this Part to a notice or other document include any form of communication in writing and references to service shall be construed accordingly.

116 Reckoning periods of time

(1) For the purposes of this Part periods of time shall be reckoned as follows.

(2) Where an act is required to be done within a specified period after or from a specified date, the period begins immediately after that date.

(3) Where the period would include Christmas Day, Good Friday or a day which under the Banking and Financial Dealings Act 1971 is a bank holiday in England and Wales or, as the case may be, in Scotland, that day shall be excluded.

117 Crown application

(1) This Part applies to a construction contract entered into by or on behalf of the Crown otherwise than by or on behalf of Her Majesty in her private capacity.

(2) This Part applies to a construction contract entered into on behalf of the Duchy of Cornwall notwithstanding any Crown interest.

(3) Where a construction contract is entered into by or on behalf of Her Majesty in right of the Duchy of Lancaster, Her Majesty shall be represented, for the purposes of any adjudication or other proceedings arising out of the contract by virtue of this Part, by the Chancellor of the Duchy or such person as he may appoint.

(4) Where a construction contract is entered into on behalf of the Duchy of Cornwall, the Duke of Cornwall or the possessor for the time being of the Duchy shall be represented, for the purposes of any adjudication or other proceedings arising out of the contract by virtue of this Part, by such person as he may appoint.

Source: The Housing Grants, Construction and Regeneration Act 1996. HMSO, London, 1996.